ADSL AND DSL TECHNOLOGIES

McGRAW-HILL
Computer Communications Titles

ADSL and DSL Technologies

Walter Goralski

Hill Associates, Inc.

McGraw-Hill
New York • San Francisco • Washington, D.C. • Auckland
Bogotá • Caracas • Lisbon • London • Madrid • Mexico City
Milan • Montreal • New Delhi • San Juan • Singapore
Sydney • Tokyo • Toronto

Library of Congress Cataloging-in-Publication Data

Goralski, Walter
    ADSL and DSL technologies / Walter Goralski.
        p.   cm.
    Includes index.
    ISBN 0-07-024679-3
    1. Telephone switching systems, Electronic.   2. Digital
    communications.   3. Telecommunication—Standards.   I. Title.
    TK6397.G65   1998
    621.387—dc21   98-5178
                                                                  CIP

McGraw-Hill

A Division of The *McGraw-Hill Companies*

6 7 8 9 0   DOC/DOC   0 3 2 1 0

ISBN 0-07-024679-3

*The sponsoring editor for this book was Steven Elliot and the production
supervisor was Sherri Souffrance. It was set in Century Schoolbook by Douglas
& Gayle, Limited.*

Printed and bound by R. R. Donnelley & Sons Company.

McGraw-Hill books are available at special quantity discounts to use as premiums
and sales promotions, or for use in corporate training programs. For more information, please write to the Director of Special Sales, McGraw-Hill, Professional Publish-
ing, Two Penn Plaza, New York, NY 10121-2298. Or contact your local bookstore.

 This book is printed on recycled, acid-free paper containing a minimum of 50%
recycled de-inked fiber.

ACKNOWLEDGMENTS

With every book I write, there seems to be an ever-increasing list of people to thank for assistance and encouragement. Whether this is a good sign or bad sign, I do not know. But this book would not exist without the following people guiding me through the whole process.

It was Dave Hill who first suggested a DSL course for Hill Associates which later grew into this book. Nigel Cole of the ADSL Forum and Orckit first made me feel as if I knew enough about ADSL to fill a book and encouraged me to write it. His review of the manuscript was one of the most thorough and complete that I have ever endured, in spite of time pressure, and the result was a much better book. Gary Kessler of Hill reviewed the first draft and made helpful comments, as always. Joe Charboneau and Mike Rude of ADC provided valuable information about HDSL and HDSL2. The ADSL Forum in general has been a valuable and unique source of information. Steve Elliot of McGraw-Hill stepped outside his role as editor to gather valuable information in his travels.

I cannot thank my family enough either. Jodi gives me space and quiet support, Christopher has a first-born's pride in my accomplishments, Alexander makes sure I keep my writing totally in perspective, and Arianna often keeps me company while I write but never bugs me (see, I put it in the book).

The McGraw-Hill Companies, Inc., gratefully acknowledges the support and contribution of Hill Associates, Inc., during the preparation of this work and for furnishing significant portions of the content and providing important editorial guidance.

CONTENTS

Contents

PREFACE

I wrote this book to do my part to promote change in the public telephone network. Although it usually comes with a price tag, change is good, and all networks must change if they are to grow and prosper. Networks that do not change will wither and struggle. It does not matter whether the network involved is the telephone network. All networks must evolve as the needs of the users change, as the technology on which the network is based changes, and as the entire economic, social, and political backdrop against which the network exists slowly transforms itself.

Consider the highway network of roads in the United States. From humble beginnings as simple "post roads" for mail delivery and stage coaches, the network adapted itself to automobiles by adding pavement and traffic signals. The system was transformed again when post-World War II prosperity encouraged automobile use for a wide variety of reasons, from commuting to the recreational Sunday drive in the country. The Interstate Highway System with limited access and guard rails better reflected this new environment.

It is not for nothing that telephone network terminology closely mirrors road system talk. Both systems have access links and bypasses. Both have tolls and interchanges. And more to the point, both have traffic and congestion. The talk about an Information Superhighway started out in discussions about an "interstate highway system for data." As will be shown, both telephone networks and highway networks share much more than terminology.

This book emphasizes one of the ways that may be used to transform the voice network in the United States (and for that matter, around the world) into a network better suited for the "automobiles" of the 1990s. If telephones are the voice network's stage coaches, then modern personal computers attached to the Internet are its luxury cars. This being the case, perhaps local access lines using *Asymmetric Digital Subscriber Line* (ADSL) technology are better suited for users than the "plain old telephone service" (POTS) access lines of the last 100 years or more.

Although many technologies can play a role in modernizing the voice network, ADSL is singled out in this text because ADSL has an edge over many others in terms of standardization, vendor activity, economics, and customer interest. The question is whether ADSL can maintain this edge. Other technologies will be mentioned in this regard, but ADSL is investigated in full. Whatever the outcome over the next 5 or 10 years, it certainly will be an interesting ride.

INTRODUCTION

All successful technologies are successful because they solve problems. The problem could be as simple as getting around faster and more conveniently (airplanes and automobiles) or as complex as finding an alternative energy source by splitting an atom in a controlled fashion (nuclear power). This book is about *Asymmetric Digital Subscriber Line* (ADSL) technology, and the problem it attempts to solve is one that is becoming more critical as people change the ways that they use networks to communicate, work, and relax. The problem is easiest to explain through a few examples.

Pacific Bell is a major local telephone company servicing much of the West Coast of the United States. Like most telephone companies, life had been good to Pacific Bell in the past. Revenues were plentiful, network usage predictable, and things pretty much revolved around the relatively routine tasks involved in supporting the typical three- to six-minute telephone call people made to order pizza, call in sick, or chat with friends. Over the past few years, however, especially since the explosion of the Internet and World Wide Web in 1993-1994, life has gotten much more interesting for Pacific Bell and other local telephone companies. There was more intense local competition, increased interest in new services, and even heightened customer awareness of perceived shortcomings of the existing Pacific Bell voice telephone network. But the biggest changes involved the interaction between the local telephone companies and the Internet. As it had done in many other fields, the Internet completely changed the rules of the voice networking game.

The Internet is not actually a network at all, but rather a worldwide collection of networks that are all interconnected, a "network of networks." This *internetwork* has been around for about 30 years, but most of those years were spent as an obscure research and educational tool beyond the consciousness of the general public. This all changed in the mid-1990s with the appearance of the World Wide Web on the Internet. The Web is technically a subset of the Internet, which means that not everything on the Internet is part of the Web. The Web is collection of computers on the Internet (the *Web site* or *Web server*) with information available to almost anyone with Internet access, and Internet access did not have to include access to the Web. A special software package known as the *Web browser* or *Web client* was needed on the home PC to enable users to get information from the Web sites on the Internet. The local telephone companies' lines were almost always used for Internet access, especially from home. This use of telephone lines changed the rules of the telephone company service game, although few noticed at the time.

But if anyone doubted that the local telephone company service game had changed for the good, these doubts were dispelled the afternoon and evening of January 6, 1997. To their credit, Pacific Bell had seen it coming, although there was little technicians or anyone else could do about it. There was little that could be done because the problems that occurred on that date were the result of design decisions and engineering choices that were made many years before.

The root cause of the problem, oddly enough, was Christmas. For their holiday gifts—which included not only traditional Christian households, but also many non-Christians who exchange gifts at that time of the year—many people, especially elementary and high school students, found personal computers (PCs) under the tree. By 1996, many had discovered that college students were not the only ones who needed PCs to do assignments and research. The rise of Internet and World Wide Web popularity in 1993-1994 had completely changed education, for better or worse. It was no longer just a matter of typing a report with the word processor software on the PC. In more and more cases, the topic was explored, the material researched, and references checked and cross-checked all from the PC itself. In fact, in some cases, the assignment itself was distributed to the class with the help of the network. Attachment of the PC to the Internet and the Web made this all possible.

So during the Christmas season of 1996, many thousands of students nationwide and in California got brand-new computers that almost universally had built-in modem hardware and connection software. This was good news for PC vendors and companies such as Microsoft, which had bundled easy Internet/Web access into their products both in anticipation of and to encourage this network trend, but Pacific Bell officials were not as happy and were downright worried.

They fretted because the telephone network that Pacific Bell and all of the other 1,300 local telephone companies in the United States had built was not designed for the PC, nor the Internet, nor the Web. How could it be? The network that evolved to connect Alexander Graham Bell's 1876 telephone invention could not have possibly anticipated later inventions such as the PC, the modem, or the Internet. The network that had evolved was designed, engineered, and built for one basic purpose: two people talking to each other for relatively short periods of time over the telephone. This network is commonly called the *public switched telephone network* (PSTN), especially by those within the telecommunications field itself.

The PSTN is not the same as the Internet. True, the Internet essentially plays a role for PC connectivity similar to the role the PSTN plays in the voice world, but there are significant differences, but as far as the PSTN is concerned, almost any device that generates the right electrical signals can use the PSTN to connect to a compatible remote device. The user devices attached to the PSTN do not have to be telephones; they might be fax machines or even PCs. Anything that makes noise can communicate over the PSTN. In fact, that is what a *modem* is for. A modem is needed on a PC to allow the PC to attach to some other remote device over the PSTN. The modem makes noise out of the digital bits that PCs and other computers understand. This noise is unlike human speech, but it is noise nonetheless. Anyone who has ever dialed a fax machine by mistake knows that such devices talk a very different language than people do.

There are more differences between the PSTN and the Internet, many more. The major components of the PSTN are called *switches* and the major components of the Internet are called *routers*, just to cite one such difference. A more systematic examination of the PSTN architecture and Internet architecture will be done in a later chapter, as well as an attempt to sort out the differences between switches and routers. It should be pointed out, however, that the entire switch/router debate is an active one. Nevertheless, some more or less firm conclusions can be drawn about these differences.

So the PSTN can be used to connect a home or business PC to the Internet. All it takes is to plug the PC (by way of the modem) into a telephone line (known as an *access line* into the PSTN), dial up any one of the more than 5,000 Internet service providers (ISPs) around the United States, and after a minimum of bookkeeping for payment options, you are on the Internet and cruising the Web in style. There are other ways to access the Internet, but you would be hard-pressed to find instances where the PSTN in one form or another is *not* used. This simple fact is both a blessing and a curse to the local telephone companies. It is nice to be so popular, but it is hard to do things in the same old way.

So the Pacific Bell people looked at the Christmas season of 1996 with some amount of fear. They were afraid of what would happen when all of those new PC users plugged in and logged on to the Internet and started surfing the Web. Because a large percentage of all the Web sites in the world are in California, they had reason to be worried, but nothing much at all happened on Christmas Day. To understand their dread, imagine that a new telephone was the prized gift of the holiday season. If everyone picked up their new phones at once and tried to make telephone calls, chaos would ensue because the PSTN is designed for a world of intermittent, independent telephone line use. There was no need to build a network where everyone used a telephone at the same time because, first of all, history had shown that people make calls more or less at random, and second, this was a basic assumption built into the very foundations of voice network engineering principles.

Apparently, there were enough other toys to play with Christmas Day to distract people, so the PSTN—the voice network—through which almost all of the analog modem calls traveled on the way to the Internet, survived. On Monday, January 6, 1997, the story was different. The students had returned to school and all seemed well, but around 3 P.M., the new PC users got home with their new homework assignments. A huge number of them turned on their computers, logged in to the Internet, and clicked their mouse, all at the same time.

Immediately, a PSTN central office (CO) where telephone lines come together to be switched in the East Bay area of San Francisco browned out. A *brownout* is not an absolute failure of the switch, which is basically nothing more than a big computer itself, but services are curtailed during a brownout, making it difficult to continue calls in progress and hard to establish new calls at all. Because both businesses and homes are serviced from the same central offices, both types of customers were affected. Throughout Contra Costa County, businesses lost their dial tone, and users heard nothing but silence when they lifted their telephones. Customers trying to dial a number heard *fast busy* signals (technically called *reorder*) that indicated too heavy a load on the PSTN trunking system, which links central office switches together. In frustration, many Pacific Bell customers had to use their cellular phones just to call up and report the service denials and outages.

The problems continued throughout the night as Pacific Bell became reacquainted with another networking truism: Congestion in any kind of network, once it occurs, is difficult to alleviate. It is much better to try to prevent congestion from occurring than it is to try to make it go away. Fortunately, things settled down, and the network more or less returned to normal. Until the next time.

And there will definitely be a next time. In has been estimated that some 10 percent of all central offices in California are chronically congested, and as California goes in many things from fashion to networks, so soon will got the nation.

The Pacific Bell problem is by no means unique. All of the regional Bell operating companies ("RBOCs": former pieces of the huge AT&T local network spun off in 1984 to go their own ways) have been dealing with problems like the one in California for a while as the Internet continues to grow in popularity, as the Web becomes a more essential part of people's lives, and as more people start or continue to work at home for a greater part of the work week. For example, a similar brownout (sometimes also called a *brown down*) occurred on the East Coast even before the Great California Brownout.

The winter of 1996 in the Northeast was harsh. Snowstorms were almost a weekly occurrence that January and February, and six inches or more of snow was not unusual for each storm. In the Washington DC area, a substantial blizzard made it nearly impossible for any government or other workers to venture in to the office. No problem: This is the 1990s. People who could not travel fired up the laptop and dialed in to the corporate network. Of course, many corporate networks are accessed today through the Internet as well, a practice sometimes called an *Intranet*, although the term implies more than just this.

As one might expect, the same problems that occurred later in California appeared, this time in Bell Atlantic territory. All over the service area affected by the storm, Bell Atlantic began to have overload and brownout problems with their switches. There were blocked 911 emergency calls, dial tone delays, and outright service denials. The effects were not as severe as in California, but the event sent a powerful message to network planners and designers.

Even if one knows nothing about networks, the Internet, or the PSTN in general, the problems posed by increased network usage—and changing network usage—are understandable. People are familiar with many systems that makes use of scarce resources. That is, the primary good that a system exists to deliver cannot be distributed to everyone at the same time. It makes no difference if the good is electricity, water, or phone calls. When workers in a downtown area hit the restaurants for lunch on a hot July day, the combination of air conditioning load and electricity in the kitchens dims lights all around town. During the Superbowl halftime, water pressure drops as previously chair-bound TV spectators rise as one and make their way to the bathroom.

However, there is one major difference between water and electricity on one hand and the new PSTN loads on the other. The difference is *timing*. The power utility knows when it is hot in July. The Superbowl's low water pressure takes no firefighter by surprise. But network engineers never know exactly when the next mass login is going to happen, so planning for it in the short term becomes impossible.

This is not to say that planning ways to alleviate PSTN congestion cannot be done in the long term. After all, any system is designed for performance maximums, whether the key activity is air handling, flushing, or Internet access. Oddly, the PSTN was built for a number of maximum traffic days. Traditionally, heavy telephone traffic take place on Mother's Day, the day after Thanksgiving, and New Year's Day. For the rest of the time, vast parts of the network more or less sit idle. The predictability of the traffic patterns is what has been shattered by the Internet and Web. The timing is off.

Another point that may strike one as add is that in some senses it does not make much of a difference whether the increased traffic comes from talking. In California, everyone logged in to the Internet at the same time, but what if everyone had simply picked up the phone to call Grandma to thank her for the new PC (as if that would ever happen)? The effect would have been the same is some respects, but different in others.

The same sorts of service denials would take place. Someone once said that country star Garth Brooks has "crashed more central office switches than everything else put together." If tickets go on sale at 9 A.M., you can bet that a couple of thousand people will call at 9 A.M. Now the point has already been made that talk is just noise and so are the signals on the PSTN from devices such as fax machines and PCs, but the digital noise from PCs is very different from the noises made by people talking. In fact, these differences form one of the major points of this book.

Because this fundamental point is so important to the book as a whole, some time should be spent exploring the differences between PC noise and people noise. Even those who have some familiarity with telecommunications may benefit from this section because this very familiarity sometimes leads to a kind of blurring of important distinctions.

Analog and Digital

There is no more profound or basic a distinction to be made in telecommunications today than the distinction between analog and digital. When two humans talk, their voices are analog signals carrying analog information content. When a user sitting at a PC employs a modem to link a PC to an Internet service provider (ISP), the link uses analog signals to carry digital information content. Consider that there are two aspects to these conversations: the information content and the signal itself. The information content is *what* is being carried, and the signal is *how* it is being carried. Either or both may be analog or digital. In other words, the information content being carried across a network may be analog or digital, and the underlying signal into or on the network itself may be analog or digital as well.

Before considering information content and signaling in detail, some words about the differences between analog and digital are in order. Even within the telecommunications industry among those who should know better, there is much debate about the correct use of the terms "analog" and "digital." For instance, sending 0s and 1s over a modem attached to the PSTN is "analog transmission," but sending exactly the same string of 0s and 1s over a satellite channel with a different type of modem (but a modem nonetheless) is "digital TV."

To develop consistent digital and analog terminology for the rest of this book, consider the information content itself. Never mind how the information gets across the network, just look at the information being sent and received. This information may itself be of an analog or digital form.

If it is analog, the information may take any value between an allowed minimum and maximum. There is no forbidden value—any value between the maximum and minimum is just as good as another. If the value of the information were plotted against time, the resulting graph would form a single continuous line with no jumps or discontinuities. Examples of analog information content would include things such as temperature, air pressure, weight, and almost all common physical measurements. The air temperature cannot go from 50 degrees to 70 degrees without passing through all of the values in between. The same is true when the weight of a person increases from 190 to 200 pounds, although it may seem that way, especially around the holidays. The point is that the human voice is an analog quantity; human speech can be described by a number of variables such as amplitude (signal strength) and frequency, all of which are analog variables that continuously vary over time between a minimum and maximum value.

The upper portion of Figure I-1 shows how the amplitude of an analog variable might be plotted against time, giving a smooth curve.

Of course, the information content does not have to be analog; it might also be digital. A variable is considered digital if it can take on only one of a limited number of discrete, or separate, values. There may be the same concept of minimum and maximum, but these limits are just the boundaries of the allowed values. Everything else outside this set of values is strictly forbidden or represents some type of error condition. The simplest example of digital information is the set of digits themselves. That is, the integers 1, 2, 3, and so on represent the only values that an integer can take. Binary digits, or bits, can take on only one of two values, 0 or 1, which is what makes them binary in the first place. A graph on digital information would show many discontinuities in value, just as one goes from "1" to "2" without worrying about values in between, such as 1.5. Modern computers are the prime example of devices that generate and consume digital information content.

The lower portion of Figure I-1 shows how the amplitude of an digital variable might be plotted against time, giving a disjointed curve. In fact, the only reason that sharp vertical lines are included between the levels is for the human eye to better interpret what is going on. Technically, there is no "jump" between one value and another. There is a time when the value is X, and a time when the value is Y, with no values between X and Y at all. As it turns out, this is hard to do in the real world because most electrical phenomena—and electricity is most often used to represent information traveling across a network—cannot be made to instantly jump from one value to another. Note also that digital values do not necessarily have to be binary. That is, there can be more than just two values for a digital signal. The figure shows at least four values.

Figure I-1
Analog (top) and
Digital (bottom)

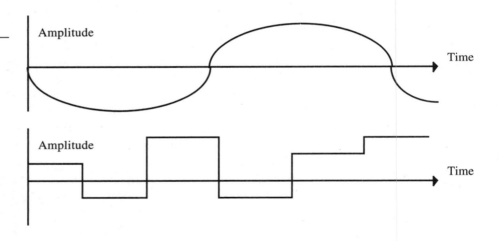

This point about the electrical representation of information content on a network brings up the next aspect of the analog/digital world of telecommunications. It is clear that the information to be sent and received across the network may be analog (voice, for example) or digital (a PC file, for example) in nature. To convey that information, some method of *signaling* must be used on the network itself. This use of the term "signaling" should not be confused with other meanings of the term in telecommunications. For instance, signaling is said to be used on the PSTN to establish, maintain, and release circuits. This use of the term refers to the way that information content is transferred and nothing else.

The same logic that was applied to information content to define analog and digital can now be used to define how that information is signaled across the network. If the actual signal generated by a sender or transmitter and received by a receiver can take on any value at all within a defined range, then the signal is analog. Note that the *signal* is analog regardless of whether the information sent is analog or digital. Sending analog information across an analog network becomes an almost trivial process, especially where analog voice is concerned. All that needs to be done is to convert the pressure waves of ordinary voice into waves of electricity using a simple transducer. The process is reversed at the other end of the network, naturally.

The underlying signaling network may also be digital in nature. In this case, as before, either analog or digital information content may be carried across the network. The digital information to digital signal process is relatively simple, but the conversion process from analog information content to digital signaling is not. Recall that an analog variable can take on any value within a range, but digital variables can only take on a limited number of values. How can one be converted into the other? In other words, if analog information can take on any value between 1 and 2, including 1.5, how is this to be represented as a digital signal which only allows the values 1 and 2? Obviously, there can be no simple one-for-one mapping of analog values to digital signals.

This dilemma has an answer, but not one that leaves everyone feeling satisfied. The representation of analog values by digital signals is known as *quantization*. It turns out that there is always some quantization *noise* that creeps in during the analog to digital conversion process. The term noise is just used here to indicate the presence of unwanted information that interferes with the desired result. In the simple example used above, the analog 1.5 value must either be represented by a 1 or 2. Perhaps an engineer could make it a rule to round off all values greater than 1.5 to 2, and all other values would round to 1. Even this simple example shows how quantization noise works. If 1.5 rounds to 2, then so will 1.6, 1.8, and even 1.555438. Now, there was obviously some meaningful difference between an analog 1.6 and 1.8, or there would be no point in generating the values in the first place, but after each has been changed to a digital 2, a 2 they must both remain. In other words, the receiver, which might want to convert the received digital signal back to analog information content, would have no way to distinguish between a 2 that started out as a 1.6 and a 2 that started out as a 1.8. Noise has been introduced into the analog-to-digital conversion process.

This quantization noise may be greater or lesser, but it never goes away. A network could try to convert analog information into 10 digital values instead of only 2, so 1.6 analog is 1.6 digital (digital values do not have to be whole numbers, just discrete numbers). Now the problem shifts to values like 1.555438, which must become either 1.5 or 1.6. The noise is ten times less, but it is still there. The question is how much time and effort should be expended to address the quantization noise issue. Quantization noise is always an issue when converting analog to digital, especially when analog voice is converted to digital signals for transfer over portions of the PSTN.

Two other points about analog and digital should be made. First, when speaking of analog, the term *bandwidth* is used to indicate the frequency range that the analog signal operates in, so it is common to say that "analog television signals occupy a 6 MHz bandwidth." The *MHz* just means "millions of cycles per second," where a cycle is the total range of analog values that the signal can take on. Here the term "bandwidth" means the same thing as *frequency range,* and this usage is common in the analog world.

But what about analog information content and analog signaling? Do not each of these have a characteristic bandwidth? Yes. If the bandwidth of the analog information (perhaps 6 MHz) is the same as or less than the bandwidth of the analog signal (perhaps 7 MHz), then there is no problem. But what if a 7 MHz analog information stream is trying to make its way onto a 6 MHz analog signaling network? Something has to give. And give it does. In this case, it is said that the information content is *passband limited* to the 6 MHz signaling limit. The term *passband* can be taken to mean the amount of analog *band*width that a analog signal will *pass* through the network. In this example, the 7 MHz of information can use the 6 MHz signaling channel by "chopping off" the frequency range between 6 and 7 MHz, chopping off the frequency range between 0 and 1 MHz, or any place in between, as long as the result is a 6 MHz bandwidth.

In the PSTN, analog local loops will passband limit analog signals in many places to the frequency range between 300 and 3300 Hz. That is, nothing below 300 Hz or above 3300 Hz will arrive at the other end of the wire. Outside of the United States, this upper limit is frequently 3400 Hz, but the point is the same. The choice of the 300 to 3300/3400 Hz passband is no accident. Some 80% of the power of the human voice is within these limits, so modest expansions of the passband has only marginal effects on perceived voice quality. The passband limit is enforced by *passband filters* in electronics gear. This analog bandwidth limiting is an important point and plays a large part in ADSL discussions to come.

But when applied to digital, a second meaning for the term "bandwidth" is common. It is commonly said that "this Internet access line has 64 kbps bandwidth." The *kbps* means "thousands of bits per second," where a bit is a binary digit and must be either a 0 or a 1.

As might be expected, there is a characteristic bandwidth associated with digital signals as well, but it is measured in bits per second rather than cycles per second. What if the bandwidth of the digital information content coming out of the back of a PC's serial port is 128 kbps and the bandwidth of the telecommunications line is only 64 kbps? Should 64 kbps just be "passband filtered" out? Well, maybe not. If the 64 kbits just discarded are part of file, the missing bits will cause a error condition at the receiving end and will probably result in the need to retransmit the whole file across the network. Why not just delay the extra bits in a special place called a *buffer*? In this context, a buffer is just a special area in memory used for communication purposes.

Why would a buffer help when PC communications are concerned? Only because of one of the major differences between using the PSTN for human-voice conversations and using the PSTN for computer communications. When people talk on the phone, a constant stream of information usually flows both ways. Even silence, which occurs when one party is listening or the other pauses to collect their thoughts, is represented by the full 3 KHz bandwidth (300 to 3300 Hz) on the network. In other words, the bandwidth assigned to a voice conversation cannot be easily taken away and given to another conversation. In fact, this process defines what is called a *circuit*. A circuit can be defined in this context as bandwidth continuously dedicated to a conversation for the duration of the conversation. The duration of the conversation in turn defines the holding time of the call, as was already alluded to above.

But data, as defined by computer-to-computer file transfers, Internet Web browsing, and the like, is different. There is no longer a constant stream of information from source to destination. Instead, information is organized into units called packets. *Packets* have a minimum and maximum size (which is usually quite small) and must conform to a standard structure and set of rules called a *protocol*. On the Internet, packets conform to the *Transmission Control Protocol/Internet Protocol* (TCP/IP) structure, now officially known as the *Internet protocol suite*.

Data is different because most data applications are *bursty* in nature, which means that when packets are sent from a source to a destination, many packet are sometimes generated in a second, but sometimes no packets at all are generated in a given second. For example, when an Internet user downloads a Web page, many packets are generated to carry text and graphics across the network to the destination PC. However, as the user mulls over the new information now presented on the PC monitor screen, no packets are sent back and forth at all. Ironically, the PC is typically linked to the Internet by way of an ordinary telephone circuit, which dedicates all of the bandwidth to the circuit all of the time. The bandwidth is still there, but it is not being used by the reflective PC user. This whole issue of "packets on circuits" is explored more fully later in the book.

So it is possible, and prudent, to buffer bursts of packets above the bandwidth limit of the communications link until there is a lull in the packet stream. The buffer can then be emptied in a more leisurely fashion without loss of packets and the resulting errors from the lost information in the packets at the receiver. Of course, it may be desirable to increase the bandwidth on the communications link if the buffer itself is constantly full or is in danger of overflowing and losing packets.

Note the tradeoff implicit in buffering versus bandwidth. More bandwidth may cost more, but it may be the only way to efficiently handle packets in the buffers, which is important because packets sitting in buffers are not going anywhere at all. Buffer time adds to the end-to-end delay that packets must endure as they make their way across the network. In this case, more bandwidth will cut down on the packet delay, make the transfer of Web information occur more rapidly, and potentially cut the time needed to access the information needed at a given Web site. This interplay between bandwidth and delay is an important one and is explored more fully later in this Introduction.

The discussion up to now has focused on analog-analog and digital-digital situations. The other possibilities should be discussed a little as well.

Because there are two alternate forms of signaling and two alternate forms of information content, it is easy to construct a matrix to show how these two aspects of communication interact with each other. Please note that whether the information content is analog or digital, or whether the signal is analog or digital, communications can still take place as long as both ends of the conversation employ the same techniques. This general rule has some exceptions, but these complications are beyond the scope of the present discussion.

When interfacing a specific form of information content to a specific form of signaling, a specialized interface device is usually necessary. If no special interface device is needed, one could say that the network is specifically *engineered* for a particular type of signal and information content. That is, the reason no special device is needed is that the network fully expects to perform with this type of information and signal.

So the analog/digital matrix would show the name of the interface device in the intersections of the matrix. If none were necessary, it would be because the network was engineered for that particular type of information content and signal. Obviously, there would only be one instance of this "default" network operation. A network cannot be engineered for two different things at the same time. Such an analog/digital matrix is shown in Table I-1.

The table shows the names of the interface devices needed for all combinations of information content and signaling, whether analog or digital. Looking at the left column, it is immediately noticeable that no special interface device is needed to send analog information content across a network using analog signaling. Actually, this is not strictly true, as was mentioned previously. A *transducer* is normally required to convert analog sound waves to analog electrical waves, but this conversion is so accepted and transparent that it is seldom mentioned in any special way. When applied to the PSTN, the equivalent statement would be "no special interface device is needed to hook up a telephone to the analog local loop." This is also just another way of saying that the PSTN is "engineered for voice." Analog voice information content is the "default" mode of operation for analog local loop signaling in the PSTN.

However, there may also be digital forms of information that need to be sent on the PSTN, wuch as would be needed to link a home PC user to an ISP over the analog local loop, as pointed out previously. In this case, the information content is digital (the PC's 0s and 1s), but the local loop remains analog. The special interface device needed in this case is the *modem*, which stands for *mo*dulator/*dem*odulator, with "modulation" being a

Table I-1

Interfaces devices for analog and digital

		Signaling	
		Analog	*Digital*
Information Content	Analog	N/A	**Codec**
	Digital	**Modem**	**DSU/CSU**

technical term for this operation. The coupled terms mean that modems must be able to both modulate digital content to analog signals and also demodulate analog signals to digital content back again. Modems may be external boxes separate from the PC or built-in on a board which is inside the PC chassis.

Continuing the matrix on the right side, a device known as a "codec" is needed to send analog information content over a network employing digital signaling. The term *codec* comes from *coder/dec*oder, and "coding" is the technical term for expressing otherwise analog information as a stream of bits. A codec device is needed when analog voice conversations are coded for transmission on a digital link using digital signaling. An example would be when analog voice from an ordinary telephone is to be sent through a corporation's or carrier's digital network.

The last device is listed as a DSU/CSU, which stands for *Data Service Unit/Channel Service Unit*, but its purpose is quite simple. Note that no special interface device was needed to go from analog information content to analog signaling in the PSTN network example using a telephone. At first glance, it would seem that the same should be true if the information content is digital and the signaling is digital, but this is not the case. The network can only be engineered for one type of signaling, and when talking about residential access, this is analog signaling, so as it turns out, there are several kinds of digital signaling methods. Digital information content can also be represented in many ways, which means that a special interface device must be used to convert between a specific form of digital information content and a specific form of digital signaling and back again. There are many forms of DSU/CSU device, as many as there are digital content forms and digital signaling formats. This book introduces some of them.

Broadband

Just as no book about ADSL can be understod without a clear definition of analog and digital, so no book about ADSL should start without a definition and discussion of "broadband." The telecommunications field is sometimes criticized—and rightfully so—for being a field so inundated by professional jargon. To be taken seriously, certain terms must be used over and over and acronyms tossed about with reckless abandon. Unfortunately, the key terms in the telecommunications field are used over and over again with slight or radical changes in meaning, depending on context, often to the dismay of those seeking to understand just what all the excitement in telecommunications is about. The term *asynchronous* is one of those overused and under-defined terms, and the term *broadband* is another.

A definition of broadband is needed because ADSL is sometimes referred to as a method of providing "broadband access" to residential users. The "access" is just the way a person at home would link up to the PSTN and is usually through an analog local loop consisting of a pair of copper wires. This is not the only form of PSTN access, of course, but ADSL primary addresses this from of access line.

In this book, the term *broadband* refers to a telecommunication link that has a "latency" less than that of a 2 Mbps (2,000 kbps) telecommunications link used for digital voice communications. The term *latency* is usually defined as a delay in the network, but there is more to latency than that. A few words have been said about the interplay between bandwidth and delay. Now is the time to explore this relationship more completely, with an eye toward defining latency and applying the definition to broadband access networks.

Both delay and bandwidth play a crucial role in networks. Today's networks must support not only traditional voice and data applications, but many applications that fall into the category of "interactive multimedia," so not only the network delay (for interactivity) but also the network bandwidth (for huge video and audio files) must be adequate for these services. Some applications on a network are inherently *delay bound*, such as voice. Assigning more bandwidth to voice does not make it better or more efficient. All that matters is adequate bandwidth and delay. People will not talk if the delay between speaking and listening is 5 seconds, or starts at 1/2 second and rises to 3 seconds.

But other applications on a network are *bandwidth bound,* such as most data applications. Assigning more bandwidth to these applications makes them run better and more efficiently. Assigning more bandwidth to a PC file transfer makes a big difference. What matters here is adequate delay and bandwidth. A file transfer does not care whether the beginning packets of the file transfer make their way across in 1/2 second and then rises to 3 seconds for the packets at the end of the file. Even a delay of 5 seconds might be okay, but sending a huge Web page over a circuit with very small bandwidth can cause delays of many minutes and will definitely make users angry.

Delays vary according to the number of bits sitting in a buffer somewhere, the current congestion load on major network components (switches or routers), or both. Most people have an almost intuitive grasp of the effects. A telecommunications network is mathematically equivalent to a *queuing* system—so are banks. It makes sense that it takes longer to cash a check either when many people are in the bank, when each teller is taking longer than usual because each person has multiple transactions, or both. A banking line is just a buffer, and a teller with a lot of transactions to sort out is congested. The delays in both situations are highly variable. A person may get in and out of the bank in five minutes in the morning, whereas someone else may languish for 30 minutes in the afternoon. Likewise, a packet may arrive 20 milliseconds after it is sent (a millisecond is one thousandth of a second) while the next packet makes its way in a leisurely 40 milliseconds. On the Internet, delay variations on the order of hundreds of milliseconds are not uncommon. On the PSTN, the delays are much lower and are generally stable.

Interestingly, the *International Telecommunication Union* (ITU), which is in charge of all international telecommunications standards, defines a term it calls *transfer time* for networks. *Transfer time* has a little different meaning than what one might expect because the ITU frequently specifies network services in terms of latency, instead of a combination of bandwidth and delay, as one might expect. Yet this latency specification is totally adequate. Because most people use the terms "latency" and "delay" interchangeably, something else must be going on here. And there is.

Figure I-1 shows how network delay and bandwidth can be calculated for a data application on a network. Information transfer takes place as a series of transmission frames (which contain the packets), which are some number of bits long (X bits in the figure). Because this is the data world, packets and frames always contain bits. The transmission path is pictured as just a passive "bit pipe" that in no way alters, stores, or converts the bits. Now, by ITU definition, the time elapsed from the moment

Figure I-2
Bandwidth and delay

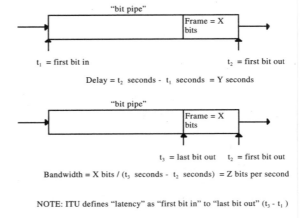

"bit pipe"

Frame = X bits

t_1 = first bit in t_2 = first bit out

Delay = t_2 seconds - t_1 seconds = Y seconds

"bit pipe"

Frame = X bits

t_3 = last bit out t_2 = first bit out

Bandwidth = X bits / (t_3 seconds - t_2 seconds) = Z bits per second

NOTE: ITU defines "latency" as "first bit in" to "last bit out" (t_3 - t_1)

the first bit of a given frame enters the network until the moment the first bit leaves the network is the delay. Note that as a frame makes it way through the network the delay may vary, have a maximum and minimum, an average value, a standard deviation, and so on. The delay could be measured end to end from sender to receiver, between two particular points on the network, or a number of different points along the way.

In contrast, bandwidth is defined in terms of bits. Note that because this concerns data applications, the definition uses the digital version of the bandwidth term given above and not the analog version of frequency range. This bandwidth is in bits per second, so digital bandwidth is defined as the number of bits in the frame divided by the time elapsed from the moment the first bit of a frame leaves the network until the moment the last bit leaves the network. Actually, this is only one possible measurement point. Note that as frames make their was from an access link to other parts of the network, and so forth, the bandwidth the frame has to work with may vary considerably.

The whole point is that the ITU defines "latency" of "transfer time" as the interval from first bit in to last bit out. As should be obvious from the figure, this concept neatly combines the effects of delay and bandwidth on frame transfer. In fact, it allows networks with lower delays to compensate somewhat for restricted bandwidths, and it makes no difference at all whether the bits of digital information content are sent with analog singling (modem) or digital signaling (DSU/CSU). The definitions of bandwidth and delay are still valid for digital information content.

Keep in mind that this example ignores the effects of errors or overhead on frame transfer time. Typically, if a frame and the packet it contains is received in error, it must be retransmitted over the network, which adds to the overall latency of the file transfer, of course. Also, some of the bits in a frame or packet are not used for information content, but for the internal use of the network itself. These are *overhead* bits. Factoring in the effects of errors and overhead would give the overall throughput of the network. The lower the error and overhead penalties, the closer the throughput is to the transfer time.

Again, it should be noted that no distinction is usually made between *delay* and *latency*. The somewhat special definitions used here are peculiar to international standards and used primarily for illustration. Many working telecommunications experts might consider this whole discussion overly "picky," but the idea here is to introduce the concepts precisely before working with them.

With firm definitions of network delay, bandwidth, and latency in hand, we can take a second look at the term *broadband*. Broadband was defined previously as "a telecommunication link that has a latency less than that of a 2 Mbps telecommunications link used for digital voice communications." It should be clear that this latency applies to both the bandwidth and delay of a 2 Mbps link. The PSTN itself, designed and optimized for voice, has minimal delays to work with in the first place, so broadband access usually refers to a bandwidth greater than 2 Mbps, purely and simply.

The current interest in broadband access lines is due to the increased use of the Internet from residential users. If someone dials in to the Internet over a modem attached to an analog local loop, chances are they are getting about 33.6 kbps of bandwidth or slightly more with some types of modems. Obviously, this is a long way from the 2,000-kbps minimum of a broadband access line. ADSL is a way to make a broadband access line out of a local loop without any major expense in re-engineering the wire in the loop or replacing it. In fact, ADSL and related techniques go far above and beyond this 2 Mbps minimum. ADSL itself can operate at 8 Mbps, and some related technologies can push bits at a rate up to 50 Mbps through the access line.

More details on the differences between analog and digital information, circuits and packets, phone calls and Web browsing, and the PSTN and the Internet are discussed in full in the first few chapters of this book. Understanding and appreciating these differences are the keys to understanding and appreciating what ADSL does, and what it is for.

This introductory chapter has introduced and defined terms that are used constantly throughout the remainder of this book. A brief glossary of some of the more important terms discussed is included at the end. This glossary in no way substitutes for the full definitions and discussions offered in the introduction, but it may be helpful for recalling the important points about each.

This might be a good place to summarize the major points examined in this section of the book. Briefly, the PSTN is designed, engineered, optimized, maintained, and operated primarily for analog voice conversations. Even though the PSTN is primarily digital today, the engineering is geared toward short time duration voice calls from human to human. But more of the information content, voice and otherwise, is being transported in the form of digital bits. Perhaps the time has come to make some accommodation to this longer lasting digital information, especially with regard to the limited bandwidth (passband) local loops, and the congested switching and trunking networks on the PSTN that these loops provide access to. This accommodation by using ADSL is the basic theme of this book.

1

Welcome to the Information Superhighway

Not long ago, it was fashionable to use the term *Information Superhighway* to describe a kind of ultimate network for residential and business use. This network would enable people at their desks at work or in their living rooms to have access to all of the information they needed to accomplish business tasks, to do their homework, or even to plan a family vacation.

It now appears that the Information Superhighway actually exists, at least in a rough, preliminary form. This would be the Internet, of course. From relatively humble beginnings as a research and education network confined to government agencies and major universities, the Internet has exploded in the mid-1990s to become an almost indispensable part of ordinary people's lives.

The desire that people have to be able to interact with network content, whether manifested as talking to a TV show or mumbling to e-mail, has never been so strongly demonstrated as with the Internet and World Wide Web. The Web is a gold mine of multimedia, and applications such as "chat rooms" form a kind of primitive video conferencing package that adds an element of interactivity to the mix.

The lure of interactive multimedia and the Internet is amply illustrated in the phenomena of "work at home"—most people are familiar with the idea of working at home by remotely accessing the Internet or Intranet at the office.

The concept of "home at work" reverses the trend: Take the whole family to the office. Why? Faster access for the Web, of course! The Web is a multimedia paradise, and many family educational and recreational needs revolve around the Web. The children can do their homework, the latest movies can be previewed and reviewed, Dad can work on his golf game with an instructional video, and Mom can plan the family vacation (and see what the view from room 422 is), all on the Web.

It is not just residential users that need adequate Web access bandwidth. The speedy lines used by large businesses are often too expensive to cost justify in small offices with just a handful of employees. This whole contingent of *small office / home office* (SOHO) workers and their families is one of the key market segments that ADSL is aimed at.

The Web has invaded almost every aspect of modern life today. Once people realized what the Web could do, they quickly came up with new and innovative applications to take advantage of the new medium. Table 1-1 shows some of the things that the Web is used for today.

It should be noted that this list is neither exhaustive nor limited in any other way. For instance, adult entertainment and related euphemisms are a big money maker on the Internet and Web. However, because many observers try to downplay the role of this segment, the significance of this market is merely noted here. All of the activities listed are current and firmly entrenched. Newer activities that have been proposed for the Web, such as television programming distribution and voice, have not been included in spite of enormous interest.

In many places around the country, mostly in California, office parking lots fill up on weekends. Employees bring the whole family to the office. Most businesses and other organizations have Web access many times faster than the analog dialup modem arrangements used at home. Whereas most analog modems operate at 33.6 kbps (or struggle to reach 56 kbps in some cases), business Internet access tends to start around 64 kbps and ranges upward toward 1.5 Mbps. Even at 1.5 Mbps, a lot of users or a lot of large Web pages and files can slow things down, but this is still

Table 1-1

Ten things to do on the Web

Activity	Advantage	Comment
School homework	No "book is out" problem at library	Web assignments routine at all educational levels
Shop at home	Online catalogs always up-to-date and largely typo and error-free	Retailers have no worries about shopping season weather
Banking	No lines at the bank	Large percentage of transactions will be done on Web soon
Bookstores	Author information, reviews available	Largest dollar volume on Web
Buy a car	Easy to get best price around	Second largest dollar volume
Product information	Cuts down on printing/ distribution costs	Potential customers can access only the information they need
News	Almost instantly updated	No wait for print journalism
Marketing	New products introduced immediately	Less need for costly ad campaigns
Research	No travel to special centers	Much historical material, scientific and classical literature is online
Software distribution	No expensive mailings needed	Convenient and simple

better than the speeds that people achieve at home. This is not yet broadband, but it is close.

Without adequate bandwidth, all of the multimedia information available on the Internet and Web does comes with a steep price—usually, a required high degree of patience. Some files available on the Internet are up to 45 megabytes (for example, a virtual reality tour of Notre Dame, complete with hunchback). It is worth remembering that computer people like files to appear small, and so measure them in bytes (8 bits), whereas network people like networks to appear fast, and so measure bandwidth in bits ($1/8^{th}$ of a byte).

So "home at work" is one way to get the bandwidth people need to access the Internet and Web. After all, it takes a lot of bits to preview a major motion picture, access a multimedia encyclopedia, and so on. Books are the number one dollar item sold over the Internet, but why stop there?

Why not put the whole book on the Web and let you read it there (for a small fee)? Text is the smallest and most compact form of information, so why not just download the whole book to your home PC? Copyright issues aside, there are problems just in terms of bandwidth in distributing even something as small and simple as book text over the Web.

Consider sending a compressed, 300-page book containing a modest 20 figures (sort of like this one). How long would it take to send this book from one site on the Web to another at various speeds? The results are shown in Table 1-2. (This ignores the slowing effects of line overhead and errors, but the comparison is what is important.)

Because the table introduces some of the characteristic speeds at which many technologies operate, including but not limited to ADSL, this is a good time to look at these technologies in general. Several of them are examined in more detail later on in this book.

The table begins with a modem that operates at 2400 bps. That's right—this modem could download a whopping 300 bytes, usually just text characters, in a single second! In the 1960s and 1970s, this was considered a *high-speed* modem. It was the size of a microwave oven and most modems the same size ran at only 110 or 300 bps. This was state of the art at the time, and fortunately not many computers that existed in those days could handle information arriving or leaving much faster than that.

Today, modems operate at 33.6 kbps, although highly publicized efforts have been made to push this speed to 56 kbps with mild success, at least in one direction. When the Web first entered the public consciousness in mid-1994, the most common modem speed was 9600 bps or 9.6 kbps. This was not bad, but downloading the Japanese national anthem (for instance) could take about 10 minutes or so. Of course, the song was a lot shorter than 10 minutes to listen to.

Table 1-2

Downloading a book from the Web

Speed	Time	Amount
2400 bps (old modem)	1.5 hours	0.0625 pages/second
33.6 kbps (new modem)	6 minutes	0.875 pages/second
64 kbps (ISDN)	3.5 minutes	1.67 pages/second
1.5 Mbps (T1, HDSL2)	7.7 seconds	0.13 books/second
6 Mbps (ADSL)	1.925 seconds	0.52 books/second
45 Mbps (T3, OC-1)	0.25 seconds	4.0 books/second
155 Mbps (B-ISDN)	0.07 seconds	14.3 books/second

Recently there has been a lot of talk about delivering higher speeds to residential users and small office users by using an *Integrated Services Digital Network* (ISDN). A full description of ISDN would fill a book by itself, and some aspects of ISDN are investigated later. For now, it is enough to point out that much of what ADSL is intended for, ISDN was intended for first—that is, a kind of early "Information Superhighway" for access to all kinds of digital information content. Unfortunately, ISDN proved very expensive to implement using 1980s hardware and software and was priced way over the heads of most users. But ISDN still used the same existing and congested switches of the PSTN. Oddly, the Web seems to incorporate many of the best features of ISDN plans. In any case, ISDN operates at a basic speed of 64 kbps, although higher speeds are routinely used.

Next in the table is something called T1 and HDSL2. A T1 is a form of digital, leased private line, which means that a company (or even a residential user with deep pockets) can lease a point-to-point circuit at a flat monthly rate from a telephone company (others offer these services today as well) and use it to their heart's content without concern for additional usage costs. This approach differs from the approach of the previous speeds listed in that the others so far have been "switched" network services; that is, a modem or ISDN user can connect to many sites over the access line, just not more than one at a time (the Internet is still just one place to these services). Although a T1 can run at 1.5 Mbps, the line can also be used in the form of 24 channels (each an individual circuit) running at 64 kbps each. Of course, all must start and end at the same physical locations, since no switching is provided—just a point-to-point link. Alternative ways of providing T1 service, such as wireless radio or fiber optic cable, have been around for some time. But the "new" way of delivering, transporting, and providing T1 is known as HDSL2 and is related to ADSL, but not closely. It should also be noted that the ISDN *Primary Rate Interface* (PRI) runs at 1.5 Mbps or 2.0 Mpbs, but no separate entry in the table is provided for this line speed since it is so very close to the T1 speed.

ADSL itself is the next speed listed and is the subject of this book. The speed listed is 6 Mbps, but ADSL runs at a whole range of speeds, and, admittedly, this listed speed is toward the high end. Much will be said about ADSL in a whole series of later chapters.

Following ADSL, the speed jumps up through the T3 and OC-1. Although it is possible to obtain speeds between a T1 and T3, these intermediate speeds are not common and are of no concern here. A T3 is a faster T1 and runs at 45 Mbps. However, like T1, a T3 is often broken down into 64 kbps channels, 672 of them to be exact. A company with a T3 running at 64 kbps may only be able to give any user 64 kbps at any one time, which can cause a problem when speedy Web access is the goal. An *Optical Carrier level 1* (OC-1) is basically a T3 running on a fiber optic cable, although a full description of OC-1 would go far beyond this simple relationship. The point is that both operate at about 45 Mbps.

The last entry runs at 155 Mbps, which is the speed for a newer and better version of ISDN known as *Broadband ISDN* (B-ISDN). One of the shortcomings of ISDN was that by the time ISDN was ready to roll out in large markets, the needs of users and capabilities of PCs had gone far beyond a mere 64 kbps. In 1988, the ITU essentially added a lot of video services to ISDN and called it B-ISDN. Popular technologies such as *Asynchronous Transfer Mode* (ATM) and *Synchronous Optical Network/Synchronous Digital Hierarchy* (SONET/SDH) were part of this 1988 effort, as well. Unfortunately, B-ISDN still suffered from many of the other shortcomings of ISDN in terms of expense and awkwardness, but many of the intended B-ISDN features have found their way onto the Web.

It seems obvious from the table that a network operating at about 1.5 Mbps or above to everyone's house would be just fine for interactive multimedia. As mentioned, ADSL can operate above even this speed. ADSL would enable people to stay home on the weekend and leave the company parking lot empty for a change.

Notice something else odd about the preceding table. At least two of the speeds, ISDN and B-ISDN, were introduced not just as a speedy network technology, although that was a key feature, but they also included a whole suite of services that users at home or at work could access. Yet the point has been made that much of what ISDN and B-ISDN are supposed to do is now available on the Internet and Web. Both ISDN and B-ISDN were the products of the ITU and the most respected networking companies in the United States and around the world. The Internet started as a research and education network funded by the United States military.

What happened? Why are people around the world clamoring for faster Internet access instead of services based on ISDN and B-ISDN? The Internet has been around in one form or another for almost 30 years now. What happened in the mid-1990s to catapult the Internet into mainstream consciousness?

The answer to all of these questions is the World Wide Web. In the middle of 1993, a new type of Internet application appeared on the Internet scene. A year later, this new Web concept was undoubtedly the most popular aspect of the Internet. Soon many people's only contact with the Internet was through the Web. Without unraveling the whole history of the

Internet, a few words about the early Internet are necessary to explain why the Web caused such a revolution when it did.

Probably the best way to understand the rapid growth of the Internet and Web and its transformation from just a giant, specialized computer network to a young Information Superhighway is to trace the development of the Web *browser* and the resulting impact on the Internet itself.

The Internet and the Web

The Internet officially began operations in 1969. While some young people living in the present gathered at Woodstock, other young people living in the future were sweating and straining to get four sites connected by the end of the year. Of course it was not called the Internet then, but rather ARPANET, named for the U.S. military *Advanced Research Project Agency*, which funded it. Throughout the 1970s, the "Internet" continued to grow and expand as it added computers from many universities and research centers.

Strangely enough, although this giant network was a decent way for computers to communicate, there were no tools specifically designed to navigate it. It was a fairly difficult environment for the users to use their various text-driven interfaces to communicate and exchange information on the network. But as awkward as it was, by the early 1980s, the ARPANET had grown so big that to better oversee its development, it was split into two parts. One formed links between the *military networks* (MILNET), and the other linked the academic networks. By the mid-1980s the academic part was overseen by the National Science Foundation and was called the NFSNET.

Today someone looking at the Internet of the late 1980s would hardly recognize it. Even with this continued growth and impressive progress the greatest use of the Internet was for e-mail, but as resources and data availability expanded as each university was added to the NFSNET, the potential for wider use grew every day. The Internet remained an untapped gold mine unless a user was familiar with their own site's system, knew specifically where to go, how to get there, and what to do once there. The problems of the early Internet can be referred to as the "messy room" syndrome. Because many of the Internet users at that time were college students, this was certainly an appropriate image. There may be valuable resources buried somewhere under that pile of clothes, or then again, there might not be. Where to start looking for anything was more a matter of luck than skill, and finding one sock did not guarantee that the other one would turn up.

This is not to say that the Internet in the 1980s shares nothing in common with the Internet or Web today. Quite the contrary. All networks have an enormous inertia that must be overcome before change can take place. The more radical the proposed change, the greater the inertia, so only small changes are possible in short periods of time. As it turned out, the rise of the Web was based on a relatively small change, but one that had enormous implications and applications. Some of the milestones in the history of the Internet are shown in Figure 1-1.

Then, as now, the Internet was based on the TCP/IP protocol. Then, as now, people could dial in for access to the Internet. Then, as now, the Internet was based on what is known as a *client-server* architecture. Because this is such an important feature of the Internet and Web, more should be said about the overall concept of a client-server architecture.

According to the client-server concept, all applications running on a computer are either client or server versions of the application. A single computer can run both client and server versions at the same time, just as a PC can run a spreadsheet and word processing application at the same time, but it is more common to use a single machine to run all client versions of an application or all server versions of an application. Not surprisingly, a PC running only client applications is called a *client*, and a PC running only server applications is called a *server*. The Internet exists basically to link clients to the servers they need to access.

All interactions between computers on the Internet follow this basic client-server model. Consider a fairly common Internet activity—transferring a copy of a file from a remote computer to a home PC. In this case, the home PC is the client and the remote computer is the server. Both computers must be running file transfer software that is compatible with the TCP/IP protocol and with each other. The easiest way to ensure this application-level compatibility is to base both client and server portions on the Internet's *File Transfer Protocol* (ftp) standard. There is an ftp client software package and an ftp server software package that embody these standards. The home PC runs the client ftp process, and the remote server runs the server ftp process. The home PC user accesses the remote

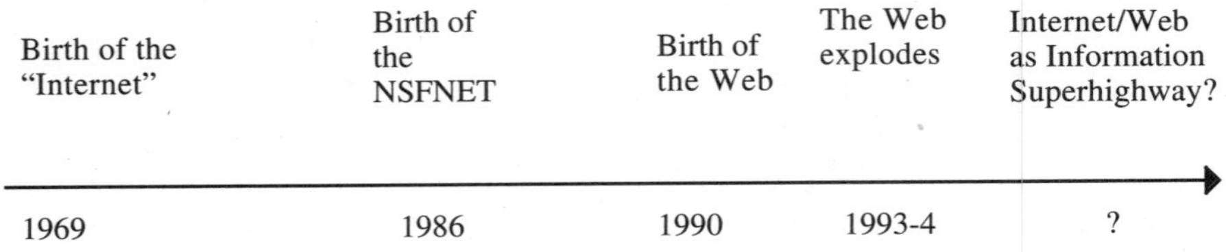

Figure 1-1
A (very) brief history
of the Internet
and Web

Birth of the "Internet"	Birth of the NSFNET	Birth of the Web	The Web explodes	Internet/Web as Information Superhighway?
1969	1986	1990	1993-4	?

ftp server over the Internet and transfers the file (actually a copy of the file; the original remains where it is). This client-server system is easy to implement and quite efficient, as long as the server and the server ftp application are up and running whenever a home PC user accesses it. A lot of effort is needed to keep heavily accessed servers up and running 24 hours a day, 7 days a week.

This very client-server model goes a long way toward explaining why the Internet was as organized as a messy room. The fact that no single company or entity ever controlled the Internet certainly helped as well: it is easier to get one child to clean their room than to try to organize three children and get them to coordinate the task. On the Internet, there eventually turned out to be thousands and then millions of children.

But with regard to client-server, suppose a user needs to send e-mail to a colleague, fetch a file from a remote server, and log in to a remote computer to run some custom application written to run under TCP/IP. Obviously, three different client applications would be needed—these are usually mail, ftp, and telnet, respectively—to enable the user to accomplish these tasks. In the old days, around 1984, each of these client applications would have to be run one at a time. In 1984, this was just as well. Few desktop computers or workstations could run more than one application at a time because the familiar Macintosh and Windows multitasking operating systems and computer systems powerful enough to run them remained in the future.

So the PC user ran the e-mail client to talk to the e-mail server, then ran the ftp client to talk to the ftp server, and finally ran the telnet client to talk to the telnet server. The whole process was awkward and slow, and by the time people got to the telnet phase, they forgot what they were looking for in the first place. Even worse, each of the client processes was *command driven*, meaning that a user had to type in a terse string of letters that had to be looked up or memorized before the server could perform the necessary task. For example, the ftp command `mget *.txt` would fetch all of the text files present in the current directory on the remote server, but `mget` was useless as a command in other applications.

Two ways were tried on the Internet to work around the twin problems of separate clients and text commands. Sometimes these approaches were known as *hierarchical* and *hypertext*. Their differences are considerable, but the important point is that the hierarchical approach was tried by an application called Gopher, and the hypertext approach was tried by an application called a Web browser. Both appeared around the same time, and the Web proposal actually pre-dated the Gopher idea. After some initial success in the early 1990s, the Gopher concept of hierarchical client and server organization pretty much disappeared, while the Web concept of hypertext clients and servers flourished. Cynics are quick to point out that the change in 1993 occurred at about the same time that the inventors of Gopher began to charge for Gopher server software while the Web server software remained free, but there was more to it than that.

The Internet Gopher was developed by the University of Minnesota and the software was released on the Internet in late 1991. Named for the school mascot and state nickname, as well as for every low-level worker who at one time had to "go fer" coffee, Gopher was designed to allow servers to be set up as information providers. Gopher servers could offer files with readable lists (instead of terse file names) which were organized hierarchically. Gopher *menus* contained locations and resources from across the Internet that enabled users to "go fer" it simply by pointing and clicking a PC mouse instead of typing a command.

By 1992 and early 1993, *graphical* (only a mouse is needed to coordinate a pointer on the screen) Gopher clients were available for both the Macintosh and Windows platforms. Because of its ease of use and installation, and availability across multiple platforms, such as UNIX computers and Windows PCs, Gopher quickly caught on and became a popular tool for navigating the Internet. The Gopher was an important early step from Internet to Information Superhighway. Today, Gopher is mainly of historical interest to all but the most dedicated Gopher site and user. Today belongs to the Web.

Early mammals lurked in the shadows of the hulking dinosaurs until changing environmental conditions killed off the dinosaurs and led to the age of mammals. In the shadow of Gopher's popularity was a piece of the Internet using a different method of data organization. In opposition to the hierarchical approach of Gopher, this method was designed to use hypertext for navigation and linking of data. The actual term "hypertext" was invented in 1981 by a computer scientist named Ted Nelson in his book called *Literary Machines*. Hypertext is based upon the idea of following a non-linear path through data and documents through a series of linked "nodes" of information.

The concept of hypertext can trace itself back to 1945 when its methods were proposed by Vannevar Bush who wanted to design a computer information database he called a *memex* (fortunately, the term never caught on). A user of a memex computer would be able to read data on a subject and from within that document be able to link to related information throughout the database. In essence, the user would follow a trail of links throughout the database that was not limited to a specific topic but allowed for connections across boundaries of classification. Just as any good scientist knows, for example, that while biology and chemistry are separate fields of study, they are in reality so interrelated that to truly understand one, a scientist must have an understanding of the other (biochemistry). Bush therefore surmised that knowledge, in any field, needed to be explored, accessed, and found by users taking different pathways in their pursuit of information. This concept became the basis for the Web.

In fairness to the Gopher approach to the two systems of Internet organization, it is important to realize that neither method is right nor wrong in any real sense. Hierarchical is based primarily on organization and classification: a place for everything and everything in its place. To counter the messy room syndrome hierarchical cleans up the room and puts everything in a drawer. If someone needs socks, they're in one drawer, pants in another, shirts in the closet, and so on. It is easy to get dressed if the person knows where to find socks, pants, shirts, and so on to complete the outfit.

On the other hand, hypertext views the *relationship* of data as the guiding factor because this method accepts that information needs the ability to cross boundaries and classifications to be useful. This method leaves the messy room messy, but allows that finding a pair of socks will likely lead to finding a pair of jeans, which leads in turn to a pile of shirts (ones that actually go with the jeans), and even though the person did not need anything else, hypertext can lead to some ties, which make the outfit perfect. Of course both hierarchical and hypertext's strengths are the weaknesses of their counterparts. In a hierarchical system a user needs to climb up and down the hierarchical tree before different but related subjects can be found.

There is one other aspect of hypertext that makes the Web a frustrating experience in some ways. Following links can be a confusing thing. Finding socks leads a person to pants and then to shirts, but in with the shirts is the disco shirt last worn for a Seventies retro party, which leads to a collection of disco records from someone's misguided youth, which leads to the entertainment cabinet, where also happens to be located the manual for programming the VCR. What does the VCR have to do with getting dressed? Maybe nothing, maybe something. But perhaps that is the important point that has lead to the success of hypertext as the navigation of choice for many Internet users.

Through hypertext a user has the flexibility to organize and link information any way they see fit (even hierarchical!) and also link information together in a way so the user can go from one concept to another without limitations. Life does not always fit into nice categories, nor can human beings be expected to think in a completely linear fashion.

Hypertext in a very real sense is the information organization idea that just would not go away, even while experts both praised and condemned its methods and implementation. Although many associate hypertext with the Internet, it would eventually have its first productive application outside the Internet. In 1987 Apple Computers released a software package called HyperCard that came bundled free with the Macintosh and its operating system. It not only allowed the linking of documents, it also abled the user to create links to sounds and images within those documents. While it was limited to restricting its links to the same system, its popularity would go a long way in setting the stage for users to adopt its methodology on the Internet. In a very real sense, the Web is just the old Mac HyperCard system with links allowed to HyperCard applications on other computers.

The first efforts to bring the concept of hypertext to the Internet began in the late 80s when Tim Berners-Lee was employed as a software engineer at CERN, the European Particle Physics Institute, in Geneva, Switzerland. The problem facing Berners-Lee was the familiar "messy room" that the Internet had become. The high-energy physics community was spread throughout various universities and industries in Europe. Information and results of various experiments and the research data it produced was spread over many computers and LANs. It was believed that there was great potential to accelerate research if a way could be found to correlate and share all this information electronically.

The problem of separate clients was discussed previously. Users might have to send e-mail to a colleague, then remotely log onto their computers to examine text, and finally transfer a file with the graphs and images, recording the results of the experiment they needed to look at. Moreover, they were forced to do each step from a different client program with different commands and usually one at a time. Berners-Lee was seeking a kind of "universal client" for the Internet that used hypertext to link files, e-mail, and graphics all in one.

By March, 1989, Berners-Lee proposed a way of linking text documents with other documents using "networked hypertext." Working with colleague Robert Cailliau, they eventually produced a design document and published it in November, 1990, as a proposal for developing a system based on their idea of networked hypertext. "Hypertext is a way to link and access information of various kinds as a web of nodes in which the user can browse at will," the document stated. "Potentially, hypertext provides a single, user-interface to many large classes of stored information, such as reports, notes, databases, computer documents, and online systems help."

The term "World Wide Web" was introduced in this document, although without the capitals. It proved to be a name that would stick, although often just shortened to "the Web" today. After passing a few documents between computers, they found that the documents needed a standard format that could be interpreted by each computer and still convey the information contained within. With this in mind, they developed a language called the *Hypertext Markup Language* (HTML) modeled after, and basically becoming a subset of, a powerful, but much more complex formatting language called *Standard Generalized Markup Language* (SGML).

The resulting Web software would be a mixture of Web *servers* (now just Web *site*) and Web *browsers* (the "universal client" terminology has become the Web browser). The browser is software that runs on the user's client computer and "talks" to the software running on the Web server to request certain files. These files could be any type representing the various resources a user needs to access. If written in HTML, the files can contain information the browser can display in addition to the names of files of related resources and their locations (the hypertext links).

By May, 1991, the World Wide Web was being used successfully at CERN. The first browser developed was a "line-mode" browser that was really just an advanced version of a telnet session, as shown in Figure 1-2. The term *line-mode* refers to the line-by-line output of this first browser. Berners-Lee then officially presented the World Wide Web to the world in December, 1991 at the Hypertext '91 conference in San Antonio, Texas. In January, 1992, the line-mode browser was made available to anyone with an Internet connection.

Figure 1-2
The original Berners-Lee Web browser

Welcome to the World-Wide Web
THE WORLD-WIDE WEB

For more information, select by number:
A list of available W3 client programs [1]
Everything about the W3 project [2]
Places to start exploring [3]
The first International WWW Conference [4]

This telnet service is provided by the WWW team
at CERN [5]
1-5 Up, Quit, or Help

In the true spirit of Internet cooperation, the ideas of the World Wide Web spread and were developed through Internet discussion groups on many newsgroups and separate conferences. In July, 1992, the University of California at Berkeley became the birthplace of the first modern-looking browser. This browser was developed by a post-graduate associate named Pei Wei. The browser, which he named Viola, was available on UNIX systems using X-Windows, which is roughly the UNIX equivalent of Microsoft Windows. It was the first browser to introduce distinctive browser features such as distinguishing hypertext links by colors and underscoring. Viola's other key feature was the capability to allow simple mouse pointing-and-clicking to activate the hypertext links. Unfortunately, Viola remained only an interesting experiment because Pei Wei had no interest in developing Viola further, not even as a commercial product.

So by early 1993, the World Wide Web had a small but solid foundation of interest throughout the Internet community, but Gopher remained king. Gopher was generating most of the Internet community interest. Gopher, with its hierarchical and menu approach, had more Internet traffic, more servers, more interest, but fewer features than the Web. During discussions within the Web community in order to try to figure out how to compete with the popular Gopher, Berners-Lee and his group of developers at CERN detailed the major limitations of the early Web.

The two major points that came out of these discussions were based on two fundamentals of the Web itself: the Web server and the Web browser. In the case of the Web server, the installation and maintenance of the server software was quite difficult even for experienced system administrators. As for the browsers, many users who accessed the Internet were going through a terminal, such as the popular VT100, attached to their company's or university's mainframe computer, and not through a PC.

The VT100 was and is a "line-mode" display device used with many DEC mini-computers. Not an intelligent device like a PC and thus unable to run its own programs, the VT100 family of devices consisted of a keyboard and a simple green-on-black monitor. The monitor was incapable of displaying anything except text, one line of 80 characters at a time. The screen was usually 25 lines high. If a line scrolled off the top of the VT100, it was lost, because the VT100 had no memory to store information from the central computer. Those who remember the DOS operating system will recognize that a DOS monitor behaved exactly like a VT100 terminal.

DEC later extended the product line to include the VT102, the VT-200, and so on. All differed in support for color and number of rows and columns on the screen, but all remained *dumb terminal* devices. Today, many PCs still run software that makes the intelligent and powerful PC behave like a VT100 dumb terminal. These *terminal emulation* packages are popular methods for accessing the Internet, especially in odd situations where Web support is not critical.

The problem with browsers and terminals was that there did not exist a reliable text-mode VT100-based browser that could deal with the many limitations of the full screen features that a VT100 client offered. More importantly, for a PC market that was already growing in leaps and bounds, a simple, full screen, color, point-and-click browser was needed to compete with the already successful Gopher clients.

The first problem with regard to improving the Web server software was immediately addressed by Berners-Lee and his group of developers at CERN, as well as others around the Web, and resulted in much improvement the next year. The second problem, improving the browsers, was once again addressed in true Internet fashion. A VT100 text-mode browser called Lynx (for following hypertext "links") was modified from an existing client software project. The Lynx parent software had been developed a year before, in 1992, by Lou Montulli for a campus information system for the University of Kansas. Lynx would develop into the standard for text-mode Web client software.

The Web Explodes

Everyone who saw Pei Wei's Viola Web browser was instantly dissatisfied with text-based browsers like Lynx. The reaction was similar to those who saw early PCs instantly recalculate spreadsheet rows and columns: "I want that!" Again, Pei Wei had no interest in developing Viola and was not interested in product development.

But others were. They quickly realized that as PC power grew, graphical Web browsers would not need the powerful UNIX operating system and X-Windows to perform adequately. Perhaps even Macintoshes and Windows-based PCs would do just as well—if not now, then quite soon. Of course, they were right.

Events were unfolding that would not only meet the requirement of providing a graphical Web browser, but would also transform the Web from a little-known corner of the Internet into being the dominant presence on the Internet and a significant presence in ordinary people's lives. This browser would also be the major catalyst resulting in truly opening the Internet and *cyberspace* to the general public.

In February, 1993, a new Web browser called Mosaic was announced for release by the *National Center for Supercomputing Applications* (NCSA), part of the University of Illinois at Urbana-Champaign. The NCSA is subsidized by the United States government to develop tools and software that provide the means for researchers to be able to work collectively and share data and resources. Berners-Lee's project at CERN and the resulting development of the World Wide Web utilizing HTML on the Internet had generated interest in the researchers at NCSA as a promising medium for the interchange of information on a national, if not global, scale.

Marc Andreessen, a graduate student at the University of Illinois at Urbana-Champaign, with the help of Eric Bina, a programmer at the NCSA, developed the new Web browser. It was originally designed to run in a UNIX environment (with X-Windows, naturally). Because of its popularity and despite being limited to a UNIX server, the MS Windows and Macintosh versions of Mosaic browser were released in the fall of 1993.

Although the Web community had specified that a *graphical user interface* (GUI) browser was important to the further development of the World Wide Web, the initial success of Mosaic was based primarily on the fact that it was easy to obtain, install, and use. Of course, this would not make any true difference if a software package was terrible, easy to use or not. Many powerful and productive software packages have failed because no one wanted to expend the effort in obtaining, learning, or just plain using them. Not only was Mosaic easy, it was flexible. It was designed as more than just an HTML browser and could handle links to other Internet services such as ftp and Gopher.

More than that, Mosaic could "learn." If Mosaic met a file it did not know how to handle, the browser allowed the user, within Mosaic itself, to set a link to an application (referred to as a *Helper Application* or *plug-in*) that could handle the file at once and then in the future. Mosaic did not stop there. Marc Andreessen wanted to expand the graphical browser and HTML beyond just text and hypertext. Mosaic was designed with the capability to display in-line images and pictures within the same document as the hypertext. Furthermore, images were taken another developmental step to exist as *image maps*. Image maps allow for different sections of the picture to operate just like a hypertext link and access another source. A user can click anywhere on the image and the browser interprets its corresponding link.

With all its new features and flexibility, Mosaic was not just a innovative Internet application, it was the tool, the "universal client," that opened the portal to the Internet for the regular person. Once there, many realized that HTML was not a terrible, scientific, complicated computer language, but rather a relatively simple format that even those who had previously only used a word processor could master in a short time. Today many word processors allow users to build pages within them and will generate the HTML code so that the users never even have to learn any of the HTML code. The users just keep using the word processor as they have been all along. With its initial success, Mosaic would be the first of many Internet browsers to come.

Marc Andreessen left the NCSA to co-found a company called Mosaic Communications, which eventually became Netscape Communications Corporation with the release of their Netscape browser, a supercharged redesign of Mosaic. Andreessen jokingly called Netscape a "combination of Godzilla and Mosaic," and the merged code-name of "Mozilla" has crept into the Netscape world in a variety of forms.

Netscape's battles with Microsoft for the world browser market have been well-documented and need not be repeated here. Microsoft even held off the release of its much anticipated Windows 95 operating system software so that it could incorporate its Internet Explorer Web browser into the package. To keep ahead of the pack, Netscape continued to offer extensions to the HTML standards to enhance the interaction and function of its Web browser. The constant race is to continue to offer a more creative, simpler, interactive medium through the browser.

This brief history of how the Web invaded the consciousness of the world could be extended to include the development of Java, cookies, and other concepts that have encouraged the Web to be used for commerce and business, as well as simple Web surfing for pleasure. But the road to Netscape and Internet Explorer was the path that needed to be blazed. In some sense, the addition of these new Web features has just been the icing on the cake.

The Rise of the Internet Service Providers

Mention has already been made of the watershed years between 1984 and 1986. In that time, the National Science Foundation found itself heir to much of the Internet infrastructure as the U.S. military sought to severe its ties to what came to be officially called the Internet. The NSF now ran the very backbone of the Internet, now called the NSFNET, which linked various regional networks together into one unified whole. There were a lot of reasons for the breakup, including international criticism and the reluctance of international standards committees to consider protocols and applications developed for a "military network" as seriously as other proposals, so for better or worse, the federal government began to funnel funds to the NSF instead of the military to operate the equipment that comprised the Internet and pay the people who kept the network up and running.

There was another watershed year in the 1990s. In 1994, about the same time that the Web was making a big splash on the Internet, the federal support dollars for the NSFNET were cut off as planned. The idea was that the federal government should provide taxpayer dollars to encourage new technologies that might improve the life of citizens (this is outlined in the Constitution), and, in any case, the Internet did not set a precedent in this regard. Samuel F.B. Morse, inventor of the telegraph, was an art professor, not a professional inventor. Only with the help of federal money was the first telegraph line strung between Washington, DC and Baltimore, Maryland. Early use of this line was free, but by the time the money ran out it was obvious that people had embraced the telegraph and money from users would support the whole industry and assure profits for all involved.

The same was true of the Internet. By the 1990s, it was obvious that the Internet formed an important part of university and college life. Withdrawing the subsidy from the NSFNET just meant that funds for continued operations had to be gathered from the users of the network, so the Internet was restructured into a small number of *Network Access Points* (NAPs) run for the most part by major telephone companies. Any part of the Internet could reach any other part of the Internet through these NAPs.

No one need think of the NSF as abandoning the Internet. When the NSF built the NSFNET, they were looking to run a network to support high-speed computing and high-speed data communications (such as they were in those days). Their intent in 1994 was the same. The result was a new backbone network and the NAP structure to make sure both backbones could communicate. In both cases, the NSF goal was always higher speeds for computing and communications.

Of course, any regional network-providing Internet service (which boiled down to access to servers on the Internet) had to pay for the lines running to the NAPs and other charges as well. How did they get the money needed to pay for NAP interconnections? They simply reached out beyond the university and college community and went public. Now calling themselves *Internet Service Providers* (ISPs), the smaller, regional portions of the Internet began advertising and signing up as many people as they could for Internet access.

It helped that there was already a precedent for such online services. In the 1980s, the availability of PCs and the appearance of these PCs in people's homes, as well as on their desktops at work, lead to the appearance of various specialized networking companies.

The appearance of the IBM PC in 1982 caused a revolution not only in business computer use, but also in home computer use. For the first time, ordinary people at home had access to the enormous computing power previously available only to a select few with home computer terminals. And these few users had no real computing power in their homes. The computing power was still in the office mainframe or minicomputer; the terminal only provided access to the office computer over a telephone line. A person with a home PC could, however, run applications and programs formerly restricted to a corporate environment. In fact, when it came to word processing, spreadsheets, and even simple graphics, the humble PC often out-performed many a mainframe and minicomputer, due to advances in technology embodied in the new PCs.

The appearance of the PC had an interesting side effect as well. Everyone could have their own personal computer, but the effect was to isolate people from one another and limit their access to hardware and software to what was installed and functioning on their local PC itself. Someone may have written a stirring piece of prose, but it could only be read by those who had direct access to the PC it was composed on. Of course, PCs had floppy drives (and most had only floppy drives back then) that someone could use to copy the writing to and distribute, if one knew who had the writing and how to get it. This "sneaker net" process was not different from Medieval scribes copying manuscripts by hand and distributing them to select libraries and repositories one at a time. What was needed was the PC equivalent of the printing press.

By the mid-1980s, the PC community had their printing press. It appeared in the form of a variety of online services, some large, some small. The smaller ones consisted of a single PC running special "server" software similar in nature to Internet server software, but not based on the same protocols in most cases. A few telephone lines connected to modems allowed others to simply dial in with their terminal emulation software packages to become "clients." This simple but powerful arrangement launched a cottage industry of *bulletin board systems* (BBSs) that lasted well into the 1990s and that still linger in portions of the networking world. These small electronic bulletin boards usually addressed the needs of a small, contained group (a telecommunications workers union local, for example) and offered meager features, such as e-mail, among users, as well as a common point to exchange files and programs. Many were staffed on a voluntary basis, and there were no charges to users beyond telephone calling time charges, which were usually minimal, except for users outside the local calling area.

The larger systems addressed the needs of the public at large for such services. Generally, the services were much the same, with the addition of some special *user groups* for those with shared interests, such as military history or house plants. These larger online services benefited from special rules that the federal government laid down in the United States in 1984, specifically to encourage the growth of such services. The larger services had many *points of presence* (POPs) throughout the United States so that local users had to constantly dial long-distance numbers and pay high rates and the consequent high bills for access to these systems. Membership was typically based on a flat monthly rate and usage charges, although some offered totally flat rate services, which users embraced wholeheartedly. Naturally, they concentrated on major metropolitan areas, but many were in almost every state in one place or another.

By the time the Web exploded onto the world in 1993 and 1994, the world of the large online service providers was filled by three companies: *CompuServe* (CSi), *America Online* (AOL), and Prodigy. This list arranges them in order of their first public service offerings.

CompuServe began in 1982, virtually at the beginning of the PC industry itself. They quickly gained a reputation as the service of choice for PC enthusiasts. That is, the more a user cared about things such as operating system optimization, writing their own utility programs, or adding their own hard drive to their system, the more the user liked CompuServe. Oddly, for all its reputed sophistication, CompuServe was one of the last to realize that the Internet and Web were important to its users and lagged behind in offering easy and integrated Internet and Web access to its members.

America Online began offering services to the public at large in 1985. Their reputation was that their members did not need a great deal of PC sophistication or have to be a computer science wizard to access their various chat rooms and sample their offerings. AOL's aggressive marketing techniques turned some people off, but certainly raised the consciousness of average PC users who at least tried the online service to see whether they would like it. Many did. AOL was among the first to see the Internet and Web not only as an adjunct service to their bulletin board-based system, but potentially as a replacement service.

Prodigy began as a joint effort between IBM and Sears in 1984 to encourage PC users to communicate online. Public service was launched in the United States in 1988, and Prodigy seemed determined to "out-AOL AOL" in terms of user friendliness and services, such as home shopping and electronic newspapers. However, Prodigy alienated some with insistent advertising from sponsors (virtually absent on other service providers' systems) and suffered from some early negative publicity.

By the early 1990s, PC users in the United States were comfortable with the idea that an online service could put them in touch with other users with similar interests and offer them access to information that would otherwise be difficult to obtain. When the ISPs were faced with a loss of government funds, many of them saw a solution in the simple repositioning. Rather than provide Internet access to colleges and universities, why not offer Internet access to anyone willing to pay the price? This involved installing TCP/IP client software on the user's PC, but software installation by this time was not the headache it once had been.

In the mid-1990s, three separate threads came together to encourage the expansion of Internet and Web access in the United States. The first was the idea that anyone with a PC could benefit from networking with other PC users through an online service. The second thread was that the ISPs were faced with having to make their operations financially self-supporting. The last thread was the Web itself, with its colorful graphics and multimedia in place of the stodgy commands of the recent past. The Internet had the Web. ISPs had Internet access. The end.

Of course, the online service providers, especially the "Big Three" of CSi, AOL, and Prodigy, saw the writing on the wall. When the Web broke onto the world, Prodigy became the first of the Big Three to offer a Web browser as an integral part of their client software package in January of 1995. AOL quickly followed and went Prodigy one better. In 1995, AOL went out and bought the assets of a company that had begun operations in 1990 as *Advanced Network & Services* (ANS). ANS was the principal architect of the NSFNET Internet backbone and was backed by IBM, MCI, and a few other organizations. Naturally, with the new Internet emphasis being to make money rather than spend government grants, this ANS purchase made sense to both AOL and ANS. ANS got money and AOL got Web access for their users. By the time CompuServe realized that their members might want access to the Internet and the Web, many had already decided not to wait and signed up with AOL or another ISP (CompuServe members generally shunned what they perceived as the blatant and crass commercialism of Prodigy).

Today, there are about 4,000 ISPs. Only about 20 or 30 have an extensive national backbone of their own (the uncertainty is due to the lack of a standard definition of "national," "backbone," and "own" when it comes to ISPs). The others basically simply link to other ISPs, who might link to other ISPs, and eventually everything is linked together at four major *Network Access Points* (NAPs). So the Internet is not one network, but more like 4,000 interconnected networks, each with a large or small number of Web servers (Web sites) maintained by their individual members or members' organizations.

Woe to the ISPs

This discussion of the Web, online service providers, and ISPs might seem to have little to do with ADSL, but it has everything to do with ADSL. The Web explosion has led to dissatisfaction with the speed and efficiency at which the Internet operates. To see why, consider the structure of the Internet at the beginning of the Third Millennium, shown in Figure 1-3.

At the bottom of the figure, millions of PCs, minicomputers, and main-frames act as either clients, servers, or both on the Internet. Although all attached computers conform to this client-server architecture, many of them are strictly Web clients (browsers) or Web servers (Web sites) as the Web continues to take over more of the form and function of the Internet at large. Only at this bottom level are the terms "client" and "server" spelled out. At the other levels, members of each ISP's network are only represented by "C" and "S." The figure ignores important details such as LANs and routers for the sake of simplicity, but it is important to realize

Figure 1-3
The Internet Today

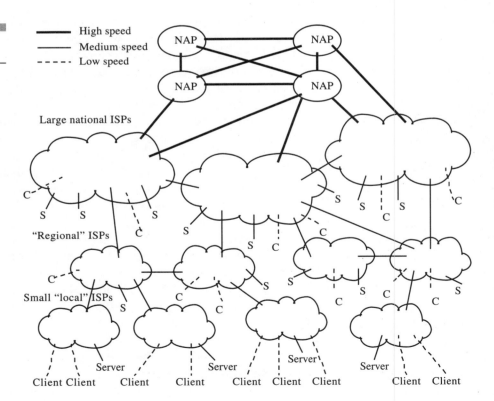

that many of the clients and servers are on LANs, and that routes are the network switching points of the Internet. The figure is a little misleading because the number of clients actually exceeds the number of servers many times over, but this is not apparent in the figure.

Moving up one level, the figure shows the thousands of ISPs that have emerged in the 1990s, especially since the Web explosion of 1993-1994. Usually, the link from the client user to the ISP is by way of a simple modem-attached, dial-up telephone line. In contrast, the link from a server to the ISP would most likely be a leased, private line, but there are important exceptions to this simplistic view. Note that there may also be a variety of Web servers within the ISP's own "cloud" network. For instance, the Web server on which an ISP's members may create and maintain their own Web pages would be located here.

Moving up again, the smaller ISPs link into the large backbone of the national ISPs. Some may link in directly, whereas other are forced for technical or financial reasons to link in "daisy chain" fashion to other ISPs, which link to other ISPs, and so on until an ISP with direct access to a NAP is reached. Note that direct links between ISPs, especially those with older Internet roots, are possible and sometimes common.

The NAPs themselves are fully mesh connected (all link directly to all other NAPs). Figure 1-3 shows the structure of the United States portion of the Internet. However, more than 50 percent of all inter-European traffic passes through the United States NAPs. Most other countries obtain Internet connectivity by linking to a NAP in the United States. Some large ISPs even link routinely to more than one NAP for redundancy purposes. The same is true of individual ISPs, except for the truly small ones, which rarely link to more than one ISP, usually for cost reasons.

Also, speeds vary greatly in different parts of the Internet. For the most part, although there are exceptions, client access was by way of low-speed dial-up telephone lines, typically at 33.6 or 56 kbps. Servers were connected by medium-speed private, leased lines, typcially in the range of 64 bkps to 1.5 Mbps. The high-speed backbone links between national ISPs ran at higher speeds still, sometimes up to 45 Mbps. On a few, and between the NAPs themselves, speeds of 155 Mbps (known as OC-3c) and 622 Mbps (OC-12c) are not unheard of. High speeds are needed both to minimize large Web site page transfers latency times and to the fact that the backbone(s) concentrate and aggregate traffic from millions of clients and servers on one network.

With the information presented in the figure, coupled with the information presented at the beginning of this chapter, the problem facing the "Information Superhighway" concept can be easily understood, which is the problem that initial implementations of ADSL have addressed. The Internet exists to provide connectivity between clients and servers. Increasingly, this translates to home PCs running Web browsers accessing remote Web sites.

But the speeds that home client PCs are dealing with are the simple dial-up modem speeds that are limited by the passband frequency ranges that these telephone circuits through the PSTN allow. Moreover, the sending and receiving of bursty packet application traffic needlessly ties up bandwidth on these circuits. The end result is that the World Wide Web is sometimes called the World Wide Wait, and the ISPs are under a barrage of criticism for not doing more to alleviate the situation.

Note that servers send much more information to clients than clients send to servers. In many cases, client output consists of simple mouse clicks and brief text strings, such as names and addresses or credit card numbers. Server output, sent in response to client requests, may be thousands or even millions of bytes (a byte is 8 bits). Yet most of the technologies used for dial-up and leased lines offer the exact same speeds in both directions. What is needed is a cost-effective, broadband solution for residential and small office access to the Internet, and this is what ADSL is for.

In fact, ADSL solves a number of tricky problems plaguing not only the ISPs, but the local telephone companies themselves. Chapter 2 takes a closer look at the structure of the PSTN, which is the cause of the problem.

2

The
Public Switched
Telephone Network

Any discussion of ADSL technology, especially considering how it is currently used in the United States, must include details not only about how the Internet and Web are structured and function, but also how the *Public Switched Telephone Network* (PSTN) in the United States is structured and functions. Only then can the major design decisions and implementations of ADSL be understood. Some of the details about the structure and functioning of the Internet and Web were explored in Chapter 1. This chapter is intended to supply the same degree of information about the PSTN itself. Although the discussion emphasizes the PSTN in the United States, there are numerous similarities with other telephone networks around the world, perhaps not in detail, but certainly in general characteristics.

It has often been observed that "network people" who work in the telecommunications industry tend to fall into two main categories: those trained in and experienced with "data networks," and those trained in and experienced with "voice networks." The structures of many organizations who must deploy and employ both types of networks reflect this distinction. This dichotomy arose at the very start of computer networking as the very different requirements between voice and data networking became obvious. However, some of the latest advances in voice technology make voice, as well as audio and video, look pretty much like any other stream of data bits in networks. The enormous implications of this simple statement will become clear later in this book.

ADSL is concerned with both traditional voice and newer data applications that essentially share a link to and from a residence, home office, or small business office. Much of the task of explaining the role and functioning of ADSL in this environment consists of explaining the PSTN to the Internet people and explaining the Internet to voice people familiar and comfortable with the PSTN. The previous chapter discussed the Internet and Web; this chapter looks at the PSTN.

A Network is a Network is a...

The whole task of investigating networks is made much easier by the simple fact that all networks look pretty much the same. That is, all networks share certain structural and architectural characteristics that make them appear quite similar to each other. This is not to say that there are no significant differences in their function and operation because there are, but all networks share characteristics that make them networks in the first place.

The discussion of the Internet introduced the familiar network "cloud." No details were offered as to the functioning of network components inside the cloud, which is why it appears as an amorphous cloud in the first place. The idea behind the cloud is that users need not concern themselves with the inner workings of the network. All users need to worry about is whether they can reach other users through the network (or internetwork of linked clouds). The rest of the task is handled by the appropriate hardware and software.

The good news is that all networks are basically simple in overall structure. They have to be or they would not work at all. After all, computers can do networking, and anyone who has had a computer foul up their bills knows how intelligent computers are in and of themselves. The bad news is that each network type has its own jargon and acronyms for network components that turn out to do much the same thing.

One word of caution: the following discussion emphasizes the similarities of many types of networks. However, this does not mean that there are not significant differences between various network types. For instance, the PSTN employs *circuit switching* and the Internet employs *packet switching*. These differences will be dealt with elsewhere.

The PSTN is a network just as much as the Internet is a network, if not more so, given the PSTN's priority in age. The PSTN is almost 100 *years* older than the Internet, and some aspects of the PSTN reflect the older technology that was state-of-the-art in its day, but is now considered aged or aging. These network discussions need not be limited to the PSTN and the Internet—the arguments extend to all networks in general. To see why, consider the network shown in Figure 2-1. This time, some of the details within the cloud are presented.

The figure shows that within the cloud are devices known as *network nodes*. Outside the cloud are other devices usually called *user devices*. This is how one knows whether they belong outside the cloud or not: network nodes go inside, user devices go outside. Usually, but not always, a user device links to one and only one network node. Network nodes, on the other hand, may have multiple links to other network nodes, but again not universally.

User devices link to network nodes by a link known as a *user-network interface* (UNI). There is nothing magical about the properties of the UNI link. It typically can run at a variety of speeds, is supported on a number

Figure 2-1
A Network

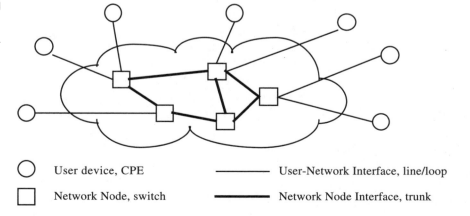

| ○ | User device, CPE | ———— | User-Network Interface, line/loop |
| □ | Network Node, switch | ▬▬▬▬ | Network Node Interface, trunk |

of different media from coaxial cable to fiber, and runs adequately up to some standardized or designed distance. Network nodes link to each other with a *network node interface* (NNI). No magic here either. These links also may vary by speed, media, and distance.

That is pretty much all there is to networking, but one exception must be added to this general network diagram. Older *local area networks* (LANs) did not conform to this generic *wide area network* (WAN) structure. Since the early 1990s, most newer LANs do indeed look like this, although the entire cloud might encompass only a single building or even a single floor.

All users can be reached through the network. Different networks play around with different hardware and software used by users, network nodes, UNIs, and NNIs, but most of the differences would be lost on those not intimately familiar with the various protocols. Remember that user devices do not send to network nodes directly, but rather to other users, and some of the network nodes do not link directly to users, but only to other network nodes. It is common to call network nodes with user interfaces *edge* devices and network nodes with links only to other network nodes *backbone* devices, but this is only by convention. The key is that these internal network nodes exist only to link network nodes to each other, not directly to users.

The network nodes, whatever they may be, typically accept traffic through an *input port*, decide where it goes by some rule or set of rules, and put the traffic out through an *output port*. The easiest way to determine where the traffic goes is to look up the destination in a table maintained and updated in the network node itself. Many variations of this theme exist, but this discussion is concerned with basic operations.

To this point in the discussion, general data network terminology has been presented. Both UNIs and NNIs are usually leased telephone lines, but there are variations. The figure also shows the terms that would be applied if the network figure were of the PSTN specifically. From the PSTN perspective, the user devices would be telephones, fax machines, or computers with modems, which all fall into the category of *Customer Premises Equipment* (CPE). In the United States, the form that the CPE must take is completely up to the customer. Any approved device by any manufacturer may be used, as long as it conforms to some basic electrical guidelines. The CPE in the United States is owned and operated by the user and is beyond the direct control of the network service provider. In other countries, the CPE can be provided and owned by the service provider under a strict set of regulations. The process of giving users more control over the CPE is part of the general movement towards *deregulation*.

This process does not concern ADSL directly, at least not yet, in many parts of the world outside of the United States.

The network node from the PSTN perspective would now be a voice *switch*. A switch is just a type of network node that functions in a particular way in terms of how traffic makes its way from an input port to an output port, the way the table is maintained, and so on. At least that is how most network people see it. There is a continuing controversy between what is meant by the term *switch*, especially between the voice and data networking communities. The controversy extends to the companion term *router*, which is another type of network node found in the Internet. Fortunately, the router-switch controversy is not a concern in a book about ADSL. Suffice it to say that a router is a type of network node found on the Internet that performs networking tasks differently than a switch (e.g. packet switching in routers and circuit switching in voice switches), which is the basic network node type of the PSTN.

Instead of the edge and backbone structure of data network nodes, the PSTN uses terms such as "local exchange," "toll exchange," or "long distance" to distinguish voice switches having links to users or not. All users link to a *local exchange*, usually called a *central office* (CO) in the United States. The local exchanges link to *toll offices*, or *tandems*, or *long distance* switches, with the actual terms varying depending on the detailed structure of that portion of the PSTN.

Moving on to the interfaces, the UNI is now the *access line* or *local loop*. Use of these terms varies, and some prefer one over the other in all cases. Some reserve the term *line* for digitized loops where analog voice is represented as a stream of bits, and others reserve the term *loop* for purely analog user interfaces. This book uses the term *analog local loop* to mean a PSTN user interface carrying only analog signals and analog information. This book uses the term *digital subscriber line* (DSL, as in ADSL) to indicate that some accommodation or optimization (in terms of handling digital information) has been performed to make the loop more efficient for high-speed digital information content. Note that this usage does not distinguish loops or lines based on analog or digital *signaling* at all. Either type of signaling may still be used. In this usage, *loops* carry low-speed digital and analog information content (modem traffic and voice), whereas *lines* carry high-speed digital information content and may also carry analog content (as ADSL does). This definition and usage may offend purists and engineers, but its usefulness will become apparent later on.

In any case, loop or line, the user interface is not normally a leased, point-to-point, private line. Rather, it is a *switched*, dial-up connection that is capable of reaching out and touching almost every other telephone in the world by dialing a simple telephone number. One such destination may be the local *internet service provider* (ISP) if the loop or line is connected to a PC with a modem or other specialized network interface device. Naturally, the ISP needs to be connected to a PSTN local exchange for this to happen at all, which is how PC users use the PSTN to access the Internet.

In the PSTN, the NNI has become the "trunk." In the PSTN, there is little to no physical difference between loops, lines, and trunks. The difference is in how their physical facilities are used. Lines and loops are used to connect users to the network. *Trunks* connect network nodes (voice switches) to each other, but suppose a link runs between the voice switch on the PSTN to the ISP router connected in turn to the Internet? Is this a line or trunk? To the PSTN, the ISP router is CPE as much as a telephone is, so the access line directly to a router is still a line or loop to the PSTN. But the router is certainly a network node to the ISP, and links between one network node (the PSTN switch) and another network node (the ISP router) are by definition trunks to the ISP. Fortunately, telephone companies and ISPs alike have started calling such PSTN-ISP links "trunks." Technically, the PSTN should consider these links to ISP routers to be lines or loops to CPE devices.

Trunks are typically high-speed links for one main reason. Trunks must aggregate a lot of traffic from thousands of users and ship it efficiently around the network to other network nodes. The same is true in general for NNIs for the same reason.

Before closing this section, it might be a good idea to summarize the differences in network terminology, not only between the PSTN and the Internet, but among many new technologies in general. These differences are shown in Table 2-1.

A few words are needed about networks other than the PSTN and Internet. X.25, frame relay, and ATM all fall into the category of packet-switching data networks, just like the Internet. Their network nodes are switches, not routers; oddly enough, however, the router can be a user device on frame relay or ATM networks. It should be noted that X.25 networks can use a special X.75 interface as a network node interface, but this is not mandated or universal. Frame relay also defines a specialized device for the network known as a *Frame Relay Access Device* (FRAD), although it is not really a user device in some senses at all. Note that of the three, only ATM defines a network node interface. (Oddly, frame relay does define an NNI acronym, but as a *network to network interface*, which handles the interface from one frame relay network to another.)

Table 2-1

Network

Network	Network Node called a	User device usually a	User-network interface is a	Network node interface is a
PSTN	Local exchange switch	Telephone	Local loop	Trunk
Internet	Router	PC client or server	Dial-up modem or leased line	Leased line
X.25 packet switching	Packet switch	Computer	X.25 interface	(undefined)
Frame relay	Frame relay switch	Router, FRAD	UNI	(undefined)
ATM	ATM switch	Router, PC	UNI	NNI
LANs since ca. 1990	Hub	Router, PC	Horizontal run	Riser, backbone

Before 1990 or so, LANs looked very different from WANs. Pre-1990 LANs were mostly shared-media, distributed networks, and some still are. Today, most LANs conform to the network node model by way of the *hub*. There are no firm equivalents for lines and trunks, however, and the "horizontal" and "riser" terminology in the table is used for convenience only.

The discussion has so far been a little abstract with regard to the PSTN. The next section explores the general structure of the PSTN in more concrete terms. It puts the PSTN in evolutionary perspective so that decisions made many years ago do not appear to be haphazard; it addresses the needs of the PSTN in the best way possible, given the technology available at the time.

The PSTN: The First Network for the People

The telephone (and PSTN behind it) has become such a part of life that like other inventions, such as televisions and airplanes, it is hard to imagine life without it. Yet such a world existed before 1876. Of course, people still had access to a network that they used to exchange information in a highly cost effective manner. It was called the postal system.

Back then, people wrote letters to communicate. Lots of letters. During the week, mail was delivered in the morning and evening (many newspapers included this "evening mail" concept in their titles because they were distributed this way), and there was another morning delivery on Saturday. People needed to communicate the same things as today, only instead of reaching for the telephone, they reached for the pen.

There was a global telecommunications network, too—the telegraph, invented and perfected by Samuel F. B. Morse in 1838. By 1876, the telegraph was in every major city and was routinely used by news agencies, government departments, and ordinary people to exchange messages deemed more urgent than those entrusted to the mail. If someone's relative was suddenly taken ill, this news could be distributed to the rest of the family by going to the nearest telegraph office, usually at the local train station (telegraph lines closely followed railroad rights-of-way from the very beginning). For a few cents a word, the telegraph operator would click the message in Morse code to another operator (humans were the "network nodes" in many cases in the telegraph network, although automated message-relaying equipment did exist in major metropolitan areas). The message was relayed to the telegraph office closest to the destination and printed out. From there, a messenger, typically on bicycle, would rush to the home of the recipient, who didn't have to wait for the mail. Letters took a day or so. Telegrams took hours.

Many of the techniques taken for granted on telecommunications networks today were first pioneered on the telegraph system. These included compression ("hw r u?"), early data terminals (which printed Morse coded messages on paper tape), and even an early form of telemarketing (Sears began this in 1886 when a railroad conductor named Richard Sears quickly disposed of a watch shipment by advertising to other railroad people over the telegraph).

The drawbacks of the telegraph system were threefold, although the benefits of the system made these drawbacks relatively minor in most cases. First, the system was closed to end users themselves. No one had a telegraph wire running directly to their home and tapped out Morse code directly. Second, although the telegraph system could distribute messages much faster than the mail, there was still a need to send really urgent messages much faster, within minutes. Finally, a telegraph wire could only handle one message at a time. There was no way to stop a message in the middle and replace it with a more urgent one. Urgent messages got in line—the telegraph form of a "buffer"—behind routine ones.

This last liability was the most damaging. In metropolitan hubs, the telegraph office backed up with messages on busy days and during busy hours of any day. Putting in more telegraph lines was an expensive solution because these lines would sit idle most of the time. Clearly, whoever could invent a way to "multiplex" several telegraph messages over a single line stood to make a lot of money.

Alexander Graham Bell was one of these dreamers who wanted some extra cash. His intended bride was quite above him in Boston society, and he thought that if he struck it rich with a multiplexed telegraph technique, he could marry. What happened instead is that Bell accidentally invented a way to expand the useful bandwidth of ordinary telegraph wire enough to actually carry an adequate amount of the voice spectrum (range of frequencies) to produce intelligible speech some distance away. Actually, it was more of a way to convert the analog voice into electrical signals (and back again) suitable for the telegraph line that Bell discovered. Only Bell's work with the deaf enabled him to realize what it was that he had found. Bell first called it the "harmonic telegraph." Fortunately, Bell eventually called it the *telephone*.

Bell used his research into the third telegraph problem (lack of multiplexing) to solve the second telegraph problem (delay). Voice communication was almost instantaneous. There was no delay for coding, relaying, or bicycling after the intended party was reached. It came as a surprise to Bell that the telephone solved the telegraph's first problem (no telegraph to the home), but this last development needs some explanation because it has implications for ADSL.

Bell saw the telephone primarily as a tool for business and similar uses. For example, factories could call suppliers instantly, and a train crash resulted in doctors being called to the scene faster than ever before. The telephone was embraced for these purposes.

Bell never could understand why ordinary people would want a telephone at home. To the day he died, Bell refused to keep a telephone in his office because he found the ringing quite distracting—people wrote letters. What Bell missed was that people can pick up inflections of voice when listening to speech and that the act of talking to someone, no matter how far apart they were physically, made the interaction that much more intimate. What Bell discovered is that there are social consequences to technology.

The early telephone "systems" were elaborate intercoms. They were point-to-point affairs where every telephone needed a direct wire to every other reachable telephone. This system worked well enough for the factory-supplier or train-crash-doctor applications of the telephone, but it would hardly do if everyone wanted a telephone at home to call whomever they pleased whenever they pleased.

To better accommodate the growing residential population using the phone, the local exchange using a *switchboard* was invented. An operator sat before a huge array of lights. When a person wanted to call someone in the community, they took the telephone off the hook and turned a crank, which generated electricity on the wire and lit a light above their circuit position on the switchboard. The operator plugged a cord into the position and asked who the person wanted to speak to. If that line had no other cord plugged into it (a "busy" line), then the connection was made by hand and the operator "cut through" the call and no longer listened in. A second light above the position showed when the conversation was over and the cords could be unplugged from both ends.

Early switchboards had no numbers, but numbers were added because new operators could not become familiar with everyone's location on the switchboard fast enough to be trained quickly. Of course, users could not be expected to remember everyone's number, so a directory of subscribers and their numbers was published and distributed.

The first such local exchange, or central office, was established in Hartford, Connecticut in 1878. The United States already had 1,000 telephones. With the exchange "switch," anyone in the local area needed only one telephone line to reach anyone else. The advantage of the central office is shown in Figure 2-2. Without the local exchange, six telephones would need 15 lines to interconnect them. The formula giving the number of point-to-point links for any N number of telephones linked in this fashion is $N(N-1)/2$, so $6(5)/2 = 15$. Once the central office was in place, the number was only N lines for N telephones. This method came just in time. By 1880, there were 50,000 telephones in the United States.

Initially, male college students were hired as operators, but this did not work out. The young men had a "relaxed" attitude about working hours, required spittoons at all work locations, and sometimes used harsh language with customers. When replaced by young women from good homes,

Figure 2-2
The local exchange
as a network node

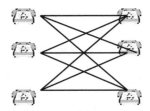

Without CO: 15 lines needed

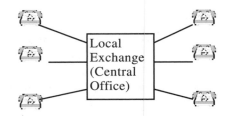

With CO: only 6 lines

all three problems went away. Mark Twain was one of the first customers and so also one of the first people to have a telephone in his home. Supposedly, Twain was one of the first people to engage in another telephone tradition: the complaint about service quality.

It is certainly understandable that anyone in those days would complain about the quality of the voice delivered through the telephone. It was pretty horrible. The standard greeting was a shouted "Ahoy!" that competed with the crosstalk of other conversations and the clicks and squeaks of adjacent telegraph signals. Users claimed that they could hear the rumblings of thunderstorms thirty miles away.

The main problem was with the local access line or local loop itself. Initially, these were single strands of iron wire, uninsulated and exposed to the elements. These loops were based on the same technology and physical plant as the telegraph system, and early telephone systems just rented spare telegraph lines from the telegraph companies. Unfortunately, iron rusted and single strands of wire acted as long antennas, picking up signals from near and far. Due to the nature of telegraph signals—simple "digital" electrical pulses representing dots or dashes—neither of the problems with rusty wire or antennas affected Morse code to any great degree. In fact, it was the other way around. The pulsing "digital" Morse code signals interfered with the analog voice signals on nearby wires when they were run together on utility poles or in trenches. Incidentally, it will be shown that this same effect has implications for ADSL implementations.

In any case, some basic changes improved voice service dramatically. In 1881, single-strand telephone wires were running the 45 miles between Providence, Rhode Island and Boston, Massachusetts. These long wires were constant sources of complaint, until one day, a technician named John Carty accidentally hooked up two wires at the same time. Suddenly the voice quality improved dramatically. Carty had invented the two-wire pair for voice *circuits*. The two wires formed a closed circuit on which the electricity could flow. Another breakthrough came in 1884 when a new method of "hard drawing" copper wire made copper affordable for telephone wiring. Everyone knew that copper had better electrical properties than iron, but up until 1884, copper had been too expensive. With copper, longer runs for telephone lines were possible, all the way to New York from Boston (292 miles), for instance.

Crosstalk from adjacent telegraph signals and even conversations in cable bundles was still a problem, however. Some early efforts were made to *shield* the wires from these effects, but this was expensive. However, even crosstalk became manageable after pairs of copper wire were twisted together, usually three or four times per foot. The new *unshielded twisted-pair copper wire* (UTP) made cables consisting of many wire pairs possible and lowered network expansion costs. Even when jammed together inside these cable bundles, the analog signals would not interfere and cause crosstalk between each other. The cables offered more environmental protection as well.

By 1915, it was possible to place a telephone call from New York to San Francisco. The call was switched by hand from switchboard to switchboard and it took 23 minutes to complete the process. The cost was $20.70. As expensive as this sounds even today, it was much more so back then when lunch cost about a nickel.

This section has positioned the local exchange or central office as the network node of the PSTN; however, each line connected to the central office has so far ended up at a telephone. Yet mention was just made about New York to San Francisco conversation being hand-switched from switchboard to switchboard across the country. How could this happen if the local exchange is only connected to local telephones? The answer to this problem is to invent *long distance*.

The trouble with the local exchange is that they connected only local users through the switchboard. The term *local* turned out to be a relative one, and in the densely populated Northeast, most users were only a few miles from the central office. In the Midwest and Far West, things were different. It could be many miles just into town, where the local exchange was usually located. Providing service was expensive and voice quality was relatively poor. For this reason, the early *Bell system* generally shunned rural areas, and it was in these areas that *independent telephone companies* first sprang up around the turn of the century. As the Bell network grew, it became common to refer to local telephone companies as *Bell Operating Companies* (BOCs). The BOCs together formed the Bell system, although this was not an official designation in any way.

So just how did a local user in one community call another hundreds of miles away? By interconnecting the local exchanges themselves. Calls within the *local calling area* could be switched by any operator. Calls outside this area had to rely on the services of the *long distance operator*. Not all operators had access to the links leading to other central offices; they did not have long distance switchboards. That was fine because few calls actually went outside the local area. But whenever someone did want to make a long distance call, the call was made to the long distance operator. The links that connected to the long distance switchboard did not actually terminate at customer telephones. In this case, there was another long distance switchboard at the other end. A call was switched from "board to board" until the local exchange servicing the called party was reached—only then did the telephone ring at the other end.

As was mentioned previously, the process was long and tedious. In most cases, the long distance operator obtained the city and local number information from the caller, and then said, "We will call you back when the call is completed." Sometimes, the call back indicated that the intended party was on the telephone themselves and could not be reached.

Of course, every local exchange in the country did not have a direct link to every other local exchange in the country. This was as impractical with central offices as with individual telephones, and the local telephone companies faced different challenges when installing and maintaining telephone lines that were not only a few miles long, but tens of miles long. And that was all these links were: regular unshielded twisted-pair copper wire, the same as that used for local loops. Special amplifiers were invented for long links to periodically boost the power of the voice so that there was something to listen to by the time the signal reached the recipient.

This long distance interconnection problem had a solution as well. The early Bell System, the largest provider of local exchange service in the country in the 1890s, founded The Long Distance Company to link local exchanges together all over the United States. While displaying a singular lack of imagination, the name was nothing else if not descriptive. For all intents and purposes, The Long Distance Company was the property of the Bell System. That is, The Long Distance Company existed to link Bell System local exchanges together and saw no reason the link any of the newly formed independents to their system. This practice generated a lot of controversy because the new and independent telephone companies could not readily afford to copy the nationwide coverage of The Long Distance Company. There were under-funded attempts made periodically, but The Long Distance Company remained the only game in town. Ultimately, The Long Distance Company did begin to connect non-Bell local exchanges, but usually only in areas where Bell had no interest in expanding their own local services.

When the long distance network was added to the network of local exchanges, it gave the telephone system a distinctive, two-tiered structure, as shown in Figure 2-3. Note that the local exchanges could be Bell-owned telephone companies or independents, which allowed nearly all of the telephones in the United States to be linked, as long as a caller knew the city and local number they were calling. Keep in mind that in these early days, all of the telephone switching was still done by hand, board-to-board, and there was nothing special about the trunks used to connect any of these switches, local or not. Back then, the *trunk* designation referred to usage, not special construction or engineering. Some long trunks needed amplifiers, but so did some spectacularly long local loops. In the rural West, these local loops sometimes used barbed wire fences to carry analog telephone signals.

The figure also shows that "local trunks" could be used to connect neighboring local exchanges. This made sense only because there were many more telephone calls made within a community than across the country.

A special problem soon arose on these trunks. Sometimes, an operator on one end would attempt to "cease" a voice circuit while an operator on the other end was also trying to do the same thing. There were other differences as well, but the whole point is that different "signaling" methods were needed on trunks as opposed to local loops. In the tradition of telecommunications terms, this use of the term "signaling" is distinct from the same term when applied to the form that information takes when traveling on a telephone line (analog signaling versus digital signaling). In this case, the term applies to a form of control information and supervisory processes that prevent the special trunk problems from occurring.

Figure 2-3
The local exchanges and the long distance backbone

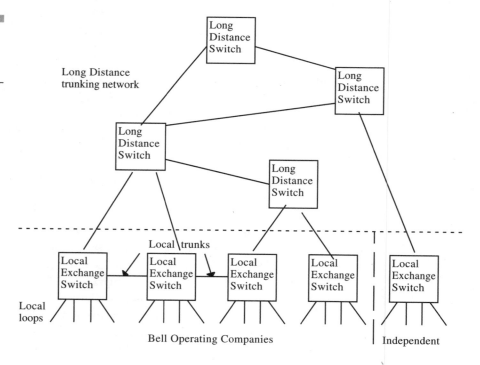

In this context, signaling means *call control procedures,* which govern the setting up, maintenance, and taking down of voice conversations on circuits. Saying that signaling on a trunk is different than signaling on a local loop is the same as saying that call control procedures on a trunk are different than call control procedures on a local loop.

The technology used in this basic local-long distance telephone network evolved rapidly, especially in the switching and signaling (both kinds) environment. A detailed discussion of the technology is the major topic of the next chapter. This chapter will continue to deal mainly with the social and political developments that make the PSTN what it is today. As concerned as ADSL is with technology, these social and political factors cannot be overlooked in any discussion of ADSL.

Long distance calling greatly enhanced the value of the telephone. Instead of waiting days for a letter to arrive, people could call each other and speak immediately. In spite of this value, telephone service in many cases was perceived to be expensive and was not available in isolated areas. Even as late as 1941, only 40 percent of Americans had telephones, but those who did generated an enormous amount of revenue for the telephone companies.

This had been true practically from the beginning, especially with regard to long distance. The Bell System basically tried to keep local service costs low to encourage the popularity of the telephone, while at the same time pricing long distance comparatively high. The result was that by the late 1890s, the parent company of the local Bell operating companies was struggling, and only the subsidiary known as The Long Distance Company was on solid financial footing. The solution was to have the subsidiary buy the parent, so the Bell System entered the 1900s as *American Telephone and Telegraph* (AT&T), with The Long Distance Company becoming a business unit called AT&T Long Lines. Some 20 or so BOCs handled local services, organized mostly along state lines (New York Telephone, Illinois Bell, and so on).

The End of the Bell System

The Bell System endured with a form of state and federal government supervision know as *regulation* from 1913 until 1984. Each state regulated telephone service quality and rate structure for calls that were initiated and terminated within the boundaries of their individual states. For interstate telephone calls—calls going between two different states—regulation was handled by the federal government. Before 1934, this was done by the Interstate Commerce Commission, but after the Telecommunications Act of 1934 was passed, control and regulation passed into the hands of the *Federal Communication Commission* (FCC).

In 1984, as a result of a decades-long battle between the FCC and AT&T, and with the new competitive long distance companies such as MCI joining in, a federal judge and the United States Department of Justice split the Bell System up into effectively AT&T Long Lines and seven newly organized *Regional Bell Operating Companies* (RBOCs). The local independents more or less continued as they were, but there were major changes in how the RBOCs and independents handled long distance calls.

The RBOCs could still carry local calls end-to-end on their own facilities. For all other "long distance" calls, the RBOCs had to "hand off" the call to a long distance carrier, which could not be an RBOC. Furthermore, the RBOCs and independents had to let their subscribers not only use AT&T for long distance service, but any of the competitive long distance carriers such as Sprint and MCI. In fact, any long distance carrier that was approved by the FCC could offer long distance services in any local service area if they had a switching office close enough.

The problem was that there was no firm definition what a "local call" was or even what "close enough" was, so the court and Department of Justice provided one. The entire United States was divided into about 240 areas with about the same number of calls within each. Calls inside these areas, known as *Local Access and Transport Areas* (LATAs), could be carried on facilities wholly owned by the RBOCs. All calls that crossed a LATA boundary had to be handed off to long distance companies, which were now called the *Interexchange Carriers* (IXCs, or sometimes IECs). The local companies, RBOCs and independents alike, were collectively called the *Local Exchange Carriers* (LECs). This whole structure neatly corresponded to the low-tier, local-long distance structure already in place.

In order to carry long distance traffic from an LEC, the IXC had to maintain a switching office within the LATA. This switching office was called the IXC *Point of Presence* (POP). The POPs formed the interface between the LECs at each end of the long distance call and the IXC switching and trunking network in between. For the most part, LATAs were contained within a single state, but there were exceptions. Any subscriber served by a LEC had to be able to route calls through the IXC of their choice, as long as the IXC maintained a POP within the originating LATA through a rule called *equal access*. If the chosen IXC did not have a POP in the destination LATA, the IXC could decline to carry the call (rarely) or hand off the call to another IXC with a POP in the destination LATA. Naturally, the second IXC charged the first for this privilege. It soon became apparent that there were just too many LATAs anyway, and as late as 1993, only AT&T had a POP in every LATA in the United States. But the system was in place, and cynics noted that the LATA structure closely mirrored AT&T Long Lines switching office distribution. With the breakup of the Bell System in 1984, it became common to speak of the entire system of telephones and switches in the United States as the PSTN.

The efficiency of the PSTN depended in large part on connecting calls as quickly as possible. There were two reasons for this concern. First, the faster calls could be put through, the more calls could be handled per hour or day, the more satisfied customers would be, and potentially more revenues could be made if the service were *metered* by number of calls and connection time. Second, the state regulators and the FCC, charged with approving telephone service rate increases, were themselves concerned with the quality of service the telephone companies provided to their customers.

Naturally, the fewer switching offices and switching steps needed from source to destination, the faster calls could go through. Also, it could hardly be expected that each local exchange could maintain a full set of trunks to each and every other local exchange office in a given area. Of course, this would just reproduce the point-to-point telephone loop problem discussed earlier on a grand scale, at the switch level. And just as the point-to-point telephone problem was solved with a switch, so too was the trunk-to-trunk switching problem solved with a switch.

In many cases the practice developed to run trunks not directly to other local exchanges (although this practice also continued based on calling patterns), but to a more centrally located local exchange. Often, this local exchange received a second switch, but one that switched only from trunk-to-trunk and not from loop-to-loop or loop-to-trunk. These trunk-switching offices were called *tandems*, and the practice of switching trunks without any loops was said to take place at a *toll office*. Keep in mind that these definitions were never spelled out in any great detail. This was just the way telephone people talked.

Usually, a call routed through a toll office was a "toll call." A *toll call* is exactly analogous to a *toll road*, which is simply a road that one must pay a fee to drive on, above and beyond the road-use taxes assessed against drivers. In the same fashion, a toll call is just a telephone call, but it costs more to make it, above and beyond whatever the subscriber pays for local service. The amount of the toll depended on the distance and duration of the call. Remember that these calls were distinct from long distance calls, which crossed a LATA boundary. A toll call stayed in the same LATA, but just cost more (there were a few odd LATA arrangements, but these need not be of concern in this general discussion).

Also, the tandem/toll office arrangement offered a convenient way for IXCs to attach POPs to the LECs' networks. Instead of running trunks from a POP to each and every local exchange, an IXC could just link to the area's tandem or toll office. Because the tandem or toll office existed to tie all of the local loops in the area together, this guaranteed that all subscribers would be able to make *inter-LATA calls* through that IXC's POP, at least on the originating end. From the IXC perspective, this preferred point of trunk connectivity was called the *serving wire center* because the POP was served from this switching office. Again, this term was used from the IXC perspective. To the LECs, a wire center was just a big cabling rack (which they called a *distribution frame* where trunks and loops connected to the switching office) in the local exchange. In other words, a wire center is nothing special to the LEC, but is quite important to the IXC. Many IXCs maintain trunks to several wire centers in a LATA, all in the name of efficiency.

The Architecture of the PSTN

Today, the PSTN has a structure similar to the one shown in Figure 2-4. The local exchanges and toll offices inside the LATA make up the first tier of the PSTN, the LEC portion. Since the Telecommunications Act of 1996, service providers may be any entity approved or "certified" by the individual states to become a LEC. Newer companies are *Competitive LECs* (CLECs) and the former service provider in a given area becomes the *Incumbent LEC* (ILEC). Terms such as *Other LEC* (OLEC) are sometimes used as well. ILECs, CLECs, OLECs, or some other exotic alphabetic combinations may still be RBOCs, independents, ISPs, or even power companies in various parts of the United States. There are some 1,300 LECs operating in the United States today, but many of these LECs are quite small and have only a few thousand subscribers in an isolated area.

The second tier of the PSTN is comprised of the IXC's networks. The IXC POP in the LATA could handle long distance calls for all subscribers in the LATA. The IXC had to have its own backbone network of switches and links, as well. The acknowledged leaders in this arena are AT&T,

MCI, and Sprint. Sprint remained an oddity for a while because Sprint was also a LEC in some parts of the country, a rare mix of local and long distance services. Some 700 IXCs are still in operation in the United States today, but most of them have POPs in only a handful of LATAs. Many still handle calls to almost anywhere that originate from people within a LATA where the particular POP appears, but frequently only by handing the call off to another IXC.

A few points about the figure should be emphasized. All lines shown on the figure are trunks, not local loops, because they connect switches rather than user devices to switches. Although shown as a single line, each trunk may carry more than one voice conversation or *channel*. In fact, many can carry hundreds or even thousands of simultaneous voice conversations. Also, IXC B, since it does not have a POP in the leftmost LATA, cannot carry long distance calls from anyone in that LATA. Of course, the same is true for IXC A in the rightmost LATA. In the center LATA, customers will be able to choose either IXC A or IXC B for long distance calls, either under equal access arrangements or *pre-subscription*, which automatically sends all calls outside the LATA to a particular IXC. Recently, the practice of deceptive IXC pre-subscription switching known as *slamming* has been universally condemned by regulators and all public-spirited IXCs. As a minor point, note that a POP need not be linked to a toll office switch. The actual trunking of the POP depends on calling traffic patterns, expense, and other factors.

At the risk of causing confusion, it should be noted that LEC B and LEC C, for instance, could both exist within the same LATA, depending on state ruling and certification. In this case, one would be the ILEC and

Figure 2-4
The PSTN today

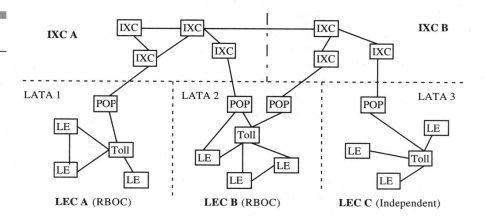

the other the CLEC (or OLEC), and both would compete for the same customer pool within the LATA for local service. Finally, there is no significance at all to the number and location of LEs, POPs, and so on, nor the links between them. The figure is for illustrative purposes only.

The PSTN, at both the LEC level and the IXC level, is what is known as a *circuit-switched* or *circuit-switching* network. Much more will be said about circuit-switching in the next chapter in order to compare and contrast this practice with *packet-switching*, but any discussion of the current PSTN architecture would be incomplete without introducing one of its most distinctive features.

All trunks and local loops are divided into *voice circuits* or *voice channels*, both usually abbreviated VC. Technically, a voice channel in each direction is needed to carry one voice circuit, but this distinction is rarely made consistently. Sometimes, the voice channel outbound and inbound shared the same frequency range; this was usually employed on the local loop and was called *full-duplex* operation in the United States. At other times, the voice channel outbound and inbound used different frequency ranges. This method was usually used on the trunks and was called *half-duplex* operation. Because many conversations consisted of local loops on each end and trunks in the middle, a means of converting from full-duplex local loops to half-duplex trunks was needed at each end of the circuit. This interface arrangement is called a *hybrid* and introduces some nagging problems with echoes on the voice circuits. Echo on voice circuits is covered more fully in subsequent chapters.

A voice circuit is just a two-way connection between two compatible end-user devices (telephones, fax machines, or modems), so that communications can take place over the PSTN. The PSTN switches, just like any other network node, exist to map the incoming voice channels to the outgoing voice channels for the *holding time*, or duration, of the call. That is, the voice switches on the PSTN switch circuits on behalf of customers, who then interact as long as they care to.

While a voice circuit is in use, the network resources, in terms of bandwidth and switching capacity, cannot be assigned by the network to anyone else. Circuit-switching is sometimes called an "all the bandwidth, all the time" approach. The network resources tied up on a voice circuit in a circuit-switching network cannot be used for other purposes. For example, even if one of the parties of a telephone call put the other on hold for three hours, and there were no voices going back and forth at all, the bandwidth that the call represented could not be used by the network for any other purpose (such as carrying Web pages).

This approach made sense in the early telephone system. People called, and if they were not talking, they were listening. When they finished talking, they hung up. The length of their interaction defined the holding time of the call. It did not even make sense to attempt to develop technology to try to use the "quiet time" while people were listening on one half of the voice circuit for other uses in those days. The periods of silence were usually quite short, and no one really had any idea how to attempt such an undertaking anyway, so circuits were assigned a voice channel in both directions, and that was that.

This concept of voice circuit holding time brings up an interesting point of paying for telephone service, both local and long distance—how to charge a telephone call to the customer's bill.

The early telephone system experimented with a variety of rate structures. By 1923, some 206 different rates for local telephone service were in place in the Bell System alone. Some locations paid three times what others paid. The actual rates were based on a combination of the community's overall wealth, historical reasons, and simple greed. Obviously, the longer the distance involved in a voice call, the more trunks and switches were tied up and dedicated to the customer for the holding time of the call, so it seemed only reasonable that the long distance calls should cost more than local calls. And naturally, the longer the call, the higher the amount charged.

But what about local service? After all, the local loop had to be there whether it was in use or not. The same applied to the switch. The same is not true of trunks, which are not dedicated to any particular customer when not in use and can carry anyone's voice circuits. In this way of thinking, after some initial installation cost for local loops and the local switch, the telephone company basically had fixed monthly costs for maintenance, salaries, and so forth, so perhaps the customers should have a *flat rate,* or fixed monthly cost, for local service. With flat rate service, the amount of a monthly telephone bill was fixed, regardless of the number of telephone calls or their duration.

The alternative to flat rate service was *metered,* or usage-based, service. A monthly telephone bill using metered service varied based on the number of calls and their duration. More technical aspects of flat rate versus metered service are detailed in the next chapter. It need only be pointed out now that customers overwhelmingly preferred flat rate service to metered service whenever such a choice was offered, and with good reason.

Early telephone service was very expensive and pretty much limited to the businesses and wealthy families in a given location. As late as the start of the 1940s, only 40 percent of Americans had telephones in their homes. As the telephone companies tried to expand services and encourage everyone to have a telephone, metered service was a problem for families of limited means or on tight budgets. This is hard to explain simply with telephone service, so perhaps an analogy would help.

Suppose the monthly payment for the family automobile was not fixed, but varied based on how many miles it was driven in the previous month. Some months, when the car was driven particularly long distances, the monthly payment might be quite high, forcing families with tight finances to scramble around to scrape up the cash needed to keep the car. The question "How much should be set aside for next month's car payment?" has no easy answer and could be far off the mark. The result is added stress and concern. People end up thinking long and hard about buying a car versus using fixed rate public transportation.

Now suppose that the automobile company says, "We know that monthly auto use varies widely, but on average, each car we sell is driven about 1,000 miles a month, so we will just average this all together and charge everyone with a car a fixed monthly payment." Now, some car owners will end up paying more per month, and some will pay less, depending on where the average is set. If the process is handled correctly, there should be no impact on the auto company's total revenues from this new "flat rate" monthly car payment system.

With flat rate payments, more families with restricted finances could figure out if they could afford a car. Payments could be budgeted consistently. The car company potentially benefits by having to perform fewer repossessions, handling fewer customer complaints over billing accuracy, having a stable and reliable revenue stream, and not having to decide whether any mercy should be granted those who cannot pay up in a particular month. There can be technical benefits, but these are explored later.

Basically, the same arguments extend to flat rate telephone service. A lot of early telephone service was flat rate simply because no one had any idea how to invent technology to track calls by distance and time. As late as the 1960s, it was still common to charge for a toll call by filling out a *toll ticket*, which was just a card that an operator filled out based on connection time and distance to the destination. These toll tickets were collected daily and used to compile the subscriber's bill (which was typed by a clerk on a typewriter).

It might be argued that by the 1960s, American families had become wealthy enough that the monthly telephone bill was a bargain and not a burden, whether based on flat rates or not. This may have been true for domestic calls, but international calls were still a problem. In the late 1960s, thousands of young people in the United States were serving in the

Armed Forces around the world, especially in Southeast Asia. When someone was wounded or—even worse—killed, it was not unusual for the monthly bill for calls to hit $700 or more. With many families at the time earning $10,000 per year or less, this was a real concern. In some tragic instances, the bill arrived on the day of the funeral. In a few cases where the family was close or well known to a telephone employee, the toll tickets would mysteriously become lost.

Flat rate local service avoided many of these social issues with respect to local calls. Once invented by the telephones companies and embraced by their customers, flat rates were mandated by many state regulators. Flat rate local service remains entrenched, in spite of perfectly valid telephone company studies that show that metered services would save a lot of people at least some money, and a few people a great deal of money (some others, of course, would pay more—much more). But in this case, the financial reasons proved to be secondary to the social reasons behind flat rate local service. There are even some who have seen Marxist overtones in the state's regulation and mandating of flat rate local service. There is no need to do this, and even Social Security was accused of being a Marxist invention at one time. There is only one fact that matters: people prefer flat rate services of all types, whether for bus fares or telephone calls.

The Pieces of the PSTN

The PSTN is a fascinating place to explore. All types of loops, trunks, switches, and equipment have been incorporated into the PSTN since 1876. Before looking at some of the more technical issues that provide a basis for ADSL in the next chapter, this might be a good place to summarize the key elements of the PSTN. The emphasis is not historical, but more contemporary.

The entire PSTN was put in place to handle basic telephone-to-telephone voice communication. The telephone just provides an inexpensive way to convert mechanical, acoustic, analog waves into electricity and back again. The electrical signal modulates the pressure waves of voice and sends them through the PSTN via copper wires. The electrical signals need to be periodically strengthened or amplified to carry them through thousands of miles.

All of the basic components of the PSTN are shown in Figure 2-5.

Figure 2-5
The basic components of the PSTN

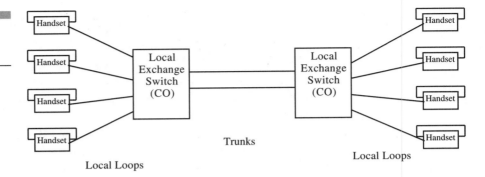

In its simplest form, the telephone *handset* contains a small microphone powered by a small charge of electricity sent through the local loop itself. For those with at least a small knowledge of basic electricity, you may know that the handset microphone is filled with carbon granules that vary their electrical resistance under the acoustic pressure of the voice. The acoustic wave compresses and releases the granules, allowing the electrical circuit to mimic the voice waves. The electricity on the local loop in turn obeys basic electrical laws regarding resistance, voltage, and current.

The electrical signal can be sent over the PSTN to another telephone handset, where the signal is sent to the speaker. This tiny speaker relies on the varying current coming in to alter the strength of an electromagnet. The magnetic variations vibrate a thin metal disk. These vibrations model the acoustic wave generated by the person talking at the other end of the voice circuit and enter the receiver's ear where it is (presumably) understood. So the handset has the transducer needed for acoustic-electrical conversion. The trick of the telephone is that although the acoustic wave is in its electrical form, it can be transmitted over great distances.

The telephone handset is connected to the local exchange or central office by a local loop. The loop is usually dedicated to one subscriber, but there were exceptions known as *party lines* (the term reflects *multi-party* use, although people could have a lot of fun with party lines). Technically, the local loop is only one form of access line from the subscriber to the PSTN. In any case, whenever a customer wants to make a call, the local loop is available unless another person at an extension (or on a party line) is using the loop. In this instance, the customer will overhear the current conversation.

In its simplest form, the local loop consists of two copper wires twisted around each other. Twisting was done to provide voice quality at a reasonable minimum and to reduce cost. With two wires on the local loop, the signals propagate in both directions. This is okay for voice conversations because usually only one person talks at a time. This two-way, but alternating, conversation is called half-duplex. However, the local loop can handle two-way simultaneous voice (try it), which is full-duplex. However, there are problems when attempting to do this with digital content, such as the output from PCs and modems. Special techniques are needed to make full-duplex transmission over two-wire local loops possible for digital information content.

At the other end of the local loop is the local exchange or central office switch. This is the PSTN network node that sets up, maintains, and terminates the temporary connections, known as calls, between any two end devices. The local exchange must terminate and switch local loops, trunks, and service circuits (for call control signaling and the like) to complete the task. In fact, a lot of effort is spent in providing *call progress* information to the user, which includes providing dial tone, accumulating the dialing digits, playing recorded announcements, generating ring back so the user hears the remote telephone ringing, and so forth.

The switch could be as elementary as a cord switchboard, as complicated as a solid-state computerized device running computer programs, or one of several varieties of *stepper*, *relay*, or *crossbar* switch in between. The switch connects loops and trunks in any combination needed to complete the call. Naturally, all of the loops, trunks, and service circuits must have *ports* on the switch itself. Some overall control is needed, which might be the human operator, but today is more likely to be a computer program running in the switch itself.

The switches are themselves connected by interoffice trunks. These can be thousands of miles long and are shared sequentially by users. Note that the user has no control at all over the trunks assigned to handle their call. Due mainly to their length and number, it was necessary and economically reasonable to provide much better voice quality on trunks than loops, so trunks use four wires (two twisted pairs) that provide a separate transmission path for each transmission direction. The main benefit is the ability to put amplifiers on the trunk that are able to operate in only one direction instead of both.

Note that in the simple network shown in the figure, there are more loops than trunks. In the PSTN, there are far fewer trunks than loops. This was possible and became an abiding principle of telephone network design because not everyone should be on the telephone at the same time. As was already mentioned, if it seems like everyone wants to use their telephone at the same time, chaos results on the trunks and within the switches. The time has come to take a closer look at this issue.

CHAPTER 3

Loops and Trunks

The previous chapter introduced the structure of the PSTN from a historical perspective, as well as its many social and political aspects, but the technology was not mentioned much at all. Loops and trunks were just "links," and the concept that humans can perform switching tasks just as well as a modern central office switch was covered.

This chapter examines the technologies behind the modern PSTN loops and trunks. Switches will be mentioned as well, but a detailed discussion of the PSTN switching issues that ADSL is meant to address is reserved until the next chapter.

The Analog PSTN

As was discussed in the previous chapter, the *Public Switched Telephone Network* (PSTN) in the United States was born about 125 years ago with the invention of the telephone. The network was "public" in that the network nodes (telephone switches, in this case) did not belong to any specific user or group of users (known as "subscribers" or "customers"); rather, all of the switches and network equipment (even the telephone handset itself, initially) belonged to the service provider. The essence of a public network of any kind is that any subscriber can reach any other subscriber. The network was "switched" in the sense that the network nodes were switches that established end-to-end connections on circuits between the source and destination telephone numbers—which are the network addresses of the PSTN, although few think of them that way. The circuits were dedicated to a given pair of subscribers for the duration of the call. This came to be known as circuit switching.

It took a while for the PSTN to take on the form shown in Figure 3-1, which is a slightly different representation than that shown in the previous chapter. By the end of the 1800s, however, the major pieces shown in the figure were in place in many metropolitan areas throughout the United States. With this architecture, subscriber telephones, which were analog, voice-only devices, fed central office switches over analog local loops, which were simply long twisted pairs of copper wire.

These analog local loops, or *access* lines, were connected by "analog" switches, initially human beings whose job it was to say things like "Number, please" and connect the source to the destination. The analog switchboard was an electromechanical device firmly under human control. Naturally, as the network grew, not all telephones could be connected to one huge central office. For this purpose, the central office switches (technically, the switchboards) were connected with special analog circuits

Figure 3-1
The analog PSTN

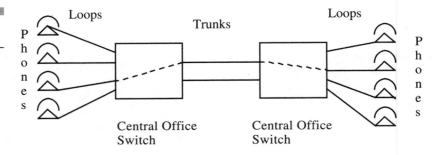

known as *trunks*. Trunks were also analog twisted-pair copper wires that connected not subscriber telephones, but switchboard operators; that way a call could be "switched" through the trunking network, usually without the caller even knowing what was going on, until the opposite party was contacted over their local loop and the operator said "Go ahead." This loop to trunk to loop switching process is shown in the figure.

So all of these central office switches were linked by trunks. The trunking network could be owned and operated by the local service provider for some calls, or the trunking network could be owned and operated by a "long distance" company, but they are all trunks. A trunk is just a common line that is shared by a number of users. Trunks are typically grouped together for use from a functional standpoint, but not necessarily a physical standpoint. A trunk may use the same physical facilities as lines, but the line coding and signaling (call control) requirements for a trunk differ a lot from those used in lines or loops. Trunks are also distinguished by a lack of user control over their use on a particular call.

Lines usually outnumber trunks, often only a couple of hundred for up to 10,000 local loops. Because there were many more local calls within the central office serving area in those days, this ratio made sense. Not all switchboards in a given central office needed, or even had access to, the limited number of trunks available. The trunks typically ended up on the long distance switchboard under the control of the long distance operator (a prized position, since at the turn of the century there was lots of free time between such calls). On older phones from the early Twentieth Century, the "0" carried a "long distance" label to direct the call to one of these switchboards.

The whole point is that the network was analog from top to bottom, with analog local loops running through analog central offices linked by analog trunks. As noted earlier in this book, the term "analog" here refers to the signaling or coding method used on the loops and trunks and inside the switches. The term does not refer to the type of information content being conveyed, which could be analog itself (voice) or digital (PC files and e-mail).

Digitizing the PSTN

Naturally, the PSTN service providers were quite concerned over the quality of the voice service they provided their customers. No real concern was paid to competition at that point because LECs were virtually guaranteed a life-long monopoly on local service with a given *franchise area* set up by the state. For the majority of people in the United States, the PSTN provider was an operating company under the AT&T or Bell System umbrella. The rest of the telephone companies were independents and operated in their own local franchise areas. The last independent company to have direct local competition with the Bell System was located in rural New Jersey and shut its doors in 1945. In short, there was little fear that the state government would revoke a local service franchise and give it to someone else.

Instead of competition, the major concern of the local telephone companies, BOCs and independents alike, was pleasing the state regulators with regard to service quality. There was one overriding reason for this concern. Subscribers who did not like the quality of their telephone service complained. Some grumbled a little, some griped all the time, but many took up the pen and wrote complaint letters to the state regulator, usually generically called the state *Public Utility Commission* (PUC), but the actual term used varied from state to state. When the state PUC periodically held public hearings about telephone service, the more disaffected tended to gather in groups and complain all at the same time.

The key to understanding the relationship between satisfying customers and satisfying state regulators is simple. In the vast majority of cases, all telephone service rate changes and increases had to be approved by the state regulator, typically after a round of public hearings. Naturally, the state PUC members tended to look unfavorably on rate increase requests when users were jamming the hearing room with signs and chants against the LEC whose rate increase was under discussion. The PUC commissioners, usually political appointees, were sensitive to situations that might become embarrassing to the administration that appointed them. Granting a rate increase to a telephone company whose service was being vilified by the public was a sure way to lose votes for all concerned. In fairness to the telephone companies, every franchise area seemed to harbor a small number of customers whose main goal in life seemed to be complaining about the quality of their telephone service. Some PUCs became fairly adept at filtering out this persistent background noise and concentrating on the major problem areas.

So the LECs got rate increases by pleasing the state regulators, who granted rate increases based on complaints remaining below a certain level for a set period of time (usually a year). Other factors were involved in this rate increase process, of course, but telephone companies with a stack of service complaints against them faced a tougher battle, without question. The bottom line was that the better the quality of service, the easier it was to get a fair hearing when a rate increase request was needed.

Voice quality depended on the quality of the electrical signal as it traveled through the PSTN. Analog systems have had one major problem when it comes to signal quality. Analog circuits are very noisy, and nearby thunderstorms and changes in electrical characteristics of the wire from rain or snow made the line hum, squeak, and pop. Shortly before World War II, a British engineer working for ITT in France became so annoyed by the quality of analog voice signals that he decided to create a way to digitize the analog voice signal.

The digitization of analog voice involves a three-step process. First, the analog waves are *sampled* at a rate adequate to represent the analog waveform accurately. For analog voice, the international standard sampling rate is 8,000 times per second. Next, the samples are *quantized*, which means that they are represented by strings of 0s and 1s. For analog voice, the international quantization standard uses 8 bits per sample.

The choice of 8,000 samples per second was established by the voice passband bandwidth. The voice passband of 300 to 3,300 Hz was basically "rounded up" to 4,000 Hz. A rule known as the Nyquist Theorem—which states that digital sampling must take place at twice the highest frequency component to accurately reconstruct an analog signal—is then applied. Twice 4,000 Hz is 8,000 and so the sampling rate was set at 8,000 times per second. The 8 bits per sample is less rigidly maintained. In fact, the only real reason that 8 bits are still used is that most early computers and processor chips handled exactly 8 bits at a time instead of the modern standard of 32 or even 64 bits; it simply made sense to generate 8 bits for efficiency and ease of processing.

Finally, the bits are *coded* for transmission over a digital transmission link. Standard line codes come in many forms, but they all must operate at 64 kbps (8,000 samples/second x 8 bits/sample = 64 kbps). The three steps of sampling, quantizing, and coding are done in hardware and together make up a unit known as a *channel bank*. In the channel bank, the sampling and quantization is done by the codec, and the line coding is performed by the DSU/CSU.

The new quantized digital signals used on digital trunks were much less susceptible to noise and other impairments. The system was much too expensive, given the state of electronics in those days, and was a little like using a Rolls Royce as a New York City taxicab. World War II soon put an end to these experiments. The technical term for the process just described is *pulse code modulation*, or PCM. Sometimes, digitized voice is just called *PCM voice*.

It is important to realize that this digital invention in no way affected the information content sent over the line. This was a change from analog signaling on a line to digital signaling on a line, and it did not even matter whether the line in question was used as a loop or trunk. This technique did not change whether the information content being sent was analog or digital. The interface device might change, of course, from a telephone transducer to a codec for voice, and from a PC modem to a CSU/DSU for e-mail, but the information was the same.

After the war, the expanding world economy, pressure on the "telcos" to provide more and better service, and advances in electronics made the digitization of the PSTN not only feasible, but almost mandatory. Even the new electronic digital computers developed during the war were being researched and experimented with. The plan was to introduce the computer into service in the central office to switch the increased number of calls faster than humans, enabling the same number of trunks and lines to handle more calls. As it turned out, this ambitious plan took many years to happen.

Although originally developed for line side (local loop) use, digital transmission was perfected and extended to increase trunk capacity and alleviate the situation in metropolitan areas, where literally not another trunk cable could be run in a conduit. To understand why, a brief discussion of the multiplexing techniques used in analog and digital system is necessary.

Multiplexing and Trunks

It is not in the least bit ironic that the telephone was invented as a result of the effort to multiplex, or combine, many telegraph messages over a single telegraph line. Multiplexing involves expanding the bandwidth available over a single medium, in this case, copper wire. Bell accidentally expanded the bandwidth so much that the wire could carry the major part of the voice bandwidth, the part containing about 80 percent of the power of the human voice. Figure 3-2 shows the bandwidths involved in human voice communication over the PSTN.

In the figure, the amplitude, or strength, of a signal is on the vertical axis, and the frequency is on the horizontal axis. Too much should not be read into the actual shape of these curves because they are intended for comparison and informational purposes only.

The human ear can generally respond to frequencies from 20 Hz (cycles per second) to about 20,000 Hz, although the sensitivity to higher fre-

Figure 3-2
Human hearing,
speech, and the
voice bandwidth

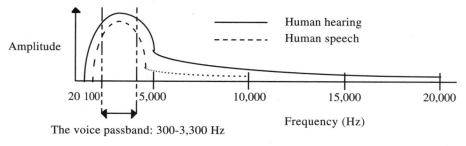

The voice passband: 300-3,300 Hz

Voice bandwidth = 3 kHz

quencies falls off with age. When it comes to speaking, human voice operates in the 100 to 10,000 range. Note that humans are capable of hearing more than the human voice can say, which is probably a result of needing to listen for the low rumblings of wild beasts and the high-pitched shrieks of many birds.

In any case, most of the power (a function of frequency and amplitude together) of the human voice occurs roughly between 300 and 3300 Hz (this extends to 3400 Hz in many places outside the United States). Any more, and telephone engineers found that transmitters and receivers became more expensive quickly. Any less, and people could not distinguish individual voices very well, so everyone tended to sound like they were speaking over an intercom at a fast food restaurant. This 300 to 3300 Hz *passband* became the Bell System standard, and most other independents also adopted it (however, there are still odd exceptions here and there where smaller passbands are used, typically at the high end (e.g., 3200 hz)).

The 300 to 3.3 kHz passband meant that 3,000 Hz (3300-300), or 3 kHz of bandwidth, was needed to transmit an analog human voice conversation adequately, which was a lot more bandwidth than was ever needed to send the simple dots and dashes of Morse code, but that was the whole point. The 3 kHz bandwidth was easily provided on twisted-pair copper wire in the PSTN.

Note that there is nothing fixed by electricity or physical laws that limited the bandwidth on the copper wire to 3 kHz. That's just where the voice was, and this is one of the keys to understanding how ADSL works. Now suppose that a given transmitter and receiver can pass not just 3 kHz, but 36 kHz along two pairs of copper wire. Obviously, this wire could be used for more than one telephone conversation. These new transmitters and receivers would cost more than the ones that just sent and received 3 kHz, but maybe that was okay.

More expensive hardware is definitely okay on a trunk. Trunks are fewer in number and much longer than local loops in most cases, so it is financially sound to spend extra money to carry as many conversations on one or two pairs of copper wire as possible. As long as the money spent on transmitters, receivers, and multiplexing equipment is less than the money spent on new trunks, it would be a wise course of action. In addition, state regulators often mandate that the local service franchise holders show that they are operating the network in the most efficient and technically advanced way possible. Multiplexing voice conversations on trunks is all part of this process.

Note that multiplexing voice conversations on a local loop is also possible. However, with little financial or regulatory incentive to do so until the Web came along, coupled with the inability of LECs to control the wiring and CPE devices in the United States, multiplexing and increasing the bandwidth available on the local loop has been an enormously expensive and technically difficult thing to do. Of course, ADSL changes all of the rules, so to speak.

It is desirable in parts of PSTN to carry more than one voice signal over the same transmission facility. The transmission facility may carry analog signals or digital signals. Both can be multiplexed, but it makes economic and technical sense to use different multiplexing techniques for analog and digital signals.

Analog signals typically use *frequency division multiplexing* (FDM). FDM gives a voice circuit "some of the bandwidth all of the time," which is done simply by dividing the passband of the transmission facility into separate frequency ranges, each corresponding to a voice channel. One voice circuit is carried in each channel, and usually one channel is provided in each direction, as is common for trunks. An FDM multiplexer device is installed on each end of the transmission facility to multiplex and demultiplex the various channels.

As an example, consider the twisted-pair copper wire with a bandwidth (technically, passband) of 36 kHz instead of 3 kHz. Suppose the entire passband was divided into 4 kHz channels. A voice conversation would comfortably fit inside each channel. For electrical reasons, the voice channels cannot sit "side by side," frequency-wise. The 4 kHz passband includes an adequate *guardband* to prevent crosstalk between the channels. Now the wire pair can carry not one, but 12 voice channels ($36 \div 4 = 12$). Another wire pair could carry the returning voice signals. The multiplexers on each end of the trunk would combine and split off the 12 voice circuits carried on the two pairs of wires. Note that if a voice channel is idle, a special "idle" signal is needed to indicate this.

FDM works best with analog signals. In the analog PSTN, FDM trunking networks were quite common. The Bell System developed many of these systems for their own use, and the technology became common even among the independents. The system just described, with 12 analog voice channels on two pairs of wires, corresponds in its basic idea (although not in detail) to something called *N-carrier*. The term *carrier* just meant that the voice circuits were multiplexed ("carried") and did not appear in their normal 300 to 3,300 Hz bandwidth.

The FDM methods used in the Bell System had their own terminology that showed both a firm grasp of simple words and a singular lack of imagination. In the FDM hierarchy, a *group* consists of 12 voice channels (as carried on N-carrier). A *supergroup* was made up of 5 groups for a total of 60 channels. Ten supergroups composed a *mastergroup* of 600 voice channels. Finally, a *jumbogroup* consisted of 6 mastergroups forming 3,600 voice channels. Frequently, the groupings were carried on coaxial cables which formed a family known as *E-carrier*.

All of these analog carrier groupings with FDM work just fine. As it turns out, however, if the signal being carried is digital and not analog (regardless of analog or digital information content), FDM becomes enormously inefficient. After the analog voice conversation was digitized, the rationale for FDM was weakened considerably.

Fortunately, another form of multiplexing was known and used for analog signals as well. *Time division multiplexing* (TDM) was quickly abandoned as a viable multiplexing technique for analog signals due to expense and annoying technical glitches. As opposed to the "some of the bandwidth all of the time" approach of FDM, TDM functions according to a "*all* of the bandwidth *some* of the time" approach.

With TDM, the transmission facilities bandwidth, perhaps 36 kHz as before, is divided into *time slots*, into which go not pieces of the analog signal (that was hard), but bits from a digital stream (that was easy). The multiplexer now loaded time slots with bits on the sending side and took them off on the receiving side. The "ownership" of bits was determined by position in the bit stream. In other words, time slot #1 always had voice channel #1's bits, time slot #2 always had voice channel #2's bits, and so on. Note that even if the voice channel is idle, the time slot must still be sent from sender to receiver. In this case, the bit pattern in the time slot says to the receiver, "ignore this, the channel is idle."

The differences between the analog FDM and the digital TDM approaches are shown in Figure 3-3. Note that the total bandwidth and numbers of channels in the figure are the same. Only the organization has changed. After the digitization of voice became possible and cost effective, it quickly became apparent that TDM was a better way to carry digital voice conversations. Of course, telephones and modems still generated only analog signals, and local loops still carried the 300 to 3,300 Hz passband only, but TDM was intended for the trunking network. TDM not only flourished there, it took off, thanks to a digital TDM technology known as *T-carrier*.

T-carrier

Although some think that digital signaling technologies for voice and data are relatively new, the T-carrier system has been around since the early 1960s. The voice quality advantage that digitization enjoyed has already been discussed. An added benefit turned out to be the increase of trunking capacity without the need to run new wires between switches. As the need for increased capacity on the voice network became evident in the 1950s and 1960s, this was a big bonus. The enormous expense of these early electronic devices, the codecs, and the CSU/DSU was offset by the savings enjoyed from fewer labor-intensive facilities installations. Thankfully, the price of electronics always goes down, while the population and wire costs always go up.

One of the most common ways to double analog trunk capacity on N-carrier links handling 12 simultaneous calls was to install the new digital T-carrier equipment on each end of the trunk. At its most basic level, this T1 could handle 24 voice channels on the same two pairs of wire. No new cable needed to be run between the central offices, and 24 simultaneous voice conversations could be carried. In another form, called T3, the T-carrier could handle 672 voice channels, or 28 T1s, on a single coaxial cable that might have been formerly been used for an analog E-carrier.

Figure 3-3
FDM and TDM

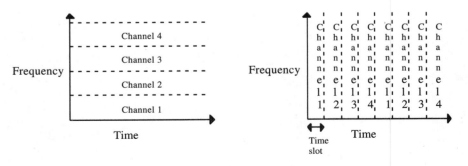

Frequency Division

Time Division Multiplexing

By the mid-1970s, digital trunks were in use in many places for quality and increased capacity purposes, while electronic computer technology was beginning to be used for switching. In 1976, the first all digital local exchanges began to appear in small, rural central offices in the United States. However, even before that time, electronic computerization techniques had made advances in the telephone system possible on a large scale. The whole point was to complete calls faster and without the intervention of a human operator. By the late 1960s, subscribers in most urban and suburban areas could dial all local calls directly, thanks to electronic systems that handled all call routing.

Oddly, it was a while before digital trunks and digital switches were used together. The smaller, rural central offices employed digital switches first for two reasons: first, the switches were new and could only handle the smaller number of lines in rural areas; second, the human operators there were often underutilized given the smaller number of calls. On the other hand, large, urban areas got the new digital T-carrier trunks first. This was done mainly to alleviate conduit congestion and handle more calls on the same facilities.

Note that the analog local loop was not considered part of the problem(s) that digitization addressed, and so was left alone. It would have been prohibitively expensive at the time to install digital equipment in everyone's home. Besides, such equipment would have been classified as *customer premises equipment* (CPE), and the telephone companies would have had as much control over its installation and use as they do with modems, which is to say, none at all.

Integrated Services Digital Network (ISDN)

This retention of the analog passband local loop had at least one consequence—the superior voice quality that digital transmission on the trunk provided was "masked" by the continuing presence of analog passband local loops. The new digital trunks did not dramatically improve voice quality, but at least they kept it from getting worse. Of course, there were still many analog trunks; because the voice quality "problem" was not the main reason for trunk and switch digitization in the first place, this approach still worked well.

It was only when long distance calling became common in the early 1960s (with the invention of area codes) that voice quality became a real issue. People objected to paying a lot more ($6-$12 a minute for a coast-to-coast call in 1961) for what was essentially the same voice service. AT&T Bell Labs responded by eventually proposing the complete and total end-to-end digitization of the telephone network. This plan formed the digital network portion of what eventually became known as *Integrated Services Digital Network* (ISDN). The whole goal of ISDN was to integrate all services reachable by a telephone connection (which included data by this time) and deliver them over an end-to-end digital network. By the mid-1980s, the digitization of switches in conjunction with digital trunks had begun. ISDN implementation basically boiled down to digitizing the analog local loops and the analog switch ports to which these loops were attached (if the switch was still an analog switch).

It should be noted that ISDN essentially digitizes the last segment of the PSTN, the local loop. No analog signals at all are carried by an ISDN line. While ISDN was a much more efficient platform for all kinds of digital information content, it required every telephone to either be converted to an ISDN digital device or use an expensive converter. Neither alternative was popular with service providers nor users.

The ISDN plan gave each wire pair running to a home the capability to carry not one but two simultaneous voice conversations, known as *bearer channels*, or B-channels. The voice calls were carried as 64 kbps digital signals (128 kbps together) along with a separate *digital signaling channel*, or D-channel, which ran at 16 kbps. Some additional bits were added to bring the total line rate up to 160 kbps, of which 144 kbps was represented by the B-channels and D-channel.

The D-channel could be used for packet data services when not needed for signaling. Some envisioned these services as being similar to what the Internet and Web provide today, but in those days this was just a dream. The B-channels could be used for almost anything: voice, data, and even video. These B-channel circuits could operate in circuit mode (for voice), packet mode (for data), or even frame mode (for frame relay "packet" services). ISDN included plans for new kinds of telephones that would have data displays and even video screens, as well as simple voice capabilities.

For its time, ISDN was an ambitious plan to funnel all forms of information down a simple pair of wires to everyone's home. Unfortunately, much of the best that ISDN had to offer is now available on the Internet and Web, and the funnel now ends not at the voice/video/data telephone, but at the voice/video/data PC.

ISDN became an international standard and could be applied not only by the Bell System, but also by the independents. A full discussion of why ISDN never took off as intended, in spite of years of publicity and deployment efforts, is well beyond the scope of this section. For now, it is enough to say that the digitization of the local loops and switches proved to be enormously expensive even in the 1980s. This meant that ISDN services needed to be priced quite high in order to allow these expenses to be recouped in the amount of time established by state regulators. Also, not many people could figure out exactly what benefits ISDN conferred, at least until the Web came along. The combined result was that as late as 1996, there was absolutely no ISDN service at all available in Alaska, Montana, most of Nevada, or New Mexico.

Analog Local Loops

Analog local loops have a number of characteristics that make any radical and comprehensive modifications (such as digitization for ISDN) difficult. Most of these characteristics are a consequence of actions that were applied strictly to improve analog passband performance and, therefore, become a hindrance in a digital environment.

Initial local loops were single wires (using something known as *ground return*) and only paired when a technician accidentally discovered that a *metallic return* improved voice quality dramatically by cutting down drastically on crosstalk. These long, parallel wires, however, suffered from signal loss (known as *attenuation*) because the wires acted as long, thin capacitors that tended to "store" the signal rather than allow it to flow freely. Twisting the pairs together was soon found to help counter the attenuation effects slightly by adding an electrical characteristic known as *mutual inductance* to offset the capacitance of the wires. *Shielding* the wire by adding an outside metallic jacket or braid would have improved analog voice quality as well, but this approach was deemed too expensive and also tended to increase the attenuation.

The *unshielded twisted pair* (UTP) was invented very early in the history of the PSTN to minimize crosstalk. It also allowed the signals to flow more freely, but only marginally. After thousands of feet, the signal was just too weak and the voice was hopelessly muffled and tinny because only so much induction could be added by twisting the wires. Furthermore, the attenuation was much worse at higher frequencies on the loop, which is true of electrical signals when the electrical loads placed on them are said to be *reactive*.

After much trial and effort, the telephone companies eventually figured out that acceptable voice quality had the following practical limits. UTP analog local loop wire of 19, 22, and 24 gauge gave acceptable voice quality up to 18,000 feet (18 kft) from the central office. These *gauge numbers* are an *American Wire Gauge* (AWG) standard for the thickness of the wire; the smaller the number, the thicker the wire. On thinner 26 gauge UTP, only 15,000 feet (15 kft) of local loop gave acceptable voice quality. Nonetheless, there were many subscribers, especially in rural areas, who were more than 18 kft (about 3.4 miles) from the central office. How were these subscribers to be reached?

Adding induction would counteract the attenuation and signal loss, but only so much induction could be added by mutual twisting. The answer was to add extra inductance to the UTP. This process of extending the analog local loop with inductance is called *loading*. The electrical components used to produce these *loaded local loops* were known as *loading coils*.

Why Loading?

A more detailed look at loading requires at least a passing knowledge of electrical circuit terminology. Each of the terms used in this section will be given an analogy or example to help those unfamiliar with the terms (or those for whom formal education was longer ago than one would care to admit). This section just reviews some of the relationships between inductance, capacitance, and attenuation.

Power in a circuit is not solely dependent on either the voltage strength of a signal or the amount of electrical current flowing. Rather, power is the mathematical product of voltage and current together. Voltage is the electrical equivalent of pressure (as in a garden hose), and current is the electrical equivalent of the amount of water flowing through the hose. It is thus easy to see that the total amount of water delivered to a barrel being filled by a hose is dependent on both the water pressure and the flow of water (that is, larger hoses deliver more water). More pressure through the same hose will fill the barrel faster, but so will a larger hose, even at the same pressure. This is true in all instances. A fire hose knocks you down not from flow by itself, nor the pressure, but from the combination.

In an electrical circuit, the capacitance changes the phase of the current relative to the voltage. This phase change means that the maximum voltage no longer occurs at the same time as the maximum current, which limits the power delivered over the path. Capacitance is the tendency for electricity to build up on a wire (or pair of wires) rather than flow freely. Interestingly enough, inductance (which is a measure of the "slipperiness" of the wire) changes the phase as well, but in the opposite direction from the phase shift due to capacitance effects.

This being the case, it is easy to see that adding inductance to high capacitance circuits such as a long local loop can affect the power delivered to a receiver. Adding inductance in just the right amounts can bring the voltage and current back into phase again, but only in a certain bandwidth range. This boosts the received power, which is, of course, the whole idea.

Note that the additional inductance cannot be added haphazardly, but must be engineered carefully to affect only the bandwidth of interest. For the analog voice passband, this is the frequency range from about 300 to 3,300 Hz. Effectively, the adding of inductance "tunes" the circuit (in this case, the analog local loop) for voice transmission. The process of adding inductance to long local loops is done with loading coils.

Loading and Attenuation

Attenuation on analog local loop UTP limits acceptable voice quality to about 18,000 feet, in most cases. Attenuation means signal loss, and if the local loop is too long, the voice is hopelessly muffled. The attenuation is caused by capacitance effects between the wires. These effects may be countered by adding inductance, which was a side benefit of twisting the wires in the first place. However, only so much inductance can be added by twisting; loading coils must also be added to the local loop.

The effects of a loaded analog local loop on attenuation are dramatic. Figure 3-4 shows the effects of the three major families of loading. The attenuation is given in *decibels* (dB) per mile, a measure of signal strength. The horizontal axis shows the effects of the three major loading families when used on 22 AWG copper wire local loops. These are the "H," "D," and "B" loading architectures. The nonloaded curve shows that the signal loss is more severe as the frequency rises.

The loaded analog local loops have much flatter signal loss profiles, which is the desired result and the key point to be made about loading effects. However, note that even with the common D and H loading schemes, attenuation rapidly increases above 3 or 4 kHz. That is okay for voice because most systems cut off local loop signals at about 3,300 Hz (the passband); but above 4 kHz or so, the attenuation of loaded loops actually exceeds the attenuation of a nonloaded loop of identical construction.

Loading posed a major problem for ISDN and for all other analog local loop digitization schemes. In order to deliver acceptable voice quality over analog local loops in excess of 18,000 feet, additional loading coils are usually necessary to add inductance to the loop. This inductance counteracts the effects of capacitance (which causes attenuation) over long distances.

Figure 3-4
Loading and
attenuation

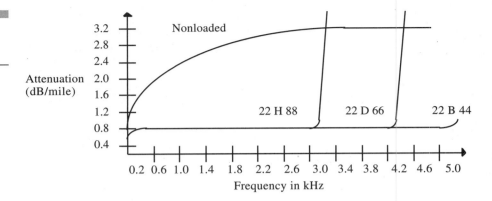

Loading coils look like iron "doughnuts," around which are wrapped each wire of the UTP loop. The inductance added is controlled by the size of the doughnut, the purity of the iron, and (most importantly) the number and spacing of the wire wrappings. The overall spacing between loading coils affects the distances at which the analog local loop operates as well. To avoid the haphazard deployment of loading coils, a series of standard "families" of loading coils and spacings evolved.

Figure 3-5 shows the general architecture of a loading system. Note that there is a central office end section, one or more spans between loading coils called the loading sections, and an outer end section. It is important to realize that the outer end section itself might extend thousands of feet. Also note that several phones might be serviced by the last loop, an arrangement usually known as a "party line." The interface between the outer end section and each individual phone in this arrangement is called a *bridged tap*. Bridged taps may still be present even when party line service is absent because the taps do not interfere with normal analog voice services.

The loading "family" is distinguished by two related parameters: first, the spacing between the loading coils, and second, the number of millihenrys of inductance the coil adds to the loop at that point. Three loading schemes are common: B-44, D-66, and H-88. These are normally referred to as B, D, or H loading. A *millihenry* is a standard unit of inductance. For some strange reason, the standard units for capacitance and inductance in electrical standards were set much too large. Using a *henry* is like trying to buy a cup of coffee with a $1,000 bill. A millihenry is one thousandth of a full henry and much more useful for common electrical situations. It is much easier to buy your coffee with a "milli-thousand-dollar bill"—a single dollar.

B-44 loading adds 44 millihenrys of inductance to the loop. The spacing is 3,000 feet. D-66 loading adds 66 millihenrys, so the spacing is longer at 4,500 feet. H-88 loading, the most common, adds 88 millihenrys and spaces coils 6,000 feet apart.

One other important point is that the central office end section is always half of the normal spacing to the first loading coil. This compensates for loaded line to loaded line connections within the office itself. Thus, for H-88 loading, this central office end section span is only 3,000 feet.

Other Analog Local Loop Features

Naturally, not all analog local loops suffered from excessive attenuation. In fact, the majority of PSTN subscribers, up to 80 percent by most accounts (although these figures are sometimes disputed), are located less than 18,000 feet from the nearest central office.

The other subscribers are reached in a variety of ways. Loading, which adds extra inductance to counteract the effects of capacitance over distance, may extend the loop to up to 30,000 feet (about 5.7 miles). Special *line extenders* (essentially amplifiers) may also be used to stretch the distance out to about 25,000 feet (about 4.7 miles). Line extenders are typically non-standard devices and so must be matched on the loop. This is not a problem as long as the same entity controls both devices (that is, one is not considered CPE). These local loop features are shown in Figure 3-6.

Figure 3-5
Loading Archtectures

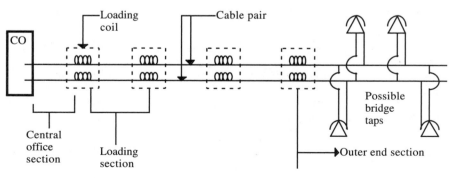

Figure 3-6

Local loop types

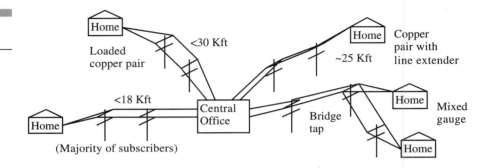

Figure 3-6
Local loop types

However, the figure also shows other analog local loop characteristics besides pure distance concerns. In many cases, the unshielded twisted-pair copper was "tapped" to service homes closer than the end of the wire, either to support party line arrangements (an older feature) or just because it was easier. In the latter case, if 100 pairs were run down a block and a new phone was installed, it was easier to "tap in" to a pair than to run new wire. Almost universally, the tapped wire was not cut because the new line service could be terminated in the future, with a new subscriber accommodated down the line at a later date. These bridged taps posed no difficulty for analog voice service.

Finally, it was not uncommon to mix wire gauges on a UTP copper local loop. Going from 24 gauge to 26 gauge was most common, but other gauges were mixed as well. For analog voice this worked just fine, although sometimes faint "echoes" occurred because signals are always reflected, as well as transmitted, whenever the electrical characteristics of a wire path change, which is exactly what mixing gauges did.

When ISDN was proposed as a method to digitize the PSTN end-to-end, most of these analog local loop features caused concern. Loading coils, line extenders, bridged taps, and mixed gauges destroyed most digital signals. Bridged taps weakened and reflected them (the unterminated ends were long antennas), mixed gauges also reflected them, and loading coils and line extenders limited the available bandwidth because they were tuned for the analog voice passband of 300 to 3,300 Hz. Bridged taps act as "delay lines" and can extend anywhere from 1,000 to 5,000 feet on their own. They also put an obvious "notch" in the line's attenuation at the frequency associated with the bridged taps' wavelength.

For the most part, the subscribers at the end of less than 18,000 feet of analog local loop were pretty much okay, as long as there were no bridged taps or mixed gauges with which to contend. In fact, ISDN specifications called for runs of 24 gauge wire, with no bridged taps, with a maximum length of 18 kft.

ISDN, Loops, and DAML

As if the different passband limited local loops variations were not enough to deal with, there is another local loop complication seen in the United States and around the world—*Digital Added Main Line* (DAML) local loops. In the United States, these lines are sometimes called to as *Digital Subscriber Single Carrier* (DSSC) links. While not common, one form or another of DAML is used in about 10 percent of the United States. The reason is simple. One of the drawbacks that quickly became apparent in the PSTN was the need to add local loops to deliver second lines to users for Internet access or other services. Obviously, if some way could be found to offer second line service on a local loop engineered for a single analog conversation, the effort would be worthwhile. DAML allows the adding of this second telephone line. The service presented to the user is still two analog voice passbands. But this technique uses the ISDN DSL structure of two B-channels (64 kbps each) to carry the PCM voice representing the analog voice over the local loop. The one D-channel (16 kbps for signaling and packets) is not used. Of course, the new "2B+0" service would not run through an ISDN switch and effectively is still just two analog voice lines on one local loop. But since few people used ISDN to make telephone calls or had ISDN equipment in their homes, this was all right.

With DAML, a telephone company could add a second line to a home with existing analog service and provide the desired analog service on the new line. Think of DAML as "analog service packaged as ISDN without the ISDN switch." DAML uses readily available ISDN components on the local loop, but of course needs no ISDN switch upgrade. But because there is no ISDN switch at the end of a DAML link, the call control messages on the D channel are simply ignored. The service is still analog. DAML equipment is also used in Europe and other places around the world. Oddly, it is seldom considered or mentioned in surveys of local loop arrangements in the United States.

Another Arrangement: CSA

Loading systems are not the only way to extend the reach of the analog local loop. In addition, loading systems still had to be engineered and "balanced" for local conditions. Too much loading (technically, inductance) would make the loops "sing" or "hum" with annoying background noise; too little loading made the voice sound muffled and just too hard to hear;

and of course, each loop had to have its own set of loading coils.

By the late 1960s and 1970s, digital technologies had matured enough to consider their use in the previously analog local loop arena. This brought the benefits of digital trunks to the local loops, namely better voice quality and higher call capacities. However, it was not a trunk anymore at all, but rather, a *carrier serving area* (CSA) that combined aspects of trunks and loops in one. The CSA architecture is shown in Figure 3-7.

The figure shows how this typically worked (there were many variations, so no generalization covers all instances). Usually, a T1 carrier was used to digitize 10 pairs of UTP from the central office to some point central to a cluster of subscribers; this central point, or "neighborhood," was the serving area itself. Normal 24 channel *time division multiplexed* (TDM) digitized voice ran on eight of these pairs (two pairs were needed for each T1). The other two pairs were for network management or other purposes. Analog CSA technologies did exist, but digital T1 systems soon became more common. When T-carrier was used for this purpose, it was not called a bunch of T1s, but rather a *digital loop carrier* (DLC). In many cases, these systems were called *pairgain* systems because they "gained pairs" back for the telephone company to use for other purposes.

There is a fundamental CSA design rule that is not emphasized in the figure. The loops must be less than 12,000 feet with 24 AWG pairs and less than 9,000 feet with 26 AWG loops.

To support the 96 voice channels (4 x 24), a device called a *central office terminal* (COT) was placed in the local exchange to interface with the 96 switch ports; a device called the *remote terminal* (RT) was placed in the field, usually in an environmentally hardened casing. This arrangement was known as *SLC-96* within the AT&T Bell System—SLC stood for *subscriber loop carrier*, which was a specific type of DLC.

The advantage was that the 96 analog local loops now only had to run

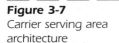

Figure 3-7
Carrier serving area
architecture

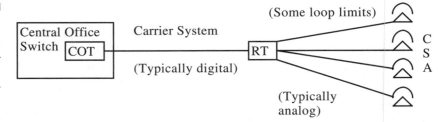

to the RT, and the same distance limits still applied at the end of the COT-RT link! So the SLC-96 (or whatever) DLC system could run 5 or even 10 miles to a housing development, and then the analog UTP local loops could extend another 12,000 feet if need be. It was even possible to load these analog sections, but this was seldom done.

The Trouble with Analog Local Loops

The progressive digitization of the PSTN to increase switching speeds and capacities proceeded in almost an "inside out" fashion. Initially, the central office switches and trunks were two targets of digitization efforts. The analog local loops, which needed neither more speed nor more capacity, were left alone.

Nevertheless, end-to-end digitization of the PSTN from phone to phone was attractive for many reasons. Once the whole network was digital, no conversion would be necessary, quality would improve, signaling (call control) would be more secure, and so forth. Even before ISDN was defined as the preferred method of digitizing the PSTN end-to-end, however, the problems inherent in analog local loop arrangements became obvious.

The main problem is that subscriber lines engineered for passband analog local loops in many cases limited the useful bandwidth available for digital signaling. This was not done on purpose, of course; it was just the end result of years of engineering for optimal passband analog voice signals.

Each of the analog local loop features, aside from basic, short-run local loops, turned out to be a concern. Loading coils cut off all the high frequency signal components required to transmit digital signals. Bridged taps cut the digital signal strength drastically because part of the signal went to each unterminated "branch" which reflected signals as well. Mixed wire gauges caused signal reflections that created echoes on the digital circuits. *Carrier Serving Area* (CSA) terminal equipment limited the bandwidth available for any one loop, typically to 4 kHz analog or 64 kbps digital, and this was true whether the outer loop was loaded or not.

All in all, these features of the analog local loop, developed over the years to solve problems of analog voice transmission, made the transition to digital local loops have a much higher price tag than anticipated.

Real World Local Loops

It is all well and good to discuss the impairments of analog local loops in terms of bridged taps and mixed gauges in the abstract, but what do local loops look like in the real world? The statistics shown in the following list were gathered from a variety of sources, including the *Institute of Electrical and Electronics Engineers* (IEEE) and Bellcore.

- 160 million access lines in the U.S., 70 percent residential, 11 kft average
- Fifteen to 20 percent of all local loops have loading coils installed
- Bridge taps and mixed gauges are almost universal
- Typically 26 AWG out to 10 kft, then 24 AWG beyond (or even 19 AWG)
- Fifteen percent are DLCs, usually 24 AWG, but 9 kft maximum
- Average loop has 22 splices, which pose corrosion and attenuation risks

It turns out that 160 million of the world's 700 million access lines (analog local loops for the most part, but a good 5 percent of the access lines in countries like Germany are digital ISDN lines) are in the United States. The figures in the above list apply to these 160 million United States local loops. About 70 percent of these run to homes, and the other 30 percent run to businesses. The average length of the local loop is about 11,000 feet, according to Bellcore. However, the larger number of shorter urban loops makes this number somewhat deceiving and lowers the impact of the much longer rural runs.

About 15 to 20 percent of these loops have loading coils, which means there are up to 32 million local loops beyond 18,000 feet. Moreover, bridged taps and mixed gauges, often presented as annoying anomalies, are just about universal (especially mixed gauges) because it was (and still is in many places) standard practice to run 26 AWG wire from the central office out to 10,000 feet. At that point, the wire was changed to 24 AWG, or even 19 AWG in rural areas, which lessened attenuation effects. Bridged taps act as "delay lines" and can extend anywhere from 1,000 to 5,000 feet on their own. They also put an obvious "notch" (more like a "ripple with notches") in the line's attenuation at the frequency associated with the bridged taps' wavelength. The bridged taps' wavelength determines the spacing between the notches.

Additionally, *digital loop carriers* (DLC) serve about 15 percent of this loop plant, or about 24 million homes and businesses. Almost 99 percent of all *carrier serving areas* (CSAs) today are serviced by DLCs instead of analog pairgain systems. These DLCs usually have 24 AWG wire extending directly to the home, but they have a 9,000-foot maximum in almost all cases.

The local loop has another impairment. Most outside plant cables, whether 200, 400, or even 3,600 pair wire, come in 500-foot reels, which means there must be a splice every 500 feet; the average loop then has 22 splices (11,000 feet). Splices tend to form corrosion "collection points" and, if they are assembled improperly, can add a lot of attenuation themselves. In order to prevent corrosion, a small amount of electricity, called a *sealing current*, was applied to prevent oxide buildup (rust). Directly relevant to digital transmission, this sealing current could not be applied when the loop was used for purely digital signaling. However, DSL allows for the presence of the sealing current.

Trunk Groups and the Law of Large Numbers

Trunks in the PSTN consist of groups of voice circuits used between switching offices. Whenever a call is made that requires a trunk circuit, an idle trunk must be found in a group of a given size. If all the trunks in a group are occupied (busy), the call is blocked and must be redialed at some point in the future. This process is called *reorder* in the PSTN. Usually, the telephone company supplies a "fast busy" tone to the caller to inform them that the destination telephone may not be busy, but all the trunk groups that were tried during the switch routing process were occupied to a level that prohibits their use. Note that some trunks are always reserved for special uses and cannot be assigned for "regular" telephone calls.

Whenever things are grouped together for any purpose whatsoever, a principle known as the Law of Large Numbers takes over. Basically, this law states that large groups are more predictable than small groups when it comes to mathematical properties. To see why, consider the average height of people in groups of 100 rather than 10. There will be a lot of variation in the average height, determining 10 measurements at a time. One group may be all over 6 feet tall, while another is under 5 feet 6, and so on. But when taken in groups of 100 at time, one would expect much less variation in the average measurements from group to group, which should be closer to the average for the general population as a whole.

Telephony traffic is no exception to this rule. In the PSTN, service providers often use something called the *Erlang B traffic tables* to predict the probability that a call will be blocked. A level of service known as B.05 will block only 5 percent of the calls attempting to use this group, or 1 call in 20. The B.1 level of service will block 10 percent of the calls attempted, or 1 in 10, and so on. These levels of service are built into tariffs and contracts and are closely watched by state regulators. Because larger trunk groups can support higher occupancies, the PSTN tends to aggregate traffic into large groups to achieve a given level of service more predictably.

So the PSTN is engineered for a given level of service when it comes to trunks. A level of service that is B.05 ("Erlang B .05 blocking") will block 1 call in every 20 (5 percent) attempting to use that trunk group. B.1 will block 10 percent, and so forth. Because the Law of Large Numbers applies, large trunk groups can be utilized closer to 100 percent at a given level of service, such as B.05.

For instance, trunk groups of 25 can be 75.9 percent occupied and only block 1 call in 20, whereas trunk groups of 200 can be 94.3 percent occupied. Larger trunk groups can be pushed closer to the edge, while smaller trunk groups need lots of slack.

However, this higher loading factor on larger trunk groups comes with a price. Because they operate closer to the edge, large trunk groups are more sensitive to traffic increases than small trunk groups. For example, a traffic increase of 14.2 percent will double the blocking to 1 call in 10 (B.1) on a trunk group of 25, but only a 7.9 percent increase will produce the same effect on a trunk group of 200.

In some cases, traffic increases have doubled the traffic loads on trunk groups. The level of service is lowered on a trunk group of 200 to B.5 service, which blocks 1 call out of every 2. Because these levels of service are built into tariffs and contracts and watched by state regulators, there are real consequences to these degraded service levels.

The impact is on the users as well as the service providers. Right after the 1993 California earthquake, for example, PSTN traffic increased dramatically. In larger cities, the added traffic pressures on the large trunk groups quickly overloaded the trunking network. Yet in smaller towns with modest trunk groupings, service degraded much less under the same increases.

The higher efficiencies of large trunk groups have resulted in large tandem switching offices (recall that tandems switch trunk to trunk) across the country. When one of these offices burned in Hinsdale, Illinois, the large trunks could not handle the traffic increases on alternate routes and quickly blocked many of the calls attempted on them.

Increasing Loads and Trunks

In the PSTN, a trunk is occupied for the duration of a call. That is, the connection is mapped onto a trunk circuit as long as the connection between two endpoints exists. Traffic engineers use complex formulas and tables to determine the number of trunks needed to support a given number of local loops (that is, access lines) at the desired level of service (usually drawn from a table known as Erlang B). This level of service may be mandated by the state, or written into service agreement contracts or state documents known as tariffs. Either way, there may be real consequences for the service provider if the service level dips below the engineered level.

The problem is that the tables were created for relatively short voice telephone calls. The characteristic holding time (the duration of a call) for a voice telephone call is only three or four minutes (some studies place this as high as six minutes) and usage is relatively constant. This only makes sense for a human voice conversation. One person talks while the other listens, and when they are done speaking to each other, they hang up. A change in these underlying assumptions of trunk traffic engineering, in terms of traffic patterns or holding times, could degrade service. Even a modest 10 percent increase can degrade trunk group service from blocking 5 calls in 100 to a possibly unacceptable 10 calls in 100.

This is precisely what is happening to the PSTN today. Families with PCs (some 17 million or so), as well as *small office/home office* (SOHO) users, have been busily ordering second lines to enable their PCs to attach to the Internet and Web. In most cases this involves a dial-up modem connection to an Internet service provider with a router (a sort of "connectionless switch") connected to the Internet.

However, data connections differ from voice connections in two significant ways. First, the holding times are much longer, typically five times longer than voice calls. The modest modem speeds available on analog local loops virtually guarantees longer holding times for the smallest data transfers today. Secondly, the modems send bursty data packets over the voice circuits. It is not unusual for data connections to send a flurry of packets initially and then sit idle for many minutes on end until another "burst" of packets occurs. Both points are important and affect the efficiency of the trunking network. The basic problem is that voice circuits have limited bandwidth, and when trunk circuits on the circuit-switched PSTN are used for bursty packet transmission to and from the packet-switched Internet and Web, a lot of otherwise useful bandwidth is wasted.

Figure 3-8 illustrates this. As more local loops are used for data connections, trunk occupancy from PC users increases to the point where voice users cannot reliably make telephone calls at all. Service levels might degrade to the point where the telephone company must make ad hoc and expensive additions to add voice circuit capacity to their trunking network. This is clearly not an ideal situation. Of course, this situation is exactly the one that ADSL is meant to address.

Actually, ADSL cannot totally rectify the situation in and of itself. But ADSL can be combined with a data "overlay" network that gets the data traffic off of the voice switches and network and more directly onto networks that were made for this purpose, such as the Internet. The next chapter shows why this is so.

Figure 3-8
Increasing loads
and trunks

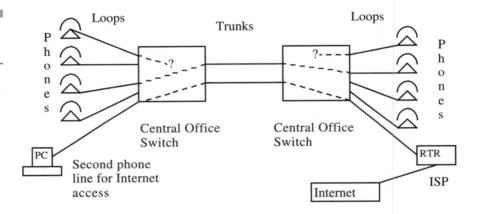

Long *holding times* exhaust trunks and switch resources
PCs will send data *packets* on the usual voice *circuits*

4

Packet Switching and Circuit Switching

This chapter examines one of the important issues concerning both the PSTN and the Internet/Web today. The issue is simply that the PSTN was designed, engineered, deployed, and optimized to be a circuit switched network and the Internet/Web was designed, engineered, deployed, and optimized to be a packet switched network. The differences, as has been pointed out earlier, are basically how the bandwidth on the trunks and lines of the network are handled and how the network nodes handle the flow of traffic from input port to output port.

A circuit switched network works best when the end devices or users have, for the most part, a constant interaction for the duration of the call. A circuit switched network dedicates "all of the bandwidth all of the time" to a particular call. If anyone makes a telephone call on the PSTN, the trunks (in most cases) will reserve 64 kbps in each direction for the bits representing the voices. This 64 kbps is usually coding "silence" in at least one direction because most people listen politely while the other is talking. This 64 kbps cannot be used by anyone else, not even the telephone service provider, while the call is in progress. The two 64 kbps channels that represent the voice circuit "belonged" to the end users for the duration of the call.

As long as the holding times were only a few minutes, this technique was fine. The trunking and switching network could still handle plenty of calls one after the other. Sometimes, with metered service, this generated revenue because more calls completed meant more money. Other times, with flat rate service, the benefit was less direct, but real nonetheless. More calls completed meant more satisfied customers, who generated fewer complaints to the state regulators who looked more kindly on LEC requests for flat rate service increases, which meant more money.

In certain social situations today, putting someone on hold is considered a particularly obvious and not very inventive form of insult, but the hold button has changed the way that people use telephones. In turn, the familiar hold button on the telephone handset, which did not exist on residential telephones until relatively recently in the history of the PSTN, changed voice circuit holding times somewhat. Instead of saying, "I'll find that information and call you back," people said "Let me put you on hold for a minute." The minutes sometimes seemed like hours, but that was a different effect.

Consider two people on the telephone discussing plans to meet somewhere at a certain time the following week. Travel plans must be coordinated and checked, perhaps with spouses who are present, but not nearby. No problem—the hold button enables one of the people to check and then pick up again. But travel plans may affect the time initially proposed. Hold again, pick up again. But the new time may trigger a better travel arrangement idea, leading to a repetition of the cycle. In all, several minutes of talk may be punctuated by longer periods of silence. Even with metered service, because the rates typically drop after three minutes or so, it may make more sense to keep one conversation going than to redial again and again.

In this case, the pattern is not connect-talk-disconnect. The pattern is connect-talk-hold-talk-hold-talk, and so on. While the call is on hold, there are no meaningful bits being sent across the 64 kbps trunk channels nor through the circuit switches, but neither the trunk channels nor the switch ports can be used to send any other meaningful information through the PSTN. The circuit is sacred in that regard. It is worth pointing out once again that there is nothing wrong or bad with the PSTN functioning in this manner. This was the most efficient way to design, engineer, deploy, and optimize the PSTN if traffic is not *bursty* or *intermittent*, as those terms were defined previously.

But wait. The whole talk-hold-talk-hold pattern appears to be quite bursty and intermittent. In fact, it is. The question is whether enough people employ their telephone in this fashion to impact the circuit switching aspects on the PSTN. Before 1994, people did not impact the performance of the PSTN appreciably, no matter how often they hit the hold button, but the Web changed that in 1994. Data sent to and from client and server PCs is nothing if not bursty—and usually extremely bursty. This similarity is shown in Figure 4-1.

Another aspect of circuits is that they connect only *one* thing to another, which is the essence of the PSTN: it exists to connect the device at the end of one access line to the device at the end of the other access line. It matters little whether the devices are telephones, modems, or fax machines, as long as they are compatible. One major aspect of the Internet and Web, however, and probably the one that fueled such explosive growth in the first place, is that the Internet exists to connect the client device at the end of one access line to *everything*. As more and more people use PSTN access lines for Internet access and not telephone calls, perhaps circuit switched networks like the PSTN are not the best way to handle this traffic. Perhaps there is a better way, such as ADSL.

On the Internet, information is sent and received in the form of *packets*. Packets are variable-length units that have maximum and minimum sizes. On the Internet, packets conform to the rules set out in the Internet Protocol Suite set of standards, which was formerly and still is usually

Figure 4-1
The hold button
and bursty data

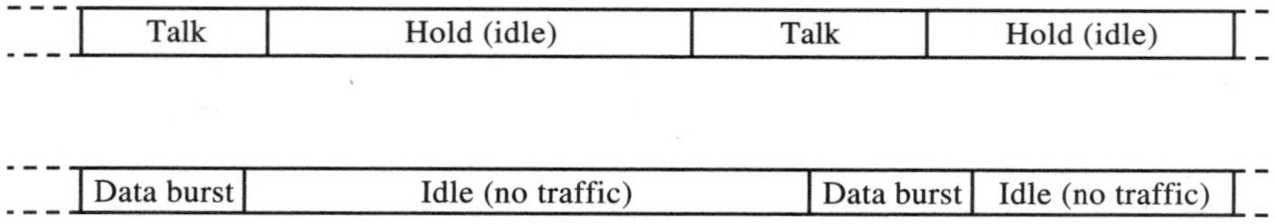

Talk	Hold (idle)	Talk	Hold (idle)

Data burst	Idle (no traffic)	Data burst	Idle (no traffic)

called TCP/IP. On the Internet and Web, all packets conform to the *Internet Protocol* (IP) rules. The information contained within packets can be almost anything at all. The contents of packet on the Internet and Web today could be a piece of a file being transferred, an e-mail message, packetized voice (known as *Voice Over IP*, or VoIP), or even packetized video.

In these last two cases, the term *packetized* means that voice and video have traditionally been handled by circuit switched networks (such as the PSTN and cable TV networks) rather than packet switched networks. Once packetized, with state-of-the-art digitization and compression techniques, this type of voice and video more closely resembles data than anything else. Packet switching also is better for bursty data.

The following plays fast and loose with "Internet" history, but the intent here is to be informative about packet switching, not precise as to the date and times that one particular event occurred. In any case, the Internet quickly became not a circuit switched network but a packet switched network for a very good reason: money. Packets were always used as basic information units on the Internet, but to connect Internet nodes (then called "gateways," not routers), long links were needed. Because these links connected Internet nodes, it made sense to call them trunks. From where did the Internet get trunks? From the PSTN, of course, in the form of point-to-point leased private lines (in some instances, they were dial-ups, but still circuits).

On course, when a trunk connected Internet Node A with Internet Node B, that was all it connected. True, a dial-up could also connect to Internet Node C later, but not at the same time that Node A was connected to Node B. It was the familiar "telephone mesh" problem all over again. The solution this time was for the gateways to route packets from one link (or trunk) onto another. So far, this was nothing more than the old central office switch trick applied to the Internet. When anyone on the early network wanted to send a file somewhere, a circuit was needed there. If someone else using the computer wanted to send e-mail at the same time, and maybe even to the same place, they were out of luck—the circuit was busy. Circuits reserve all of the bandwidth all of the time for *one* thing. Naturally, these long trunks were paid for by the mile, but it was very expensive to require one circuit for each potential user. Of course, the applications that people were running remained as bursty and intermittent as ever. A lot of expensive bandwidth was tied up on these circuits while people examined and scratched their heads over experiment results and the like.

To save money and make more efficient use of these long and expensive trunks, the Internet people made one big leap beyond basic switching to bring packet switching to the Internet. In this way, individual packets could be switched from trunk to trunk, not based on what circuit the packet represented, but for which end application the packet was carrying information (which is how a single link on a packet switched network can connect one end device to everything). The packet switch, now called a router on the Internet, could send a packet literally anywhere on the network, based on the individual address carried in the packet header.

So instead of a client PC needing a separate circuit for sending e-mail to an e-mail server, another for transferring files to another PC, and another for accessing a Web site, only one link was needed for all these activities. The packets just had different addresses in them that allowed the routers as packet switch network nodes to distribute them properly. This link could still be a *circuit*, but make the distinction between this link and a circuit because packets still must flow on something. This "something" may even be a PSTN dial-up circuit, but there are other forms of links that might be used. In fact, many of these alternate forms are detailed in the next chapter.

This individual address information attached to every packet gives another perspective for distinguishing circuit switching from packet switching. Circuit switches, such as PSTN local exchanges, switch the entire bandwidth of the circuits (all the bandwidth all the time) from one place to another. Packet switches, such as Internet routers, switch the individual packets that form the content of the bandwidth. If a PSTN circuit switch was a packet switch for voice, it would be possible to establish one call to the switch itself and then say something like, "This next sentence is for Person A," and then, "This next sentence is for Person B," and so on, all on one telephone call. The sentences would be voice "packets," and the "next sentence" statements would contain the "addressing information" for the "packet."

In short, packets can be mixed to a whole host of destinations on the same link, which may still be a PSTN circuit, dialed or leased. In ISDN, which shares many features in common with the Internet today, this link for ISDN services would be the *Digital Subscriber Line* (DSL) B-channel, which operates in either *circuit mode* or *packet mode*, but not both at the same time.

The current discussion has become quite complex. The whole point is that sending packets on circuits, especially passband-limited voice circuits, may not be the wisest thing to do. It is time to look at a real example of how the transfer of Internet packets on the circuits of the PSTN impacts the performance of the PSTN.

Circuit Use and Packet Use

Figure 4-2 shows the basic differences between circuit switched networks (the PSTN) and packet switched networks (the Internet and Web). There are others, of course, but the figure shows the major issue discussed in this book.

In a circuit network, a local exchange has trunks to a tandem switch, which in turn connects to an *interexchange character*, or IXC, for long distance calling, and the ISP, for Internet access. In the simple network shown in the figure, only two trunks lead from the central office to the tandem switch, and there are only four devices, but the effect is the same, regardless of absolute numbers. Usually, 10,000 potential users are served by as few as 600 trunks. Note that the trunk itself, which may have a lot of bandwidth in and of itself, is always divided into 64 kbps bandwidth circuits (naturally). For example, a T-3 has 45 Mbps in bandwidth, but only has 672 channels of 64 kbps each. No user can ever get a circuit with more than 64 kbps.

The point of the figure is that if the two home users have both telephones and PCs, they may want to make both calls (circuits) and access

Figure 4-2
Switching circuits
and switching
packets

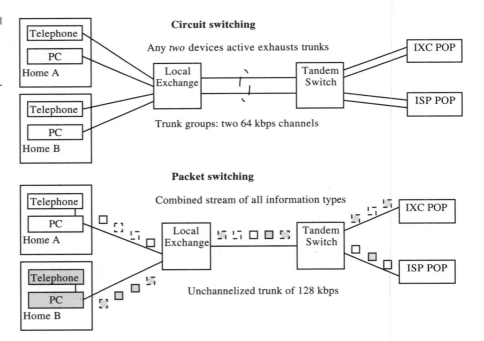

the Web (packets). However, the packet equipment (routers) is in the ISP office. The only way to get there is through a circuit (local loop to trunks). Because circuits are assigned for the duration of the call, it is easy to see that any two connections, voice or data, from telephone or PC, will exhaust the trunking capacity between central office and tandem switch. If two PCs are currently accessing the ISP router ports, no long distance calls can be made. As more and more homes obtain second local loops for PC access, it is easy to see that the number of trunks must increase, as well as the number of switch ports to service these extra access lines and trunks.

The figure also shows the result when local loops, trunks, and switches are "packetized." When packets are used for data and compressed digital voice, it makes no sense to have circuits anywhere at all. Now a combined digital stream of voice and data (and perhaps video!) packets can easily share the unchannelized loops and trunks to the IEC or ISP. No dedicated bandwidth exists to eat up the PSTN resources.

The figure shows the effects of hooking both packetized voice equipment and a PC to a single "packetized" local loop. A voice call no longer consumes all of the bandwidth on the loop. Voice compression and silence suppression (the twin features of voice packetization techniques) leave plenty of bandwidth for data packets (or any other type of packets, such as video). The figure codes the voice packets with a distinctive border and shades the packets from one home. Note that the packets share the loop and trunk bandwidth, but still arrive at the IXC or ISP switch correctly. The local exchange and tandem switches must now switch on *content*, not just bandwidth. Of course, the issue of whether this is still the PSTN or not is open to debate.

The "packetization" of local loops and trunks has many technical solutions. ISDN was and is a possible solution, but only when used as a packet mode connection, not in pure circuit mode. B-channel circuits are still 64 kbps circuits. ADSL, or any xDSL technology can be used, and there are other alternatives, such as ATM. The big questions for today are: who gets to do it, and who pays for it?

The Local Exchange Switch and the Internet Router

Much has been said so far that a PSTN switch and an Internet router are just different types of network nodes, but to look at them, one would never guess that they had much in common at all. The PSTN switch is housed in a huge building called the local exchange, or central office. The Internet router is housed in an ISP office which may be—believe it or not—someone's bedroom. How can two things be at once so different and so alike?

Part of the answer is that the PSTN switching office is the culmination of over 100 years of backward compatibility and technological evolution. The Internet has been around for about 30 years. Another part of the answer is that the PSTN switch might service 10,000 local loops and 1,000 or so trunks. ISPs that are housed in bedrooms rarely handle more than 30 lines. If an ISP does handle 10,000 lines, it is often as big as a local exchange building (well, maybe not quite so big).

Now is the time to look inside the local exchange switching office and the ISP site to see just what is in there that is generating so much excitement today.

The typical structure of a PSTN local exchange is shown in Figure 4-3.

In the figure, access lines (local loops) and trunks (to other switches) enter the local exchange or central office below ground level. It does not matter whether the access lines run on utility poles or underground for most of their length, the entrance facility to the local exchange is below ground, with rare exceptions. This sub-basement is known as the *cable vault*, and technicians sometimes dread having to work there, depending on the lighting and drainage conditions. In some major metropolitan areas, there may be 40,000 lines and trunks all snaking their way from underground conduits into the cable vault.

The access lines and trunks make their way through the cable vault to the *main distribution frame* (MDF), which is the facility commonly called a *wire center* by the LECs (to the IXCs, it usually means the actual building where the trunks to the LEC terminate). The MDF is essentially one huge patch panel. Every trunk and access line appears on the MDF. There may be other frames in the local exchange just for this group of access

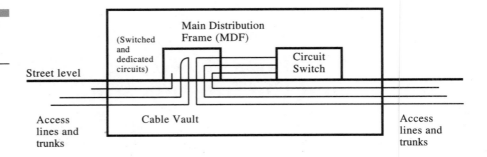

Figure 4-3
Inside the local exchange

lines or trunks, but everything sooner or later ends up on the MDF. Each cable on the MDF is identified by a vertical and horizontal number assigned to it. V=365 and H=4533 is the way a subscriber's access line or trunk to an IXC POP might be identified within the local exchange itself.

For switched services, the access lines and trunks must be patched through the MDF to a circuit switch port. Naturally, these switch sizes and capacities range from small in rural areas (a few hundred ports) to gigantic in major metropolitan areas (up to 40,000 ports). Some are modular and digital; others are barely upgradable at all and still analog; some use computers only to control the switching process; and others are nothing more than big computers in their own right. Because there are some 13,000 local exchange switches in the United States today, some installed last week and some installed in 1932, it is hard to make valid generalizations about the switch itself without some qualifying statements.

Generally, there are two ways for information to make its way through the local exchange or central office. Circuits may be switched through the local switch, directed by the telephone number dialed by the originating telephone equipment. Alternatively, circuits may be "nailed up" as leased private lines that always lead from point A to point B. These were called "nailed up" because early implementation permanently connected the wires forming the circuit, which made modifications difficult and awkward. No call control is needed on the private line circuits, so a line appearing on the main distribution frame may go directly into the PSTN switch or into another piece of equipment known as a *digital cross connect* (DACS or sometimes DCS). The DACS equipment got around the need for physically "nailed up" private lines by allowing the input and output connections to be made in a software configurable "mini-switch," but calling a DACS a true switch is probably a gross overstatement of its capabilities. Whether routed through the DACS for private lines services of through the PSTN switch for switched services, the line eventually made its way into the trunk transport system, unless the other end of the circuit was served by the same local exchange. This detail of DACS/PSTN switch alternatives is shown in Figure 4-4. In either case, the switching is always port to port, bandwidth to bandwidth, and is transparent to any of the bits or packets traveling on the circuit. Note that the number of links leading between the components is not intended to imply relative numbers, but to show connectivity.

Another point is that the DACS equipment operates on digital access lines, not analog local loops. That is, if an analog access line was to be converted to private line use and it was necessary to put this private line through a DACS, the access line had to be digitized first. There were a number of ways of doing this, which are beyond the scope of this discussion. The emphasis here is on switched services as opposed to private line services.

The point is that the circuit switch is needed in all cases for *switched services*. Switched services are those where the endpoint is determined by some call control procedures (signaling), such as a person dialing a call to another telephone or a PC speed-dialing an ISP POP. Obviously, the switch capacity can be exceeded in terms of the number of ports currently supported, especially as more and more people order second lines to attach to the Internet. The only thing to do is to install more switch ports, which is sometimes easier said than done. Only so many switch ports can be added before a total overhaul or upgrade to the entire switch is needed. This can cost anywhere from a few hundred thousand to a million dollars.

So every circuit that needs to be switched must have an input and output port on the circuit switch. Whenever a home PC user dials in to their ISP, the access line must be switched to a trunk leading to the ISP POP. Note that once this is done, the PSTN switch adds absolutely nothing to the circuit. The packets are invisible to the circuit switch.

Switched services are not the only type of services that the local exchange can offer, however. These are all *dedicated services*, such as those represented by leased private lines. Because the MDF is basically one huge patch panel, it is easy to provide these dedicated circuits. All that needs to be done is to "patch" the vertical and horizontal position of one access line on the MDF to another vertical and horizontal position on the MDF. No switch port is needed because no switching needs to be done on the circuit. In the PC user to ISP world, this is the equivalent of using the second telephone line as a dedicated connection to the ISP POP. However, the user could never use this line for anything other than ISP access, and the ISP POP presently cannot support any other user on this router port.

Figure 4-4
Switching and cross
connecting

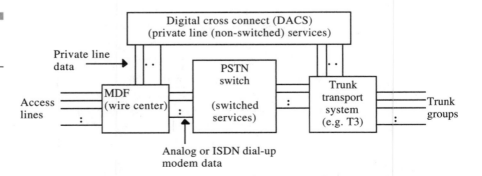

Dedicated services are point to point, which saves having to run the circuits through the PSTN switch.

A local exchange has many other features, of course. The emphasis here has been on the handling of access lines and trunks, and the role of the MDF and circuit switch.

The typical structure of an ISP office is a little tougher to categorize. About 4,000 ISPs are in business today, and some have as few as 100 customers, whereas others have millions of customers. Of course, after the number of customers grows past a few thousand, it is nearly impossible to service them all from a single ISP site. Nevertheless, small or large, most ISP offices have many features in common. The typical structure of a medium-sized ISP is shown in Figure 4-5.

Figure 4-5 has two distinct sets of outside lines. The first set consists of the trunks that lead to the PSTN switch ports, which is how the ISP handles users dialing in from home with their PC modems. It is not unusual for the ratio of users to trunks to be on the order of 10 to 1. In other words, if an ISP has 1,000 customers, the number of trunks could be as small as 100, which means that only 10 percent of the customers could ever be attached to the ISP at one time. All the others would get a busy signal. This practice started before the Web became popular in 1994 and simply reflected the fact that not everybody wanted to be on the Internet at the same time. If customers complained loudly enough, the ISP could add trunks, but this added to the ISP's monthly expenses without necessarily increasing revenues, which were usually billed at a flat rate. ISPs tended to resist adding trunks until absolutely necessary. Today, even a

Figure 4-5
Inside the ISP

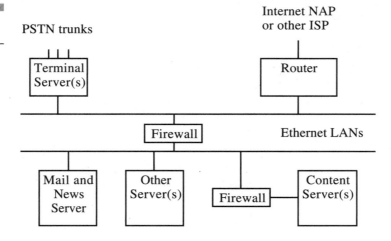

3:1 user-to-trunk ratio might not prevent user dissatisfaction and complaints. The lines terminate with a series of modems in a *terminal server*, a term that reflects older technology when this device handled calls from dumb terminals and not PCs, but the term seems to have stuck.

The second set of outside lines is usually just a single link. This leads to the Internet itself. The link may lead to a major Internet NAP (only about 20 ISPs link directly to a NAP) or, more likely, to another ISP's router. Note that only one link is needed because the router is essentially a packet switch and not a circuit switch. A T1, if used as this Internet link, can be used not as 24 channels of 64 kbps, but rather as one unchannelized link running at a full 1.5 Mbps.

User traffic need not be isolated by circuits because each user's traffic is distinguished by the addressing information in the packet—an important point. It means that if there is only one active user, that user's packet gets to use all 1.5 Mbps for packets, instead of just a 64 kbps channel. Even if 25 users are active, they will gracefully share the T1 bandwidth, although, in this case, things would be slower than having a 64 kbps channel of their own. But statistically, with bursty traffic, this should not happen often. With circuits, supporting the twenty-fifth user would be absolutely impossible.

The ISP exists to shuttle packets to and from users dialed in on the trunks and the Internet itself. In between, there is usually an Ethernet LAN running at 10 or 100 Mbps. A special *firewall* device separates the terminal server and router from the rest of the ISP equipment to prevent unauthorized users from gaining access to the site, through either the Internet or one of the dialup ports.

Behind the main firewall is another Ethernet LAN. Authorized ISP customers can access a variety of servers here, at least one of which will be the e-mail and Internet newsgroup server. This server stores e-mail sent to the ISP's customers until the user logs on and retrieves the messages. Newgroups are a special application of e-mail where like-minded users can exchange e-mail messages among themselves. Typically, users will belong to or monitor several newsgroups.

Other, more specialized servers may exist, depending on the size and focus of the ISP. Some ISPs try to attract scientists, others are geared toward commercial artists, and so forth. These servers contain the specialized services that the ISP is trying to promote. Today, many ISPs that are experimenting with voice and video over the Internet employ these servers here as well.

Behind another firewall lies the heart of the ISP itself. This, the *content server*, may also be comprised of more than one server. The content server stores the ISPs *home page*, the Web page that users first see when they dial in to the ISP. Users may store their own pages on these servers, although many ISPs limit the size of user pages. Note that nothing stops a user from having a Web site on their home PC other than the fact that others on the Internet can only access the page(s) when the user is dialed in and attached to the ISP.

Smaller ISP may skimp on some of the firewalls and number of servers. Even a large ISP may have several sites around the country that appear the same as this medium-sized ISP's office, so the figure is a good representation. The actual equipment is very small. There are seldom more than seven separate "boxes" that need to be housed, each about the size of a PC. A modest office in a small building will do. Of course, the ISP can handle only a few hundred simultaneous users with this arrangement, whereas the LEC switch might handle several thousand, but no ISP needs to worry about switches installed in 1932 nor access lines installed in 1919.

Please note that many LECs and IXCs are also ISPs today. In that case, the ISP equipment is usually tucked away in a corner of the local exchange. This is certainly deregulation in action.

Internet Service Providers (ISPs)

As millions of U.S. households and every organization worth its salt hooks up to the Internet and World Wide Web, more and more Internet service providers are appearing all over the United States to satisfy this hunger for Web access. In most cases, these ISPs supply Internet access on switched local loops and trunks, and then onto leased trunks from carriers to ISP *points of presence* (POP); the ISPs are generating more and more traffic on the public switched telephone network.

However, ISPs are quick to claim that they really generate no traffic at all on the PSTN! The ISP is just a way to connect clients (user PCs in most cases) to servers (Web sites in most cases) on the Internet. Technically, the users and servers generate the traffic. All the ISP does is deliver it. The point is that the traffic the ISP adds to the PSTN would not be there without the ISP.

Unfortunately, the PSTN is engineered for three-minute voice calls. Actually, for engineering purposes, call lengths can vary from three to seven minutes, but the three-minute residential call (that's when metered rates went down, right at the end of most calls) is enshrined in the hallowed halls of PSTN traffic engineering guidelines everywhere. This has been the guideline and general rule of thumb for close to 100 years.

The problem is that Internet access and Web browsing sessions last much longer. Thirty minutes seems to be the current average, and much longer sessions are not uncommon. One oft-repeated tale speaks of a two week call that was manually terminated by the *local exchange carrier* (LEC), only to get a call five minutes later: "What happened to my Internet access?" Flat rate monthly charges from most ISPs (as low as $15 in some cases) give many people little incentive to ever hang up at all! Why not just attach to the Internet the minute the PC comes up in the morning? Then it's there if you need it.

As a result, other calls may be blocked on the PSTN. An access line ties up at least two switch ports (the loop and trunk) and an outbound trunk to the ISP itself (usually a 64 kbps T-carrier channel). Tales have gone around of 911 calls being blocked (no switch capacity or outbound trunk) due to increased Internet usage in a neighborhood. Now, these 911 calls usually have priority in the trunk group, but only so many trunks are set aside for priority purposes. The rest can be claimed by any call. The problem is that when 100 drivers see an accident, or 100 civic-minded citizens spot a fire, there are instantly 100 calls to 911. The reserved trunks are overwhelmed, and if the "regular" trunks are all full, the 911 call is blocked. If one of these 100 callers is trying to report a robbery, they are out of luck.

In one sense, the problem is obvious. Using a series of switched 64 kbps circuits (with their dedicated bandwidth all the time) for a long holding time data call is an inefficient use of PSTN resources.

To make matters worse, from the carrier perspective, the ISPs are explicitly exempt from *access charges*, which are paid to the LECs at each end of the call by the IXC for completing the local calls to and from an IXC POP. This exemption was put in to reflect the fact that the ISPs actually add value to the basic telephone service. Also, the federal government explicitly wanted to encourage data services by keeping financial burdens to a minimum when the whole system was set up for divestiture in 1984. All an ISP needs to do is lease a trunk (usually at what are known as wholesale, *reseller* rates) from switch to ISP POP, and then on to the Internet network access point (or other ISP), and they are in business. This flat rate monthly cost makes it easy for ISPs to offer flat rate monthly service to their customers. The only initial concern is whether the number of customers for a given ISP can cover the fixed costs of the ISP.

The "Packets on Circuits" Load

In a very real sense, all packets flow on circuits anyway. Sometimes, the packets are said to flow on "virtual circuits" when the packet flow follows a *connection*, as defined by international packet switching standards. The Internet protocol suite even defines a concept known as a *flow*, which for all intents and purposes is nothing more than a virtual circuit on the Internet.

So although technicians talk loosely of "packets on circuits," the real issue is more one of "PSTN circuit switching versus Internet packet switching," as was covered in detail previously.

In any case, when bursty packets from long, but intermittently used, Internet connections are used across dedicated circuits built for short voice connections, the impact on PSTN resources can be considerable. This impact is shown in Table 4-1.

The table shows the results of an extensive study done by Bell Atlantic in the Summer of 1996. It compares the traffic load from ISPs' access (measured in traditional voice traffic units, hundreds of call seconds, or hundred call seconds per hour) to the traffic load from traditional telephony use.

The first two lines show ISPs' load over a 1-MB line (one line, measured business) and through an ISDN *Primary Rate Interface*, or PRI (circuit mode, not packet mode). The last two lines compare a business with a *multiline hunt group* (MLHG), common for businesses and the average for the entire local exchange, with voice residential users factored in. The units are in *hundreds of call seconds per hour* (CCS), a telephone traffic engineering concept. The precise definition of a CCS is not necessary here. All you need to know is that 36 CCS equals 100 percent utilization, the point at which no more calls can be made.

The impact is amazing. The ISP use of 26-28 CCS approaches the maximum on the circuits (that is, 36 CCS). Voice use peaks at a modest 12

Table 4-1

Impact of Packets on Circuits

Customer Type:	Average Busy Hour CCS:	Busy Hour:
ISPs on 1MB lines	26 CCS	11:00 PM
ISPs on ISDN (PRI)	28 CCS	10:00 PM
Business with MLHG	12 CCS	5:00 PM
Local exchange average	3 CCS	4:00 PM

CCS and averages only 3 CCS. Note also that the busy hours (the study calls them "peak hours") have shifted from late afternoon (voice) to late evening (data).

This timing affects the trunking and overall capacity of the central office, which is engineered, built, and provisioned around 3-12 CCS with a late afternoon peak. The LEC claimed that in order to prevent voice call blocking (like that of 911), an additional $17.8 million had to be spent to increase trunking capacity (that is, 44 percent trunk growth for ISPs versus 9 percent "normal" voice growth). Dial tone delays grew as well, impacting state service requirements, and added personnel were needed to service additional trunks.

The plea of local exchange carriers is clear. Why should ISPs be immune to access charges when it is the ISPs who are killing the PSTN? Why should ISPs be treated as "special cases" with reduced interconnection charges (wholesale), when clearly they are special cases, but in another sense entirely?

Of course, the ISPs and others concerned about inexpensive Internet access have not taken all of this lying down. The Bell Atlantic study, in particular, has been dismissed by consulting firms such as Puttre, Incorporated, as "silly" and criticized for using obscure and "obsolete" units, such as CCS, to make their point. This criticism goes to show how little understanding, sympathy, and respect many of the technical consulting firms and Internet service providers have for the basic engineering concepts employed in the PSTN.

Flat Rate Service

The assessment of rates charged to subscribers by telephone companies for their services fall into two broad categories: metered service and flat rate service. With *metered service*, charges accumulate based on how much of the service is used. With *flat rate service*, a fixed monthly charge is assessed, regardless of how much of the service is used. In the telephony world, metered charges are usually assessed by connection time (the duration of the call); however, this requires that all calls be monitored by the service provider for duration, which is a very complex process and one that tempts users to cheat.

To understand why, consider someone who rents a log splitter for a day. The rental agency could assess charges on a metered basis—the cost would increase for every log split. How does one determine the exact number of logs, short of looking over the person's shoulder all day? Cheating by undercounting the logs actually split would yield the benefit of lower cost to the user. If the rental amount charged is a flat rate for the day, however, the renter does not have to watch the user and no one cares if the device is used or not.

In the United States, a number of regulatory decisions have resulted in LECs offering flat rate local service. Customers seem to prefer this billing method. Most local telephone companies have adopted flat rate for their basic telephone services. The ISPs have followed right along because flat rate users are easier to bill, track, and administer. Revenues are stable and not as dependent on fickle user patterns. In any case, few ISPs have the software or hardware needed to meter usage.

Most calls to an ISP are just flat rate local calls through the LEC switching and trunking network. The ISP charge is flat rate as well, generally $20 a month or so for unlimited access. The cost is the same whether the ISP session lasts three minutes, three hours, or three days!

Also, some home office users and telecommuters have begun hosting their own Web sites on their home PCs. After all, the days are past when a powerful, $20,000 SUN SPARCstation was needed to do so—any $1,000 PC will do today.

The question is, with flat rate local service, why should someone ever hang up? Why not just connect to the ISP when my machine boots up and stay there forever?

Typical Residential Network Usage

The limitations imposed by the continued presence of the analog local loop and its restricted bandwidth and, to some extent, the trunking network and its persistent circuit connections, are most apparent when one considers what people do at home today. Nowadays, networks and service providers do not merely provide simple voice connectivity; they can supply a wide range of services, from home shopping to telecommuting, as well as Internet access.

Table 4-2 shows how all these new services differ in terms of bandwidth and holding time. Note that most of them are *asymmetric* in nature, which means that the characteristics in terms of bandwidth and holding time differ in the upstream (out of the home) and downstream (into the home) directions. Sometimes the differences are drastic (broadcast-quality digital TV, for example). All the differences reflect the traffic patterns of client

Table 4-2

Typical residential network usage

Type of Service	Minimum Bandwidth (Minutes)	Holding Time (Minutes)	Downstream Minimum	Upstream Holding Time Bandwidth
Video on Demand	3.0 Mbps	110	64 kbps	0.1
Teleshopping	384 kbps	7	64 kbps	0.7
Broadcast TV	3.0 Mbps	120	64 kbps	0.1
Near VOD	3.0 Mbps	110	64 kbps	0.1
Delayed broadcast	3.0 Mbps	30	64 kbps	0.1
Video games	384 kbps	60	64 kbps	60
Telecommuting	384 kbps	60	384 kbps	60
Web audio/video	3 Mbps	20	128 kbps	20

Source: ANSI/TIA-1558B (1995)

to server (upstream: not much) and server to client (downstream: a lot). The low holding times upstream for interactive services reflect the brief messages needed for start, stop, reset, and so forth.

Some of the services may be unfamiliar. Near Video on Demand means that the movie is not started at the instant a user requests it, but within a specific time interval, usually the nearest 15 minutes. Delayed broadcast services can start a TV show (for example) that is normally broadcast at 9:00 PM on a Wednesday and show it at 10:00 PM on a Friday.

Naturally, the more bandwidth the better. These figures are minimum requirements. It should further be noted that there are some 17 million home PC users. According to most ISPs, the average home Web user spends 6 hours per week online, in 30 minute sessions, normally between 6 P.M. and midnight, and sends and receives 9 e-mail messages per week.

Current ISP Traffic Woes

The rise of flat rate ISP service, coupled with flat rate telephone company local service, has caused real troubles for ISPs, as well as for the telephone companies. The links to the ISPs suffer from occupancy overload effects. The equipment at the ISP must still answer the call and connect the user to the Internet. More processing is required to validate the user login request.

Heavy loads all conspire to defeat these simple steps at one point or another. A study done by Inverse Networks in March of 1997 showed that users failed to get onto America Online a whopping 60 percent of the time between 6 P.M. and 12 A.M. in March of 1997. The problems were either no answer (no port process available to handle the request), line busy, connect fail (the overloaded equipment drops the user), or login fail (the ISP database process is too busy to handle one more request for validation).

Even a respected ISP like Sprint had an 11 percent failure rate, while the top-rated ISPs in the survey, IBM and CompuServe (now CSi and part of America Online), had a 5 percent failure rate.

The line busy failures are the direct result of trunk pressures either within the PSTN or to the ISP. Limited speed, analog local loops feeding trunking networks chopped up into voice bandwidth circuits lead directly to longer holding times and higher trunk group occupancy. It is immediately apparent from the visual that if something is done to increase the bandwidth available to ISP users and move them off of the voice circuit switch and onto a packet switch, the absence of the "line busy" problem will improve service dramatically.

By the way, CompuServe is one of the few ISPs that did not offer flat rate service at the time the survey was done! There are no busy lines due to users sitting on trunks forever.

Switch Blocking

The pressures brought on by the rise of Internet packet usage on telephone circuits extend not only from the analog local loop (in terms of speed) and the trunking network (in terms of occupancy or holding time), but also to the very heart of the PSTN itself—the central office switch.

Switch blocking occurs when there is no path available from an input port to an output port. The call request to set up a connection through this switch from loop to trunk (or loop to loop, or trunk to trunk) is now blocked. After all, switches can only handle so many circuits and connections at one time. Even the most powerful processor has limits in terms of memory and data bus capacity. Furthermore, the vast majority of central office switches cannot be replaced fast enough to keep up with the rapid advances in computing, making these switch computers less powerful than some desktop devices.

The problem becomes most apparent when a significant number of connections are used for data packets rather than voice. The long hold time associated with data usage can overload switch resources, causing blockage to occur in the switch, even if adequate trunk resources are available.

Under extreme conditions, essential calls to services such as 911 or fire departments may experience dial tone delays. Not only is the impact on the service provider negative in this case (in terms of state regulation or contracts), there is obviously a negative social impact as well ("I have cheap Internet access, but my house just burned down").

Is this the Information Superhighway?

Many people, including high government officials, computer hardware and software company executives, and computer network visionaries, have a glorious dream of an up-and-coming information superhighway. This superhighway would link homes and businesses through high-speed digital access lines (local loops) to state-of-the-art switches and routers linked together with higher speed digital trunks (nonchannelized to reflect increasing packet use and ease the restrictions and inefficiencies caused by trunks installed for voice circuit bandwidths). People would have fast and easy access to all types of information and even entertainment.

Unfortunately, for many people, this dream seems more like a nightmare. The lack of access charges assessed to Internet service providers for linking to the local network encourages ISPs to set up shop almost anywhere and charge unbelievably low, flat monthly rates for unlimited Internet access. The flat rate service encourages long (or seemingly never-ending) "phone calls" to these ISPs. Furthermore, the lack of adequate bandwidth on the analog local loop only makes the problem worse. Building more circuits in the form of trunks may turn out to be self-defeating as these facilities quickly are pressed into service for ISP access. On November 25, 1996, the *New York Times* reported on a central office in California with 2,000 trunks running to a single ISP.

Perhaps segregating data on the local loop will help. With bandwidth available for data packets, data could be routed directly to a packet switch network.

The Trouble with Loops Today

The primary trouble with local loops in today's environment of long data "calls" to the Internet and Web, as well as a need to provide enhanced services (video, for example), is that the local loop firmly remains analog, with few exceptions.

The passband (sometimes called "narrowband") analog local loop was not considered to be an issue when digital switches and trunks were introduced, and so were not addressed by this initial wave of digitization. When digitization arrived with ISDN, several stubborn characteristics of analog local loops confounded attempts at quick and cost-effective digitization.

Loading coils, which add inductance to counteract the attenuation caused by capacitance over long distances, are a big obstacle. Each coil must be painstakingly removed to digitize the loop; and even then, the possible presence of bridged taps and mixed UTP wire gauges only makes the situation worse.

ISDN deployment methods did not resolve these situations cost effectively for many reasons. A major issue was that ISDN, as an international standard, was a "package deal" that required not just the digitization of the local loop, but major changes to the central office switches that cost about $500,000. This cost left little to gracefully digitize the analog loops.

The relatively low-speed modems that analog local loops support encourage longer holding times for data sessions. This contributes to trunk group "exhaustion" and lower service levels for one and all.

Finally, the continued use of analog modems means that data traffic goes through the switch the same as voice traffic (that is, the data "sounds" just like voice to the switch), but the use of circuits for Internet packet usage merely ties up switch resources and leads to blockage all around.

Why should the PSTN worry about rebuilding the voice network for packets? Because packets will not go away, and, in fact, there will be more packets on the PSTN, not fewer. Even voice will be packetized as the new international standards for silence suppression and compression become common in the next few years. In late 1997, Reed Hundt, chairman of the *Federal Communications Commission* (FCC) said during a speech that "packet-switched networks will soon carry most of the country's bits, and that will change the economics, the structure, and just about everything else about the telecommunications industry. We are not sure how it will happen or how long it will take, but all indications are that it will happen."

What technologies exist both to increase the bandwidths available on local loops and relieve some the pressure on congested trunks and circuit switches? ADSL is one of them, but there are a host of others, most of which are introduced in the next chapter.

CHAPTER 5

Possible Solutions

Many solutions are possible for the congestion caused by the limited bandwidth and modern data traffic patterns on analog local loops. In fact, the issues extend to the trunking network and even the central office switch itself. One of the key solutions is ADSL, of course, but there are others. This chapter reviews them all and concludes that the family of technologies that includes ADSL might offer a number of distinct advantages over the others.

One possible solution is faster modems using the new—but incompatible —technology from a number of modem vendors, which allows for connections at 56 kbps (technically, the speed is limited to 53 kbps and often operates more in the 40 kbps range). This speed was formerly thought of as impossible, given analog local loop passband bandwidth. The catch is that one end must be digital (usually the remote end!) and that the 56 kbps is only one-way (that is, "downstream"). As has been mentioned previously, current regulations limit the actual speed to 53 kbps in many places. In addition, marginally faster downloading of Web pages may not reduce holding time. Users may simply look at more pages. Hold time, not data rate, might be the real culprit in this speed range.

Cable TV companies have been playing around with "cable modems" for a while now. These devices essentially bypass the local loop altogether and use the CATV network (usually itself analog) to provide high-speed Internet connections directly to a router or terminal server.

In the wireless arena, services such as *multipoint multichannel distribution services* (MMDS) and *local multichannel distribution services* (LMDS) have attracted attention as forms of wireless cable TV. In the case of MMDS, however, the data services still require a local loop connection for upstream traffic.

Satellite service providers have also considered data and interactive video services, notably DirecPC and newer systems, such as Teledesic, which have their own limitations.

The technology that has generated the most amount of interest in terms of solving the local loop issues (as well as most of the trunk/switch issues) once and for all is known as *Asymmetric Digital Subscriber Line* (ADSL). ADSL is a particular flavor of a whole family of technologies known as xDSL, or simply DSL, technologies.

The technologies introduced as potential solutions to the local loop, switch, and trunk capacity problems are shown in Table 5-1. The rest of

Table 5-1

Possible solutions

Possible Solution	Comment
56 kbps modems	No breakthrough technology, but good use of what is there
Cable modems	Plenty of bandwidth downstream on cable TV network
MMDS	Sometimes called "wireless cable TV"
LMDS	Sometimes called "cellular cable TV", but much more
Satellite systems	DirecPC is good example
ADSL	One of the xDSL family of DSL technologies

this chapter surveys them briefly and points out some of their advantages and drawbacks.

Overview of 56k Modems

The technology used in analog modems has increased in power and decreased in price dramatically over the past few years. Analog modems quickly went from 9600 bps to 14.4 kbps to 28.8 kbps to 33.6 kbps in a short time. However, things tended to stall as theoretical limits and physical impairments seemed to establish 33.6 kbps as the upper speed limit for full-duplex operation on the 3 kHz (or so) analog local loop bandwidth.

This is a good place to briefly review the history of modems in general and try to appreciate just how difficult it is to create modem devices that operate at higher speeds.

As was mentioned previously, the term "modem" is a sort of acronym that stands for *MOdulation/DEModulation*. A *modem* enables two devices generating digital information (such as computers, but not *always* computers) to communicate by using the public switched telephone network. The PSTN is optimized and limited to the frequencies used for human voice and can only carry sounds. Modems need to translate the computer's digital information format into a series of sounds that can be transported over the phone lines. When the sounds arrive at the destination, they are demodulated and turned back into digital information for the receiving computer or digital device.

All modern modems also use some form of compression and error correction. Compression algorithms enable the modem throughput speed to be enhanced from two to four times over normal transmission rates. Error correction examines the incoming digital data stream for absence of errors and requests retransmission of a frame when it detects a problem.

Modems in the early days of telecommunications networks—the 1950s —were all proprietary. They used simple techniques to operate at 300 to 2400 bps. These simple modems either used or were built on technology borrowed from radio frequency techniques developed during World War II and then applied to wireline telecommunications.

Prior to 1969, all modems had to be made by AT&T and installed by the Bell System. Customers played no role at all in the modem selection or installation process. The modem was as much a part of the PSTN as the telephone itself in those days. Naturally, this practice limited the spread of modems, as absolute control over technology usually does.

The rise of the modern market for analog modems can be directly traced to July 1968. In a landmark ruling (part of the famous Carterfone decision), the FCC decided that "the provisions prohibiting the use of customer-provided interconnecting devices were unreasonable." On January 1, 1969, AT&T had to revise their tariffs to allow the attachment of customer-provided devices (such as modems, answering machines, and so forth) to the public switched network. This was subject to three important conditions that AT&T imposed:

- The customer-provided equipment was restricted to certain output power and energy levels, which guaranteed that the customer device would not interfere with or harm the telephone network.

- The interconnection to the public switched network had to be made through a telephone-company-provided protective device, which in those days was sometimes referred to as a *data access arrangement* (DAA).

- All network control supervision and signaling, such as dialing, busy signals, and so on, had to be performed with telephone company equipment at the interconnection point.

The biggest problem turned out to be the protection device, which had to be purchased from the Bell System, was expensive, and was often not available. By 1976, the FCC recommended a plan in which the current protective devices would be phased out in favor of a type of registration plan. Registration would allow for the direct PSTN electrical connection of equipment that had been inspected, certified, and registered by an independent agency, such as the FCC, as technically safe for use on the PSTN.

In 1948, Claude Shannon, a Bell System scientist, wrote a paper that established for the first time firm theoretical limits on the speed that a modem could operate. The paper considered the top speed for a power limited and bandwidth channel hampered by noise. In other words, the typical analog telephone channel. The paper did not explain just how to reach this upper speed limit; it simply stated that this channel capacity limit could be approached by using the proper techniques.

Approaching this upper limit did not become an issue for years, but as more customers started buying and using modems, speed (and reliability) became important issues. Each individual modem vendor tried to get as close to "Shannon's limit," as expressed by "Shannon's Law," as they could.

By the 1960s, networks has become important enough to try to increase modem speeds. In the 1964 session of the ITU (then the CCITT), the first modem Recommendation, known as V.21 (1964), established an international standard for a 200 bps modem (which is now 300 bps). Oddly, V.21 is still used in the very 1990s V.34 (and V.8) modem *handshake*, which establishes basic compatibility between modems. Significant advances in line coding techniques came in 1968 and 1984 (V.22bis).

Also in 1984, a major advancement in modem technology and standards came in Recommendation V.32 with the addition of echo cancellation and trellis coding. Trellis codes were a major breakthrough because they paved the way for providing some degree of *forward error correction* (FEC) techniques to modems. Shortly thereafter, ITU Recommendation V.32bis improved on this use of trellis coding and increased the data rates to 14.4 kbps.

These breakthroughs only made technicians eager to achieve higher operational modem speeds. Work on the V.34 standard turned serious in 1989 and 1990. Acknowledging improvements in basic telephone network equipment such as switches, the initial V.34 goal of 19.2 kbps was moved to 24.0 kbps and then on up to 28.8 kbps. The latest version of V.34, from 1996, supports modem speeds up to 33.6 kbps. These latest modems achieve 10 bits per Hertz of bandwidth, which closely approaches the theoretical limits established by Shannon.

It came as somewhat of a surprise when several modem makers, most notably US Robotics, announced plans to market a modem that would operate at speeds of 56 kbps (often called "56k modems" and not "56 kbps modems," for some reason). Now, the "normal" T-carrier digitization technique for voice quantizes and encodes speech into 8-bit *pulse code modulation* (PCM) "words" 8,000 times per second, giving 64 kbps operation (8 bits x 8000/second). 56k modems use a similar trick, but because these modems operate over nonconditioned pairs (*conditioning* tunes the electrical characteristics to support 64 kbps), not all PCM words could be used, so the modem makers dropped a bit, leaving 128 signal levels instead of a the full 256 in 64 kbps operation. Technically, they use only 7 of the 8 companding regions for PCM voice. *Companding* is part of the voice digitization quantization process. This translates to 56 kbps operation (7 bits x 8,000/second).

Now, if it were this simple, surely such a step would have been taken before. As it turns out, it is not simple at all, and 56k modems are not just a trivial upgrade to devices at each end of a link (although this upgrade is still a requirement). 56k modems require one end of the link to be essentially a DS-0 running with 56/64 kbps digital line signaling and an internal digital trunking and switching network. Fortunately, these conditions are quite common.

Furthermore, there can be only one digital-to-analog conversion along the path from end to end. Usually, this is at the interface to the analog local loop in the downstream direction to the home. In the downstream direction, this digital-to-analog conversion is not susceptible to quantization noise and makes 56k operation in this downstream direction possible. The operation in this direction is straightforward: the 128 PCM word generates a tone (just as in voice PCM) that can be recognized at the far end so that the 128 PCM word can be reconstructed at the 56k modem attached to the user's PC.

Upstream is another matter. Here, analog-to-digital conversion noise (quantization noise) can be fairly severe, and so 56k modems are limited to "regular" 28.8 or 33.6 kbps operation in this upstream direction (where analog-to-digital conversion is necessary to enter the digital trunking/switching network).

Oddly, 56k modems are a good example of "asymmetric" operation, where the speed in one direction is not equal to the speed in the opposite direction. The general idea of 56k modem operation is shown in Figure 5-1.

Current federal regulations (FCC Part 68) limit the maximum voltage that can be applied to a telecommunications link to limit the possibility that the link might broadcast signals like a small radio transmitter and interfere with surrounding signals. In many cases, this limits the 56k modem operation to 53 kbps downstream. Upstream operation, at 28.8 or 33.6 kbps, is not a problem.

Pros and Cons of 56k Modems

New technologies are invented all the time. Some flourish (VHS and cassette tapes, for example) whereas others fade away to occupy niches or become footnotes (BetaMax and 8-track). All technologies are born with a set of advantages and disadvantages that follow them throughout their technical lifetimes. Sometimes these seem obvious, but they usually be-

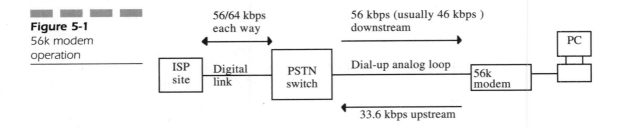

Figure 5-1
56k modem
operation

come apparent only in retrospect. Also, there is no official list of pros and cons, agreed upon by everyone, that characterizes a particular technology. Therefore, what follows is, of necessity, at best subjective and at worst all wrong.

Some of the pros and cons of 56k modems are summarized in Table 5-2.

In its favor, 56k modem technology requires no changes to premises wiring. If it worked with another analog modem, it should work with 56k. This is worth pointing out because many of the proposed residential solutions to "save" the PSTN require extensive premises wiring changes, and these costs are borne by the customer (of course).

Another nice feature is that the 56k modem can fall back to "regular" V.34 operation at 33.6 kbps when faced with an internal network without all digital trunks and switches, or with another analog local loop at the other end. Backward compatibility and flexibility are always strong technology selling points.

Also, the fact that this is "just a modem" is a big plus. People are familiar with and understand modems well—not just engineers, but many users. Familiarity has done wonders for Ethernet.

On the other hand, 56k modems are currently based on competing technologies without set standards for guidance. The two contenders are the X2 technology from US Robotics (now owned by 3COM) and Texas Instruments; and the k56flex technology from Lucent Technologies (a part of AT&T), Motorola, and Rockwell. This sounds like a common drawback for many technologies, and it is. But having two competing 56k modem technologies is worse because modems have earned a reputation for compatibility and interoperability. Also, changes are needed at both ends of the link. That is, you must have compatible 56k modems at each end because interoperability is impossible at this point.

Finally, 56k operation relies on a digital access link at one end. If two sites want to communicate, and both are currently analog local loops, who

Table 5-2

Pros and Cons of
56k Modems

Pros	Cons
No premises wiring concerns	Competing technologies and slow standardization
Falls back to "regular" modem speeds in a pinch	Changes needed at both ends of the link
Modems are familiar and well understood	Relies on digital access line at one end

gets to bear the expense of going to digital? All in all, the 56k modem is not an extensible technology (there is nowhere else to upgrade to from here), but 56k modems do maximize use of existing infrastructures with minimal changes. That said, the fact remains that 56k modems are not really 56 kbps, are not really modems, and are not even truly full duplex!

Overview of Cable Modems

To the providers of cable TV service, the battle over increasing the 4 kHz bandwidth on the analog local loop to residential dwellings seems silly. After all, the cable TV networks have delivered hundreds of MHz for years, and some systems now operate at 1 GHz (1,000 MHz). Of course, these signals are almost universally analog video, but this is no longer true in all cases.

In some cable TV networks, cable modems are used to convert digital data signals for transmission over coaxial cable. Cable modems are true modems in that they convert digital information content input and output for transmission as analog signals over an analog path. Cable modems carry data packets on *radio frequency* (RF) cable channels, the same as analog video. It should be noted that even cable TV systems with fiber installed (that is, *hybrid fiber/coax* (HFC) systems) still operate, almost universally, in an analog fashion. Only in these newer HFC systems can two-way communication take place easily. Only HFC systems support traditional two-way, analog voice services over the cable TV systems.

Cable modems take advantage of the enormous bandwidths that cable TV networks enjoy. Even a single "Internet channel" can run at 6 Mbps or better. Moreover, all cable TV systems have upstream channels reserved for interactive services. The FCC set these aside in the 5 to 50 MHz range for interactive services years ago. Less than 5 percent of cable TV networks can use these, however, because few devices in the home can generate upstream signals other than noise. In other words, cable amplifiers are usually one way only (to the home) and interpret anything in the opposite direction (from the home) as noise to be ignored or even eliminated. This makes sense given the one-way, broadcast nature of current cable TV networks. Also, the upstream channels must be shared by potentially large numbers of users.

The use of cable modems neatly sidesteps all of the issues regarding analog local loops, trunking capacity, and switching resources. Cable modem data traffic can be fed directly into an Internet router at the cable

TV head end. If voice channels and services are provided, this is the only traffic actually fed into the PSTN. In fact, the voice switch may even be in the cable TV head end site itself, especially if the telco is also the cable TV company. The general idea behind cable modem operation is shown in Figure 5-2

The nice thing about cable modems is that they can integrate both TV and PC functions into either device. Both exist as "Internet TV" (a TV with PC capabilities) or "PC TV" (a PC with a cable TV board and coaxial cable connector). Generally, PCs with cable TV video boards can watch 16 channels at once (but only listen to one), and these can even be iconized. The progression to allowing data communication seems a natural one. In fact, cable modems actually encourage this type of "convergence."

Pros and Cons of Cable Modems

This section covers the pros and cons of cable modems that are listed in Table 5-3.

On the plus side, cable modems take advantage of existing cable TV networks. No brand-new infrastructure is necessary, although changes to the cable TV network could be inevitable.

In addition, the radio frequency components required for cable modem operation are now inexpensive and plentiful. Many chipsets (the components of circuit boards that perform the communications tasks) are available at competitive prices and interoperability is quite high.

It should not be overlooked that in many ways, cable modems fulfill the early promise of interactive cable TV systems and services. Instead of building the functions into the set top box, however, the PC has taken its place.

Figure 5-2
Cable modem operation on HFC cable TV network

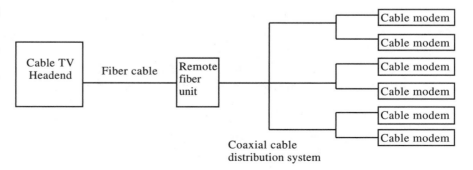

Table 5-3

Pros and cons of
cable modems

Pros	Cons
Runs on existing cable TV networks	May require extensive premises and drop cable rewiring
RF components relatively inexpensive and plentiful	Operates in upstream "garbage" bandwidths
Fulfills early promises for interactive cable TV systems	Shared upstream channels liable to easily overload

On the minus side, most PCs are not located anywhere near a TV. The TV is a distraction, and PC monitors and bus frequencies can distort television pictures. Extensive premises rewiring may be needed. However, where new premises cabling is needed, it is either coaxial cable or un-*shielded twisted pair* (UTP), both of which are inexpensive and well understood.

Cable modems may also require reinstallation of 90 percent of all drop cables that deliver services to the premises. Drop cables are often hastily and poorly installed, and lead to what is known as *ingress noise*. Another major problem is homeowner-installed splitters that hook up several TVs to one cable TV coaxial cable feed into the home. These also are often poorly installed.

A more serious drawback is the fact that the upstream 5 to 50 MHz bandwidth picks up a lot of "garbage" and noise from other home devices. Freezers, refrigerators, and a host of other devices radiate signals in this bandwidth, which is concentrated as it makes its way upstream through the branching bus network to the head end. In self defense, many cable TV networks filter out this "noise." Only 5 percent of cable TV networks can still use these bandwidths. In one-way cases, a normal PSTN modem must be used for upstream communication.

In addition, upstream amplifiers are typically nonexistent. Virtually all cable TV service providers use the PSTN to order pay-per-view events. Cable TV service providers cannot even handle the modest bandwidths needed upstream in order to support these services, let alone support the megabit rates needed for advanced services.

Finally, because the upstream channels must be shared by many users (perhaps thousands), a real danger of traffic overloads exist as all of these data packets converge on the single head end site. This has already been identified as a potential trouble spot by the cable industry, and various techniques for dealing with this issue have been proposed.

Multichannel, Multipoint Distribution System MMDS

Multichannel, multipoint distribution system (MMDS) technology is a strange blend of various video services and bandwidths. One component is the *Instructional Television Fixed Service* (ITFS). These twenty channels are used for the distribution of educational materials. Educational institutions such as colleges and universities must use at least 20 hours of air time per week in order to qualify as an ITFS licensee. In order to utilize the ITFS channels in building a full system, the MMDS service provider must use a technique called *channel mapping*. In this approach, when an ITFS channel being used by MMDS is required by an educational provider, the signal being carried is switched to an available ITFS channel. Of course, this switch must be transparent to the subscriber, so some reasonably sophisticated signaling between head end and set top is required to provide the service smoothly.

Eleven other channels are assigned to MMDS services, and two more are drawn from something called *Multipoint Distribution Services* (MDS). All together, MMDS could have 33 channels for video services or high-speed Internet access along the lines of cable modems.

The central transmitter used in MMDS costs about $1.5 million and the cost per subscriber for antenna, wiring, and set top is on the order of $400. Using these numbers, the Wireless Cable Association estimates that the break even point occurs at 10,000 subscribers serviced by a single MMDS transmitter. The basic MMDS architecture is shown in Figure 5-3.

Although the MMDS service has over one million subscribers on 73 systems in the U.S. and is being actively studied and employed by major U.S. providers, the real growth in this system is overseas where the penetra-

Figure 5-3
MMDS in operation

Set top

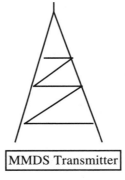

20-30 mile range of operation

MMDS Transmitter

Set top

tion of "wired cable" systems is considerably smaller than it is in the U.S. Indeed, Scientific Atlanta reports that they are selling more wireless cable (that is, MMDS) converters overseas than they are selling conventional systems. Internationally, 90 service providers supply 5 million subscribers with MMDS.

Pros and Cons of MMDS

Some of the pros and cons of MMDS are listed in Table 5-4 and then discussed.

A distinct advantage of MMDS is that the future could belong to wireless. The acceptance of cordless and cellular phones provides ample precedent. Restricting people to certain locations has long been considered a punishment of sorts. Wireless shows great promise in removing these restrictions.

Also, wireless bandwidth and speeds have been growing in leaps and bounds. On most MMDS channels, 54 Mbps downstream is achievable with state-of-the-art equipment (this equipment is admittedly expensive, however).

A further plus is that the FCC has begun issuing two-way licenses, which is necessary because the FCC is firmly in charge of which signals and strengths can be transmitted legally. Obviously, two-way wireless transmission requires a sending unit in everyone's home, though the FCC seems willing to grant such licenses on a routine basis now.

However, an important issue that remains is whether users will embrace what is, at heart, simply another TV package. The history of such "me too" services has been spotty. Without a clearly new service to offer or drastic price advantage, MMDS offers little to entice customers in huge numbers. Also, early MMDS trials have been disappointments in the

Table 5-4

Pros and cons of MMDS

Pros	Cons
The future could belong to wireless	Will users embrace another TV package?
Bandwidth is becoming plentiful (54 Mbps possible)	Early trials have been disappointing
Two-way licenses from FCC are beginning	Yet another totally new technology and system

United States. Signal strength has been erratic, even in the most carefully engineered systems.

Although not listed in the table as a separate issue, at MMDS wavelengths, a pine tree is like a stone wall to the signals, so line-of-sight from the receiver dish to transmitter is necessary. But this limits coverage, especially in tree-rich areas such as the Northeastern United States.

Finally, MMDS utilizes totally new technologies and systems. They might take a while to mature, and a lot of money will be needed to deploy them. In comparison, consider that after some 15 years, cellular phone service is still not universal in the United States.

Local Multipoint Distribution Service (LMDS)

Local multipoint distribution service is a multicell, point to multipoint, wireless distribution system, originally operating in the frequency band from 27.5 to 29.5 GHz. LMDS, sometimes known as "cellular cable TV," was invented by Bernard Bossard under experimental license from the FCC. The company that he formed, CellularVision, was originally funded by a New Jersey real estate developer. Bell Atlantic subsequently bought a minority interest and will operate and market CellularVision's services.

CellularVision (in theory) covers an area with multiple cells, thus avoiding many of the line-of-sight problems associated with MMDS. Shadow areas are projected to be covered by repeaters or passive reflectors. Adjacent cells use the same frequencies, but at different polarization. Although the general topic of polarization is beyond the scope of this material, some idea of the concept is presented here because it is of importance in understanding the technology.

All electromagnetic radiation (including light and TV signals) propagates from a source to a receiver as a wave, where the electric and magnetic fields associated with the radiation undulate like ocean waves. The direction of the electric and magnetic fields in a general signal vary in essentially all directions perpendicular to the direction of the propagation of the wave. Suppose that some device could be built that would block any electric field that did not point in a specific direction. The electromagnetic wave would continue to propagate through this device, but some energy would be lost, and the wave would contain a single electric field direction. Such a wave is said to be *polarized*. A common polarizing event for visible light is reflection off water. Light that reflects off water is horizontally polarized (that is, the water absorbs any electric field that does not lie in the horizontal plane). Polaroid sunglass lenses pass only light that is vertically polarized, which is how they work to remove the glare associated with light reflected off the water; they simply block anything with a direction of polarization that is not vertical.

CellularVision works the same way. Two adjacent cells use different polarizations of signal, and the antennas are like polaroid lenses (they reject signals from the adjacent cell and accept signals from the proximate cell). Coupling polarization with a small frequency difference between the same channel in adjacent cells results in a reduction of 65 dB in the signal strength from an adjacent cell at a receiver. This reduction is enough to prevent a phenomena known as *multipath fade* from taking over.

The concept of polarization also provides a reverse channel capability. One can signal "upstream" using the opposite polarity for the signal. No one else's receiver will accept the upstream signal; only the "head end" will hear it. Figure 5-4 shows the general LMDS architecture, which is quite similar to MMDS.

To accommodate current and proposed uses for the spectrum, the FCC devised a peculiar frequency plan for commercial deployment. Two licenses will be auctioned per geographic area. Block A (1.15 GHz) will comprise segments 27.5-28.35 GHz, 29.1-29.25 GHz, and 31.075-31.225 GHz; Block B (150 MHz) will be composed of 31.0-31.075 GHz and 31.225-31.3 GHz segments! Although potential services are not specified (not restricted), large incumbent cable and telephone operators are prohibited from license ownership until the year 2000.

Figure 5-4
LMDS in operation

Set top

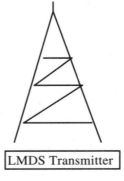

LMDS Transmitter

2-3 mile range of operation

Set top

Pros and Cons of LMDS

Table 5-5 shows some of the pros and cons of LMDS for high-speed digital services.

Most of the advantages and disadvantages of LMDS closely mirror MMDS and need not be repeated. The wireless advantage, 56 Mbps channels, and two-way licenses are the same. A very good sign for LMDS is that a large amount of interest exists, although much of it has been in Canada.

If TV services form the heart of LMDS, however, the issue is still whether users will embrace what is, at heart, just another TV package. Oddly, LMDS seems destined to form a new LAN interconnection package for urban businesses. Video services may be an afterthought. Also, both MMDS and LMDS lack an easy way to add more capacity. This is not so much of a problem in one-way broadcast systems where everyone shares outbound channels, but for inbound traffic, MMDS systems especially might be overwhelmed with no easy way to increase the licensed bandwidth to add capacity. A similar concern is present in the cellular telephone network.

With LMDS, the smaller transmitter size and coverage area (cell) can keep deployment costs manageable, especially in dense urban areas. However, this same small cell size could be a problem in suburban and rural areas. Suburban residents are notorious for opposing "unsightly" towers, and rural areas would need a lot of two-mile-across transmitters for any kind of penetration. Is it really feasible to give every farm its own LMDS tower?

Finally, like MMDS, LMDS is a totally new technology and system.

Table 5-5

Pros and cons of LMDS

Pros	Cons
The future could belong to wireless	Will users embrace another TV package?
Bandwidth is becoming plentiful (54 Mbps possible)	Small coverage area could be a problem in suburban and rural areas
Two-way licenses from FCC are beginning	Yet another totally new technology and system
Plenty of interest and manageable costs	

Overview of Satellite Systems

Direct broadcast satellite (DBS) systems already deliver high-quality TV pictures to many homes. Most deliver digital TV pictures (although to the same old analog TV for now) by using *Motion Pictures Experts Group II* (MPEG II) encoding. Audio (sometimes a kind of "jukebox" service) is generally Dolby Surround Sound delivered with MPEG I Layer II audio encoding. This is CD-quality stereo sound, not the tinny voice service or TV audio that most are used to.

Driven by equipment vendors and satellite system operators faced with delay problems for two-way voice and data services, DBS systems have found a ready market for their one-way, all-digital TV services. Most use some form of MPEG compression and carry multiple channels to provide a "near" video-on-demand service (that is, movies begin every 15 minutes). DBS systems have been hampered by problems carrying local broadcast channels, delays due to compression processing (sports events are delayed about one second), and lack of two-way operation (although some subscriber-to-satellite systems do exist). Most DBS systems rely on a telephone modem connection for upstream communication.

A single DBS channel can easily operate at 23 Mbps for data services. However, the most common speed is about 400 kbps (DirecPC, for instance). Current systems use *geosynchronous earth orbits* (GEO) at 22,500 miles altitude (about one tenth the distance to the moon). This increases delays for two-way, interactive services.

However, the next generation of satellites, such as Iridium and Teledesic, will use *low earth orbits* (LEO) systems, which will have minimal delays and require lower power transmitters. Once dismissed as a dream for the future, these systems are happening now.

One indication of the commitment service providers have in satellite systems is their cost. In the LEO arena, Iridium planned to have satellites costing a total of $5 billion. The number of satellites needed has already dropped once, from 77 to 66. In fact, the very name, "Iridium," comes from the fact that the element Iridium has 77 electrons circling its atomic nucleus (like 77 satellites orbiting the earth). The number of satellites dropped, but the name remains, probably because element 66 is Dysprosium, which does not have as nice a ring to it. However, Iridium is mainly targeted for premium cellular telephony. On the other hand, the Teledesic service, called "Internet-in-the-Sky," leaves no doubt as to who is being targeted. Teledesic's 288 satellites (!) will cost about $9 billion. There is also Globalstar, a child of Loral and Qualcomm, that will have 48 satellites costing $2.6 billion, but which is aimed at traditional data, telephony, and fax applications. Skybridge, planned for the near future, will launch 64 satellites for $3.5 billion.

The GEO systems are not about to go away after the LEOs come along, however. DirecPC, run by Hughes Communications, is the biggest (and only) satellite Internet operator right now. The 200- to 400-kbps downstream rate will be enhanced by the planned Hughes Spaceway GEO system of 8 satellites running at up to 6 Mbps downstream and costing $3 billion. On its heels is Hughes' planned Expressway system of 14 satellites for about $4 billion, delivering Internet server content at 1.5 Mbps and beyond. And we cannot forget CyberStar, a product of Loral (the same LEO people) and Alcatel, with a planned 3 satellites costing $ 1.6 billion. Cyberstar is planned to be truly two-way, with 2.5 Mbps upstream to go with the 6.5 Mbps downstream speed.

Motorola, a company with a 21-percent stake in Iridium, also has planned Celestri, an interesting mix of 9 GEO and 63 LEO satellites, costing $12.9 billion. Motorola plans to combine broadcast and multicast GEO strengths with the attraction of LEO systems for low delays for interactive Internet services. The action in the satellite world will continue to heat up for some time to come. Figure 5-5 shows the overall architecture of a satellite system.

From time to time, even more exotic plans for sky-based systems are proposed. For instance, a company recently proposed flying blimps over metropolitan areas with wireless equipment. Others have turned to tethered hot air balloons over Los Angeles and other cities. Blimps are one thing, but last time tethered balloons were tried in the 1970s, they had a disturbing habit of shearing the wings off aircraft that wandered into the guy wires. Such failures will not stop some from trying.

Figure 5-5
Satellite system in operation

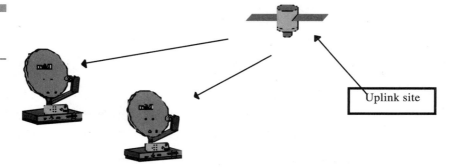

Pros and Cons of Satellite Systems

Some of the pros and cons of using satellite systems for high-speed digital services are listed in Table 5-6 and discussed here.

The biggest advantage, especially for geosynchronous earth orbit systems, is the wide ground coverage. Satellites of any type maximize the area that a given transmitter can reach for service coverage—an ideal situation (although there is some fading at the edges of the area).

Along the same lines, satellites give absolute terrain independence, and function in valleys, mountains, and at sea with equal ease. The services that satellites offer are totally insensitive to distance as well. So, a satellite serving sites thousands of miles apart is just as efficient as a satellite serving sites in the same neighborhood, as long as the sites are not in these fringe areas.

Another nice touch is that the satellite systems leverage the combined use of digital voice, video, and data technologies. After all, the very first satellites sent and received data; voice service was not added until years later.

The last advantage may seem silly, but the satellite people have played this card to maximum advantage for years. Where do you think the cable TV signal comes from? From satellites, of course! All that satellite systems do, the sales pitch goes, is eliminate the cable TV company in the middle.

On the other hand, satellites suffer from a number of serious and annoying technical problems. Satellite services may be eclipsed by planes and even lower orbit satellites. Solar flares may adversely affect signals, and they must routinely be shut down for a while when the sun passes behind them on its annual journey between the solstices. Some solar effects cause the atmosphere to literally "swell up," and low-orbiting satellites must be constantly moved to avoid the cumulative effects of atmospheric drag. The lifetime of a satellite is also determined by the amount

Table 5-6

Pros and cons of satellite systems

Pros	Cons
Maximum potential service area coverage	Solar effects and outages
Terrain independence, distance insensitivity, many others	Capacity and licensing could be an issue
Where do you think the cable TV signal comes from?	Crowding and sharing may become a real problem

of fuel it carries. Small gravitational effects must be compensated for by periodic rocket firings. When the fuel is exhausted, an otherwise functional satellite is pretty much useless.

The delays due to distance and digital compression, complex signal handoffs needed with LEO systems, and a need for true two-way service could become issues quickly. Ironically, some satellite transmitters are designed for home use, but they presently cost about $4,000.

In addition, naturally, the more two-way there is, the more important are issues of capacity and licensing. The FCC rules the airwaves, but a number of agencies, including NASA and the ITU-R, decide what goes into orbit and where.

A final drawback is that crowding of prime orbital positions may become a problem. The most desirable GEO positions have been occupied for years, and the same may happen with LEO orbits, although not for years to come. And of course the sharing of limited bandwidths among truly huge numbers of users could prove to be a problem.

Switch-Based Solutions

It seems clear that a lot of the congestion problems of switches and trunks in the PSTN could be relieved by identifying long holding time digital information streams and redirecting this traffic directly onto a dedicated network built for this purpose, such as the Internet. The question is now one of finding the best or most cost-effective solution for doing so. Solutions such as cable modems or wireless systems avoid the PSTN switch and voice trunking network entirely for this traffic by shifting it off the local loop altogether. Perhaps this traffic could still be identified as it arrives from the local loop and redirected before entering the full voice switch and trunking network. This is sort of a "make a left at the switch" approach. Several vendors have developed equipment to do this. A potential problem is that many of the solutions are tied to a particular vendor's switch or architecture. In other words, these are hardly standards-based solutions at the present time. There are five major vendors using this approach:

Lucent Technologies' *Access Interface Unit* (AIU) and Access Gateway are software-based methods used to identify long holding time traffic, usually at the ISP-attached central office. The software enables a service provider to move this traffic onto a more cost-effective part of the digital switch or to the AIU itself. The AIU provides a "virtually nonblocking" (that is, it rarely blocks) path through the switch and offers high capacities. The AIU only works with Lucent (AT&T) digital switches, but the Gateway boxes can reroute traffic arriving on the voice network for transmission over a dedicated data network.

Premisys' Interlude operates in two modes: for directly dialed or ISDN traffic, or in a remote hub configuration. In direct dial, the box just shunts aside data or ISDN traffic from the central office switch. As a remote hub, the Interlude gathers traffic, and directs voice to the PSTN and data somewhere else (usually the Internet).

DSC Communications' Intelligent Internet Solution combines hardware and software methods already on the market. They use the PSTN's signaling network (based on a protocol called "SS7") to redirect online service and ISP traffic away from the PSTN switch. Other signaling software detects these data connections and hands them off directly to a digital cross-connect (from DSC, of course) to make a connection to the data service provider.

NORTEL's Internet Thruway product is naturally based on NORTEL equipment. A NORTEL AccessNode, a *digital loop carrier* (DLC) device, looks at the dialed number and routes the call to the voice or data network as appropriate. Voice goes through the DLC to the switch, but data goes over a separate trunk to a NORTEL Rapport dial-up switch and from there to a packet-based data network such as the Internet. This package is very popular and NORTEL plans higher speed support, "always-on" connections, and data-over-voice support.

Finally, Telco Systems' Intelligent Call Routing employs an "access server" that intercepts incoming calls from sites with ISP/PSTN service and sends them to the CO switch or onto a packet network as appropriate.

ADSL and Other Technologies

All x-type *Digital Subscriber Line* (xDSL) technologies, of which ADSL is an important member, represent just one effort to solve the problem of broadband residential access for advanced services. Other methods exist, including direct broadcast satellite, MMDS/LMDS wireless CATV systems, cable modems for cable TV systems, ISDN (the original DSL, with two B-channels running at 64 kbps each and a D-channel running at 16 kbps), and even *digital data services* (DDS), which are leased lines running at 64 kbps. The only technology that is not really new is the 56k modem, for reasons already discussed. All the others are compared according to the four criteria shown in Table 5-7.

Table 5-7

DSL and other
technologies

Technology	High Initial Deployment Cost	Low-speed Return	Major Wired Infrastructure Change	No Analog Voice Support
DBS	X	X		X
MMDS		X (typical)		X
LMDS	X (but will diminish)			
Cable modems on HFC	X		X	
Cable modems		X		X
ISDN	X		X	
ADSL				

The table compares these other technologies to the xDSL family. The table contains four criteria, explained here:

- **High Initial Deployment Cost.** Some of the other technologies will require the investment of huge sums of money just to get off the ground. xDSLs are incremental costs—that is, analog local loops can be converted to xDSL almost on a home-by-home basis. (Of course, the services also have to be put in place, but these can be done incrementally as well.)

- **Low-speed Return.** In this context, low-speed does not mean asymmetric, but rather that some of the other technologies are inherently one-way at present, and so must use a "normal" analog modem for the upstream path. Ironically, maybe xDSL will help these technologies in this area! The slowest xDSL runs at 64 kbps upstream, and many will operate at much higher speeds.

- **Major Wired Infrastructure Change.** This refers to the fact that some technologies, although not requiring an entirely new infrastructure to be put into place before services can be offered to even one customer, will still require extensive changes to the way such systems currently operate. With xDSL, the intent is to maximize reuse of the existing infrastructure.

- **No Analog Voice Support.** This refers to the fact that some of the other technologies make no provision for older analog phones. A word of caution is in order here: although ISDN supports analog telephones, a conversion device (a terminal adapter, or TA) is needed. Also, some xDSL schemes (such as HDSL) make no real provision for analog phone support at all.

Note that the ADSL row in the table is blank. ADSL, and indeed all xDSLs, suffer from none of these limitations. All in all, xDSL offers more advantages and fewer disadvantages than any other technology. The time has come to see how.

CHAPTER 6

Introducing the xDSL Family

ADSL is more than just a technology that allows for broadband access from the residence or small office to a network service provider, ISP, or not. ADSL is one of a number of access technologies that can be used to convert the access line into a high-speed digital link and to avoid overloading the circuit-switched PSTN. These technologies form a family loosely called *x-type digital subscriber line* (xDSL) technologies, where the *x* stands for one of several letters of the alphabet. It is important to note that some of these technologies are based on modems. That is, some of the xDSL family use analog signaling methods to transport analog or digital information content across the access line or local loop; they have much in common with other modem technologies, of course. Other members of the xDSL family use true CSU/DSU arrangements. These members use true digital signaling to transport digital information content (seldom analog information) across the access line or local loop. They have much in common with T-carrier arrangements.

This chapter details the operation of each of the DSLs and provides additional information about them, as well. The focus shall be on ADSL and its close relative, RADSL (pronounced *RAD-sil*). The common "thread" among them is that they are all based on existing pairs of copper wires installed as local loops, as opposed to most of the alternative solutions examined so far, which rely on totally new networks and technology infrastructures for the most part. Once the frequencies above 4000 Hz can been used, higher speeds can be attained on the local loop. The xDSL family is a set of copper-based solutions.

Copper-Based Solutions

There exist many possible solutions to the problems of overloading the *public switched telephone network* (PSTN) voice network with packetized data and interactive broadband services. Some involve building entirely new systems based on wireless and satellite networks, and there is certainly nothing wrong with this. However, it may be better to start with something that already exists and builds upon or improves the operational capabilities of the copper-based analog local loop. The only solution based on using the copper local loop not considered here is the new 56k modems because these do nothing to avoid problems with the circuit switches in the PSTN. Indeed, they may make matters worse by encouraging even longer holding times on voice switches.

This approach is not meant to detract from cable modem or other solutions. It is simply a more practical approach. For instance, this is just an acknowledgment that because more than 90 percent of cable TV systems run as strictly one-way downstream operations, adapting these systems for cable modems would still amount to rebuilding the entire infrastructure. A more realistic and cost-effective solution would be the following:

1. Maximize reuse of existing analog local loops
2. Include some provision for backward compatibility with existing voice telephony equipment (that is, the analog handset)

For the time being, it seems that only copper-based solutions satisfy these two criteria.

DSL's full name, Digital Subscriber Line, began with the *Integrated Services Digital Network* (ISDN), which was created to foster the total digitization of the PSTN end-to-end, from user device (handset, PC, and so on) to user device. ISDN was the first DSL service, and its position as the first of the DSLs should never be forgotten or minimized. Many of the advanced features of ADSL became possible only through the experience gathered with ISDN DSL methods.

For residential services, the ISDN DSL takes the form of the Basic Rate Interface (BRI). The BRI operates at 144 kbps full-duplex, organized into 2 Bearer (B-) channels running at 64 kbps and one D-channel for signaling and data running at 16 kbps. The two B-channels may be bonded to yield 128 kbps in most circumstances, although not universally. The ISDN switch (called the Local Exchange (LE) in ISDN) must allow this bonding to occur through the assignment and use of Service Provider Identification numbers (SPIDs). Some uses of SPIDs by the LE makes B-channel bonding impossible in certain ISDN LEs.

Newer DSL technologies are more interesting and promising. As was stated previously, it is common to list these as xDSL, where the x represents any one of a number of letter designations. Some xDSL technologies are sometimes called *duplex*, in the sense that the speeds are identical in both directions. Note that this use of the term *duplex* differs from the usual sense in the United States of "both directions." When applied to DSL it means "both directions *at the same speed*," and so contrasts with "asymmetrical." In spite of this use of the *duplex* term, it is far more common to describe DSL speeds as simply "symmetric" (same speed in both directions) or "asymmetric" (different speeds in each direction).

However, many broadband residential services are distinctly asymmetrical, such as video-on-demand or Internet Web access. That is, the amount of traffic sent upstream from a home or client PC is much less than the traffic sent downstream to a home or from a server Web site. In this case, it makes more sense to allow for higher speeds downstream (into the home) than upstream (out of the home). Indeed, newer versions of xDSL (ADSL, RADSL, and VDSL, for example) are inherently asymmetrical.

High-bit-rate DSL and HDSL2 (a more up-to-date version of HDSL) is a duplex technology (again, in the sense of "symmetric speeds"). The speed upstream and downstream is either 1.5 Mbps in the United States or 2.0 Mbps in most other areas around the world. This aligns HDSL with the existing T-carrier DS-1 speed in the United States and the existing E-carrier E1 speed elsewhere. In fact, HDSL and HDSL2 are intended for transporting a DS-1 over copper lines and are most often deployed as a more cost effective way to deploy DS-1 services. The customer still sees and buys a DS-1, but it is provisioned as HDSL or HDSL2 within the network. As just a "newer and better" DS-1, HDSL and HDSL2 are most often used for the same purposes, namely carrier services for feeder plants such as DLC pairgain systems, or to the customer (as a DS-1) for LAN interconnection or leased-line WAN access.

Next to be considered is *Symmetric* (sometimes seen as "single-line") DSL (SDSL). For a while, SDSL seemed to be a promising variation on HDSL, intended for all that HDSL could do, and more. For instance, SDSL was capable of both 1.5 Mbps and 2.0 Mbps operation in both directions, but only to limited distances. However, it now appears that in its most common form, SDSL will refer to solutions delivering speeds below T1.

Some xDSLs are asymmetrical in nature. The speeds upstream are typically much less than the speeds downstream. Given the extremely asymmetric nature of most client-server interactions, however, especially on the Web, this should not be a drawback in most cases. A possible exception is when a home PC user or SOHO business wishes to run a Web server in their home. In this case, naturally, it is desirable to have at least symmetric (that is, duplex) speeds. Perhaps HDSL or HDSL2 would be more suitable for such home Web site arrangements.

Asymmetric DSL (ADSL) and its close relative Rate Adaptive DSL (RADSL) do not differ much, if at all, in terms of speed and distances. In fact, since most ADSLs are now rate adaptive, it makes less and less sense to distinguish ADSL and RADSL, but this distinction is still made here for historical and educational reasons. Both function between 1.5 Mbps to about 8 Mbps downstream and 16 kbps to about 640 kbps upstream, but these are only common figures. Both have a variety of uses, all centered on interactive multimedia applications. For simple Internet or Web access, either ADSL or RADSL will do just fine. And even for video-on-demand services, of simplex (one-way "broadcast" quality) video TV services, either ADSL or RADSL is more than adequate at higher downstream speeds. For remote LAN access for telecommuters, ADSL and RADSL will be a key service, as well. Of course, both ADSL and RADSL allow for the continued use of existing analog telephones, a real plus.

ISDN DSL (IDSL) sounds odd because ISDN already employs a DSL, but this combination actually makes sense. IDSL supports the ISDN 2B+D BRI structure running at 144 kbps in both directions. The problem in part is that the ISDN BRI is used in large measure for fast Internet and Web access. ISDL gets the ISDN BRI off of the circuit switch when the line is used simply for Internet and Web access, alleviating a great deal of the switch congestion and allowing more "real" ISDN users to be supported through the switch. The D-channel in the BRI can no longer be used to set up ISDN voice connections on the PSTN, but if the BRI is used exclusively for ISP access, this is not a major consideration anyway.

Finally, *Very High Speed* DSL (VDSL) is at once promising and ambitious. Usually considered asymmetric, the VDSL specification calls for an optional symmetric configuration. VDSL speeds cannot be achieved wholly on the longest lengths of local loop copper and must employ fiber

DLCs for at least half of the access line distance to the switching office. Speeds are an amazing 13 Mbps to 52 Mbps downstream, and, a not an inconsiderable, 1.5 Mbps to 6.0 Mbps upstream. The applications supported include all that ADSL/RADSL is intended for, plus *High Definition TV* (HDTV) digital television services. VDSL is usually seen as the ultimate evolutionary goal of all xDSL technologies.

Please note that supported xDSL speeds vary in both directions, depending on the physical characteristics of the analog local loops on which they are deployed. Although one home may be able to use ADSL at 1.5 Mbps downstream, another home located nearby may enjoy only 768 kbps transmission speeds downstream. In most cases, though, speeds greatly exceed those available with "regular" modems or ISDN BRI.

Another point is that one should treat all xDSL speeds and distance limits with caution. In most cases, these are merely *design parameters* and not set in concrete. Vendors are always pushing the xDSL envelope of speed and distances. But it is always true that higher speeds must be balanced by lower distances.

The xDSL Family in Detail

The major characteristics of the current x-type Digital Subscriber Line family of technologies are shown in Table 6-1. The emphasis here is more on technical operation than applications. Newer variations on the xDSL theme, such as Multispeed DSL (MDSL), exist, but these are relatively new, championed by one particular vendor or another, and difficult to distinguish as members of the family in their own right at this point. The order is roughly by age of the technology. That is, *High-bit-rate* DSL (HDSL) essentially came first and *Very High-speed* DSL (VDSL) is the newest, with the exception of *Consumer* DSL (CDSL), for reasons discussed later.

Remember to treat all xDSL speeds and distance limits with caution; consider them design parameters, not hard and fast rules. Vendors are always pushing the xDSL envelope, so in many cases today, it is possible to see HDSL products based on *discrete multitone* (DMT) modulation (rarely), full 1.5 Mbps ADSL out to 18 thousand feet (kft), and Symmetric DSL with analog support via splitters (just to mention a few of the variations).

The xDSL family members listed in Table 6-1 are explained here:

■ **HDSL/HDSL2—High-bit-rate DSL.** As was mentioned previously, HDSL runs at 1.544 Mbps (T1 speeds) in the United States

Table 6-1

The xDSL Family in particular

Name	Meaning	Data Rate	Mode	Comment
HDSL/HDSL2	High data rate DSL	1.544 Mbps	Symmetric	Used two pairs;
		2.048 Mbps	Symmetric	HDSL2 uses one wire pair
SDSL	Single Line DSL	768 kbps	Symmetric	Uses one wire pair;
ADSL	Asymmetric DSL	1.5 to 8 Mbps 16 to 640 Kbps	Down Up	Uses one wire pair; 18 Kft max
RADSL	Rate Adaptive DSL	1.5 to 8 Mbps 16 to 640 Kbps	Down Up	Uses one wire pair, but can adapt data rates to line conditions
CDSL	Consumer DSL	Up to 1 Mbps 16 to 128 kbps	Down Up	Uses one wire pair, but does not need remote equipment in home
ISDL	ISDN DSL	Same as ISDN BRI	Symmetric	Uses one wire pair; sometimes called "BRI without the switch"
VDSL	Very high data rate DSL	13 to 52 Mbps 1.5 to 6.0 Mbps	Down Up	High data rates; 1 to 4.5 Kft max. Needs fiber feeder to function, and ATM

(really throughout North America) and at 2.048 Mbps (E1 speeds) almost everywhere else. Both speeds are symmetric (the same in both directions). The original HDSL at 1.544 Mbps used two-wire pairs and extended to 15,000 feet (about 2.8 miles). HDSL at 2.048 Mbps needed three wire pairs for the same distance (but no longer). The latest versions of HDSL, known as HDSL2, employ only one pair of wires and are expected to be much more standardized to enable vendor interoperability.

■ **SDSL—Symmetric (or Single pair) DSL.** If the goal of xDSL technology is to reuse analog local loops, then perhaps it would be better to employ only a single wire pair, which is what analog loops are. SDSL uses only one wire pair but tops out at 10,000 feet (less than two miles), at least in its design specifications. Nevertheless, the speeds are the same as with HDSL. SDSL is typically provisioned at 768 kbps using single pair HDSL. Because it seems likely that HDSL2 will do all that SDSL can do and more, it is expected that SDSL will be "cannabalized" by HDSL2.

- **ADSL—Asymmetric DSL.** SDSL used only one wire pair, but the need to support duplex speeds limited distance. ADSL acknowledges the asymmetrical nature of many broadband services and at the same time extends the reach to 18,000 feet (about 3.4 miles).

- **RADSL—Rate-adaptive DSL.** Typically, it is assumed when equipment is installed that some minimum criteria for line conditions are met to allow for operation at a given speed. At least this has been true of former digital technologies, such as T-carrier or ISDN. However, what if line conditions vary or operational speeds make equipment sensitive to small environmental changes? RADSL, which is an inherent property of ADSL using *Discrete Multitone* (DMT) coding, can actually adapt to changing line conditions and adjust speeds each way to maximize the speed on each individual line.

- **CDSL—Consumer DSL.** Although closely related to ADSL and RADSL, CDSL is sufficiently different to rate an entry of its own. CDSL is generally more modest in terms of speed and distances compared to ADSL/RADSL, but it has a unique advantage. With CDSL, there is no need to worry about remote devices known as *splitters* on the customer premises. The function of the splitter on the premises is to allow existing analog voice telephone and others types of equipment, such as fax machines, to continue to operate as before. The splitter needed with ADSL and RADSL is discussed more fully later in this chapter. The point is the CDSL needs no remote splitter and it's associated wiring.

- **IDSL—ISDN DSL.** This technique takes the normal 2B+D channels of the *ISDN Basic Rate Interface* (BRI), which runs at 144 kbps (two 64 kbps B-channels and one 16 kbps D-channel), and runs the BRI not into the ISDN voice switch, but into the xDSL equipment. IDSL also runs on one pair of wires and extends up to 18 kft, exactly the same as an ISDN DSL.

- **VDSL—Very High-speed DSL.** The newest member of the family, VDSL is seen by some as the "ultimate goal" of DSL technology. Speeds are the highest possible, but only over 1,000 to 4,500 feet (less than a mile) of twisted-pair copper wire. This is not a problem for VDSL. VDSL expects to pick up a fiber feeder at this 1,000 to 4,500 foot point, and it is also intended to carry asynchronous transfer mode (ATM) cells, not as an option, but as a recommendation. This aspect of VDSL is explored more fully in a later chapter.

HDSL and T1

Before there was DSL in any form, there was T1 (T-carrier, level 1 multiplexing). T1 started as a local loop digital system to aggregate voice channels and reduce analog noise, but quickly came to be used as a digital trunking system intended to carry many voice calls between switching offices. Because many DSLs are based on at least some components of T1 circuits, a brief review of the basic T1 components is in order.

Because it was designed as a trunking, or "carrier," system, T1 expects some multiplexing to take place. In 1984, after years of internal use in the PSTN, T1 was made available for customer installation. This multiplexed access arrangement is shown in Figure 6-1.

When used to carry analog voice conversations, a channel bank was installed as customer premises equipment (CPE) at the customer site, which was typically owned and controlled by the customer. Up to 24 analog voice inputs (and outputs) could be attached to the channel bank, which digitized each one into 64 kbps full-duplex. There was then a *Channel Service Unit* (CSU) and *Digital Service Unit* (DSU) to multiplex the conversations and encode the bits for transmission. The CSU/DSU was attached to a *network interface unit* (NIU), which essentially formed the demarcation point between the service provider's network and the CPE.

At the other end of the link, a similar CSU/DSU was attached to the central office channel bank, which, in turn, fed the digitized voice channels into a variety of switching or alternate arrangements.

Figure 6-1
T1 components

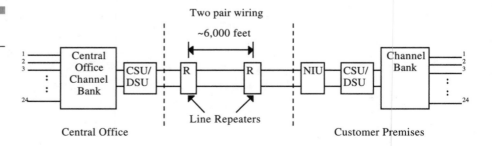

In between each end of the link, the T1 was carried on two pairs of copper wire. One pair was for sending, and the other was for receiving. Transmission took place at 1.544 Mbps (1.536 Mbps for user traffic) in both directions. Naturally, the T1 might extend for miles from the customer premises to the central office. In that case, special devices known as line repeaters, which "cleaned up" and repeated the digital signal, were placed at intervals along the line. The most common spacing was 6,000 feet, which made sense because analog loops typically had H-88 loading coils at 6,000 foot intervals anyway. Therefore, conversion from two analog wire pairs to a single T1 was easy. An outside plant crew removed the loading coils and installed line repeaters. Capacity on the two pairs now went from only two analog conversations to 24 digital conversations. Bridge taps also had to be removed, if present.

T1 was wildly successful for digital access to the PSTN. In fact, the popularity of T1 led directly to the adaptation of HDSL as a way for service providers to deploy T1 circuits more quickly and cost effectively for both internal use and to deliver to customers.

HDSL as "Repeaterless T1/E1"

The T1 multiplexed trunking system formed the basis of the Integrated Services Digital Network. As the ISDN *Primary Rate Interface* (PRI), a T1 provided a business customer with 1.536 Mbps of access to an ISDN switch. In most cases, however, this 1.536 Mbps was provided as exactly twenty-four 64 kbps channels. As the ISDN Basic Rate Interface, conceptually, a "piece" of a T1 was provided to a residential customer (or small office/home office (SOHO) user) with 144 kbps of access to an ISDN switch. However, this 144 kbps was in most cases provided as two 64 kbps channels and one 16 kbps signaling/data channel. Usually, the two 64 kbps channels could be bonded to yield 128 kbps.

So far, so good; but as analog modem speeds grew closer and closer to 64 kbps on their own, ISDN speeds—once blazingly fast—seemed relatively slow for many services. Even PRI was most often channelized into 64 kbps chunks.

Outside of North America and Japan, the PRI was supplied on an E1 and not a T1. The E1 is similar to the T1 in that there are 64 kbps channels present. However, where the United States T1 supplied 24 channels, the E1 supplied 32 channels, and 30 of these could be used for customer information. The E1 line rate was 2.048 Mbps.

Additionally, a T1/E1 for PRI required two pairs of wires and active line repeaters (which "cleaned up" and repeated the digital signal) every 6,000 feet. This made widespread use of T1 for residential services too expensive since each circuit installation had to be carefully engineered. However, if a T1 could be made to use one pair of wires or to function without repeaters for longer distances, perhaps T1 (and E1) would have new life as a residential DSL technology. HDSL was designed specifically to meet these conditions.

Bellcore proposed HDSL in the mid-1980s to address ISDN channel rate limits and the physical restrictions of T1s for PRI (and BRI, for that matter). HDSL was called a "non-repeatered T1/E1 replacement model" at the time. What Bellcore adapted was a scheme that still used the same 2B1Q (two binary, one quaternary) line code from the ISDN BRI DSL, but was now defined to attain up to 784 kbps on one pair up to 12,000 feet without repeaters. (As a word of caution, some HDSL products use other line codes, such as CAP or even DMT. Details on the operation of and differences between these line coding techniques are discussed later.) The 784 kbps represents twelve 64 kbps "channels," along with the normal BRI 16 kbps signaling/data channel.

When the original HDSL was duplexed onto two separate pairs, HDSL could attain 1.544 Mbps in each direction, the same as a T1, but without the 6,000-foot spaced repeaters. Even some bridge taps were tolerated, as long as they were not too long. The 12,000 feet also applied to loops beyond the *remote terminal* (RT) in a *carrier serving area* (CSA) on 24 gauge wire, or 9,000 feet on 26 gauge wire. However, the HDSL access line still needed two pairs of copper wire, which limited the applicability of HDSL to home or SOHO situations, where a second pair of wires was harder to come by. HDSL2 will address this limitation.

Initial versions of HDSL at 2.0 Mbps (E1 speeds) just re-used the available 784 kbps chipsets and therefore needed a third pair to deliver the E1 line rate of 2.048 Mbps. Because the two pairs in HDSL ran at 784 kbps, the third pair was needed to move beyond this speed. Newer versions of HDSL products can actually run at 1.168 Mbps on each pair, making it possible to support E1 speeds. Therefore, HDSL needed no special engineering of the loop with repeaters for distances up to 12,000 feet, and the speed is still equal to T1. Confusingly, a technology called "half-duplex HDSL," but really a form of SDSL, attains a speed of 384 kbps on distances up to 18,000 feet.

So HDSL was intended as a way to furnish T1 speeds in a duplex fashion to bandwidth-hungry customers. The HDSL architecture is shown in Figure 6-2.

Figure 6-2
The HDSL
Architecture

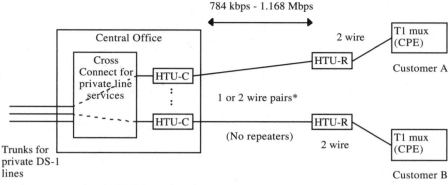

784 kbps - 1.168 Mbps

Central Office

Cross Connect for private line services

HTU-C

2 wire

HTU-R

T1 mux (CPE)

Customer A

1 or 2 wire pairs*

HTU-C

HTU-R

(No repeaters)

2 wire

T1 mux (CPE)

Customer B

Trunks for private DS-1 lines

* early E1 2.048 Mbps HDSL required 3 pairs

In the figure, a typical HDSL system is used to provide DS-1 private line services at 1.544 Mbps to two customers. Before HDSL, this would have required the use of two pairs of wires and repeaters every 6,000 feet to provide the classical T1 architecture. With HDSL, the service provider purchases two units, which currently must be from the same vendor, because interoperability has not been a goal or concern before the latest version of HDSL, known as HDSL2. The two units are the HDSL Termination Unit (HTU) in the service provider's central office (the HTU-C) and the HTU-R, which is the remote unit placed as close to the customer's premises as possible. The HTU-C is typically a rack-mounted series of units placed close to the central office's wire center or main distribution frame. The lines from the HTU-C are cross-connected to provide channelized or unchannelized T1 service as usual.

The HTU-R unit connects to the HTU-C over a single pair of wires in most cases. Beyond a certain distance, which varies from HTU vendor to vendor, two pairs of wires are still needed. Even so, the advantage of HDSL is fact that no repeaters are needed. With some older HTU equipment, a third pair of wires was required to provide E1 service at 2.048 Mbps outside of the United States. Newer HTU equipment provides E1 speeds over two pairs, or even one pair over very limited distances in some cases.

The HTU-R is still very much service provider equipment and not customer premises equipment. From the HTU-R to the customer premises, two pairs of wires are still needed. The customer premises equipment is still a common and relatively inexpensive T1 multiplexer (technically, the endpoint of the HDSL link is the CSU). The two pair interface to the customer is still retained mainly for backward compatibility. The current HTU-Rs are not standardized enough yet to allow customers to purchase just any vendor's HTU-R and expect it to work with the service provider's

HTU-C unless the two HTUs are from the same vendor. This will change with HDSL2, but for now this is the case.

The main advantage of HDSL is that it allows the service provider to provision T1 service more quickly and cost effectively. Almost any local loop wire pair will do. No repeaters or special engineering is needed on the line, and one pair of wires should always be available. This simplicity allows service providers to lower the monthly costs of what is essentially still T1 service, but internally provisioned on HDSL.

Note that HDSL does not run to the PSTN switch. HDSL is just a point-to-point private line solution. The other end of an HDSL link is another access line used for private line service. This access line could be a "real" T1 or, in many cases, another HDSL link. It is still a leased line service, not a switched service. The HDSL HTU-Cs and HTU-R are not modems—they are properly DSU arrangements; the same as for T1 lines, which HDSL basically mimics. No analog signals exist on a digital line using 2B1Q unless an extraordinary effort in made to do so. Therefore, a local loop employed for HDSL cannot be used with analog telephones at the same time, except in rare HTU arrangements.

No provision at all in HDSL is made for backward compatibility with existing analog telephone handsets. Presumably, a special digital coding unit would be attached to these telephones so that they could utilize one of the 64 kbps digital channels. Either that or the analog telephones would be removed and regular office-type digital telephone used in their place. Often, a small user *private branch exchange* (PBX) or *key telephone system* (KTS) would be installed.

Sometimes, especially recently, the single pair basic 784 kbps "version" of HDSL has used at this lower speed (sometimes called SDSL or even *Medium-speed DSL* (MDSL) by various vendors), which extends up to 22,000 feet. However, early trials show the 22,000-foot-distance reliable speed to be more like 272 kbps in most cases. HDSL distances are improving with each new wave of products. All this may change, however, with the introduction and standardization of HDSL2. More details on HDSL and HDSL2 are presented in the following chapter.

2B1Q and ISDN

It might be a good idea to take a closer look at the relationship between 2B1Q line encoding and ISDN. The 2B1Q (two binary, one quarternary) line encoding was intended for the use of the ISDN DSL. 2B1Q is a four-level line code (so known as quaternary) that represents two binary bits (2B) as one quaternary symbol (1Q). The 2B1Q line coding was seen as a major enhancement over the original T1 line coding, which was something called *bipolar alternate mark inversion* (bipolar AMI), because 2B1Q encoded two bits instead of just one with every signaling state (baud). This also means that a modern ISDN link running the BRI at 160 kbps (two 64 kbps B channels plus a 16 kbps D channel plus another 16 kbps overhead) operates at 80 thousand *symbols* per second (80 kbaud) instead of 160 kbaud.

2B1Q line encoding was intended to deliver ISDN BRI speeds (144 kbps, plus the line overhead) through local loops up to 18,000 feet. This was done on only one pair of wires, though, and basically gave 144 kbps full-duplex in each direction using the same frequency range.

So what's the point about 2B1Q and xDSL? As it turned out, 2B1Q not sophisticated enough to achieve multimegabit speeds at long distances, and, although 2B1Q required less bandwidth than bipolar AMI, 2B1Q still used the frequency range that analog voice would normally use on a purely analog local loop. Figure 6-3 illustrates this. Note that carrierless amplitude/phase modulation-based (CAP-based) HDSL uses much less of the available spectrum on twisted pair loops than either bipolar AMI or 2B1Q. Also, CAP, as any other passband modulation method, at least holds out a chance to preserve the 300 to 3300 Hz passband for analog voice service on the same wires (a later chapter covers CAP in more detail). Not only that, the 2B1Q technology was becoming dated by the mid-1990s. Perhaps for newer DSL technologies, 2B1Q is no longer the best way to go.

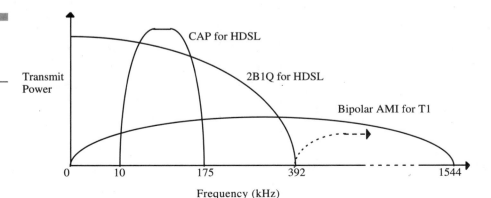

Figure 6-3
Bipolar AMI, 2B1Q, and CAP frequency ranges

Now may be a good time to discuss various line coding techniques for DSL technologies. The familiar 2B1Q code for ISDN is a member of a family of line codes known as *Pulse Amplitude Modulation* (PAM) codes. To make PAM codes more suitable and efficient for DSL uses, however, it is common to try to "optimize" the PAM code with a technique known as *spectral shaping*. All this means is that the PAM code should not "go all the way down to 0 Hz frequency," or add a direct current component to the access line. It is important to realize that a PAM code, whether optimized or not, is not a CAP or QAM, but something different. PAM is still very much a "baseband" code, although a formal definition of "baseband" in this context is sometimes difficult to put in non-technical terms.

Every time a new "flavor" of xDSL technology is proposed, a great deal of debate revolves around the use of one type or another of PAM, CAP, or QAM as a line code. The important point is that these debates concern three main factors:

1. Whether the code naturally uses all of the available bandwidth.

2. How efficient the code is in terms of speeds and distances.

3. Whether the code is susceptible to outside interference from other line code schemes used in nearby wire pairs.

To make things even more complicated, because the ISDN BRI used one pair of wires, a special hybrid arrangement was necessary because the rest of ISDN essentially decreed two-pair wire operation. This operation necessitated the use of special echo canceller devices within the DSL device. (Generally, whenever full-duplex operation is needed with the same frequency range over the same wires, echo cancellation is required.) Note that echo cancellation is required for full-duplex, long distance voice conversations, as well as full-duplex, shared frequency digital links—this applies equally to analog voice, ISDN, or xDSL, which is a key point. Whenever the same frequency range is to be used for signals in both directions at the same time, some form of echo cancellation must be used, whether the signal is analog or digital.

Moreover, all modems since the early 1980s have used their own echo cancellation techniques to achieve full duplex operation on the same passband voice frequency range (300 to 3300 Hz) across the single pair analog local loop.

In any case, it is no longer universally true that HDSL products use 2B1Q line coding exclusively. Other codes, such as carrierless amplitude/phase modulation (CAP) or dual multitone modulation (DMT), are used as well, although such use is extremely rare.

"Single Pair" HDSL: SDSL

SDSL is usually defined as Symmetric DSL, but because HDSL (and several other xDSL variations) are also symmetric, this definition loses something and is less than helpful. SDSL started out be taking half of a two-pair HDSL system. So SDSL ran at 784 kbps on a single pair of wires.

Lately, SDSL has begun to be defined as "single pair" HDSL, which is better because SDSL shares much in common with HDSL, and yet, functions on a single pair of wires. (The original HDSL requires two pairs in most cases, and originally needed three pairs in some other cases, such as when supporting E1 speeds of 2.048 Mbps.)

Of course, using multiple pairs of wires for residential service is not an ideal situation. Digitizing the analog local loop would be better if a DSL scheme used the existing single pair of wires—hence, SDSL.

Once the SDSL idea caught on, vendors started to get very creative. Some vendors made versions that operated faster (i.e. 1.5 and 2.0 Mbps), but with limited transmission distance. Other vendors made versions that operated at lower speeds (i.e. 384 kbps), intended to support longer distances. This lower speed SDSL version is sometimes known as MDSL, but this is not a generic DSL at all.

In spite of the popularity of HDSL and the much anticipated HDSL2 standard, SDSL might still have a place in the DSL world. Figure 6-4 shows an SDSL system used in a pairgain system.

Figure 6-4
The SDSL architecture in a pairgain system

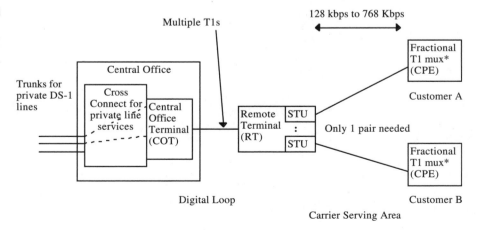

* modified CSU needed to accept one pair interface

In the figure, a Central Office Terminal and Remote Terminal are connected by multiple T1 links to form the common Carrier Serving Area architecture. Two pairs of wires normally would need to be run from the RT to the customer to support the T1 speed of 1.544 Mbps. the most common RTs, however, only support four T1s themselves, so the architecture is limited if the whole idea is to deliver full T1 rates to each customer. But in a SOHO or residential neighborhood, perhaps just a fraction of a T1 is okay, as long as the price is a fraction also. Maybe the customer could make do with 256 kbps (four 64 kbps channels) or even 128 kbps (two 64 kbps channels)—the T1s servicing the RT would not be exhausted so readily. This is what SDSL and it variations are for.

Note that there is really no such thing as an "STU," at least not as a standard device. Whatever it is called, the STU is housed at the RT. Only one pair of wires is needed to deliver services to the customer site, which needs a modified fractional T1 multiplexer (technically, a modified CSU) to accept a one-pair interface. Also, "half-duplex" translates to 128 kbps in some cases, and up to 768 kbps in others, over only one pair of wires (hence the term "single pair"). The 128 kbps represents the bandwidth of two 64 kbps channels. The speed of 768 kbps represents the bandwidth of twelve 64 kbps channels.

SDSL allows a service provider to provision DSL service based on three basic parameters: the cost, reach, and speed of the service. Based on performance needs, distance from the local exchange, and budget considerations, customers can choose from a number of the SDSL options. Service providers typically price different services on a staggered scale.

The currently supported maximum speeds and distances of SDSL are shown in Table 6-2

SDSL could very well continue as a common variation of HDSL, even as HDSL2 appears.

Table 6-2

SDSL speeds and distances

SDSL Data Rate	Maximum Distance
128 kbps	22,000 feet (6.71 km)
256 kbps	21,500 feet (6.56 km)
384 kbps	14,500 feet (4.42 km)
768 kbps	13,000 feet (3.97 km)
1.024 Mbps	11,500 feet (3.51 km)

Enter Asymmetric DSL (ADSL)

Asymmetric DSL addresses some of the limitations that HDSL, HDSL2, and their variations imposed on the newly digitized local loops.

First and foremost, HDSL, SDSL and the others rarely made allowance for analog voice. (However, nothing prevents vendors today from supporting analog voice, along with techniques like SDSL, especially with line codes other than 2B1Q, and in fact, some do.) Most people had, and have, analog telephones in their homes, yet HDSL, HDSL2, and its variations only carry digital signals, so pure HDSL or SDSL required users either to purchase special conversion units (known as terminal adapters, or TAs) or to purchase digital telephones. However, neither alternative appealed to users or to the telephone companies. Maybe some way could be found to allow the continued use of analog phones on the newly digitized loop with less expense.

Secondly, a concerted effort was made on the part of the telephone companies in 1992 to deliver digitized video (and the accompanying audio) to the home. Many technologies were explored, and the attraction of ADSL at the time was the promise of delivering these services over the same loop used for analog voice, so ADSL could form a basis for video-on-demand services and so-called "video dial tone" systems that were heavily promoted at the time. Such video service required large amounts of bandwidth downstream (that is, inbound to the home), yet not much bandwidth was needed upstream (that is, outbound from the home). After all, brief commands to start, stop, fast forward, or freeze a video stream were small data packets in the first place.

It also turned out that many home-based activities followed this asymmetric model. Internet Web server access, normal client/office server actions, and even home shopping are inherently asymmetric in the same fashion. Any DSL technique that supports larger bandwidth in one direction and smaller bandwidth in the other is, by definition, an Asymmetric DSL—ADSL is just that.

For a while, it seemed that ADSL was an "umbrella" term for a variety of asymmetric xDSL techniques, including RADSL and VDSL. However, the terminology seems to be moving back toward a more uniform and specific approach to individual xDSLs. The ADSL Forum Web site (**www.adsl.com**) has information about all xDSL technologies, not just ADSL.

More detailed information on ADSL is presented at length later in this work. For now, it is enough to position ADSL within the xDSL technology family. Figure 6-5 shows the general ADSL architecture.

Figure 6-5
The ADSL architecture in general

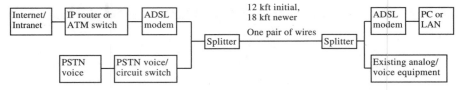

Downstream speed defined to 8.192 Mbps

In the figure, the two distinguishing features of ADSL compared to other xDSLs are apparent. The *splitter* is a device that comes between the local exchange and the customer premises; its function is twofold. First, the splitter allows existing analog voice telephone and other equipment, such as fax machines, to continue to operate as before on the customer's premises. Second, the splitter allows the long holding time data traffic to be rerouted around the PSTN voice switch (where it is carried on circuits) onto an IP router or ATM switch network (where this data traffic is carried in packets). This alleviates pressure on the PSTN and cuts down on user costs because not all customer equipment needs to be changed or interfaced with special adapters (as with HDSL and others). The routers or ATM switches carry user traffic to servers presumably located on the Internet or a corporate Intranet, although many other variations are allowed and envisioned.

Next, ADSL is asymmetrical. The downstream speed is much greater —sometimes ten times greater—than the upstream speed. A maximum speed downstream of 8.192 Mbps is defined for ADSL. However, this speed may be quite difficult to achieve in practice, not due to ADSL limitations, but due more the limitations in throughput given the current Internet architecture and backbones. This would be a little like having a 100 mile per hour on-ramp to a 55 mile per hour freeway. So for the foreseeable future, most equipment vendors and service providers would settle for 4 Mbps to 6 Mbps as a maximum.

Also note that the ADSL link is *not* switched in and of itself. In other words, the ADSL link forms another type of leased private line from a user's PC or LAN to one other place in the world, but while the ADSL link is not circuit switched, the *content* of the ADSL link is packet or ATM cell switched. That is, if the service-provider end of the ADSL link ends at an IP router or ATM switch connected to the Internet, the traffic on the ADSL link may still be able to find its way almost anywhere that the switched telephone network does today.

Rate Adaptive DSL (RADSL)

Rate Adaptive DSL addresses a possible limitation of some early ADSL devices, especially those based on carrierless amplitude modulation/phase modulation (CAP). Basically, once some initial ADSL equipment is in place on a formerly voice bandwidth analog local loop and connections are made, the newly digitally enabled line operated at a fixed speed upstream and downstream for the duration of the link's lifetime. The downstream speed especially might vary from location to location and wire pair to wire pair, typically in increments of 32 or 64 kbps. One of these limited ADSL lines on a block could achieve 640 kbps, for example, but a neighbor's might be limited to 608 kbps or even 576 kbps.

The problem is that line conditions on local loops vary all the time. Line conditions may improve or deteriorate, usually depending on rain conditions, or even as the solar radiation on, and overall temperature of, the wire rises and falls seasonally or from day to night.

RADSL is theoretically able to adapt to these changing conditions on the fly, even during active sessions, although there is currently no provision to do so in current RADSL products. For example, users might receive traffic that starts at 576 kbps in the morning, goes to 640 kbps in the afternoon, and then drops to 608 kbps in the evening.

The whole concept is similar to the idea of "self-equalizing" modems. Modems of 20 years ago needed line conditioning to maximize performance, which was relatively easy to do for leased private lines, but next to impossible to guarantee on switched dial-up connections. A key parameter in line conditioning was *equalization*, which balanced attenuation over the whole frequency spectrum used. When newer modems could perform self-equalization, it was possible for modems to "drop back" to lower speed operation if line conditions were poor (which is why even new 33.6 kbps modems still connect at 28.8 or 14.4 kbps). After the 14.4 kbps connection was made, however, that is where it stayed, even if the connection lasted hours and line conditions improved. RADSL speeds might someday be able to vary on the fly, moment by moment.

All of the other properties of RADSL essentially mimic ADSL in terms of maximum speeds and distances. RADSL is a natural progression of ADSL, and all ADSL equipment should evolve into RADSL in the future. For equipment based on Discrete Multitone (DMT), RADSL is an inherent capability that appears as a result of the way the technique functions. Although RADSL operation is not impossible in CAP-based ADSL, such operation is difficult to achieve and adds a lot of circuitry and procedural overhead to the CAP devices. The biggest difference is that with CAP, the signal spectrum changes. Nevertheless, several CAP-based ADSL vendors have introduced RADSL equipment.

The basic architecture of a RADSL link is exactly the same as the basic ADSL architecture.

CDSL: ADSL/RADSL without the Splitter

Early trials with ADSL and RADSL uncovered a rather serious situation, regarding the customer's premises. ADSL/RADSL required the installation and maintenance of a remote device—the splitter introduced above. The premise's splitter's main function was to allow the continued use of existing analog telephony and faxing devices in the home or SOHO location. However, besides introducing complexity, the presence of the splitter also raised issues about the premise's wiring and configuration, issues that are detailed in a later chapter. For now, it is enough to point out that there was a real concern about the continued care and feeding of the remote splitter device.

The first concern involved the need for the service provider to make an appointment with the customer for splitter and possibly wiring installation. This may not sound like much of a concern, but in today's highly mobile world and everybody-works-who-can environment, finding someone at home during business hours can be a chore. A customer often had to take a day off from work, and missed installation appointments lead to complaints about missed work time on the part of the customer, and so forth. A further concern was the need to dispatch a truck and technician, which added considerable cost to service initiation and slowed deployment. In many cases, the remote splitter was provided as part of the service, adding to the service provider's capital costs. The associated wiring issues only added to the service delay, expense, and complexity. Clearly, if a way could be found to install and configure ADSL or RADSL speeds and distances, and at the same time support existing analog devices without the need for the remote splitter, this would be a very attractive alternative to pure ADSL/RADSL.

Near the end of 1997, Rockwell Semiconductor Systems introduced an xDSL variation it called *Consumer DSL* (CDSL) to address these concerns and limitations. Rockwell also proposed its method for standardization before the ITU, as something called "G.adsl lite," which shows the close relationship between ADSL/RADSL and CDSL. In fact, the only significant difference between ADSL/RADSL and CDSL besides the absence of the premises splitter and wiring concerns is a restricted operating speed range (most importantly, 1 Mbps downstream as opposed to about 8 Mbps with ADSL). By the end of 1997, Nortel, Microsoft, Compaq, and Intel had all made CDSL support announcements. The future of CDSL appears not only assured, but bright.

The local exchange (central office) side of the link is unchanged. That is, a splitter is still needed there to separate high-speed data packets from voice conversations. The local exchange splitters are housed in the same type of equipment as in the ADSL/RADSL architecture. The big change is on the premises. The overall architecture of CDSL in shown in Figure 6-6. The remote splitter is eliminated.

CDSL has a number of characteristics that make it attractive both to customers and service providers. The major ones follow:

1. **Adequate access speeds for most users.** Faster Internet access might have the effect of slowing down other portions of the network. This "moving bottleneck" could be a serious problem with many users expecting data at 6 to 8 Mbps downstream. Can servers and the Internet backbone keep up? CDSL adds only 1 Mbps at a time to the load, which would still be more than enough

Figure 6-6
The CDSL architecture in general

Figure 6.6: The CDSL architecture: ADSL/RADSL without the splitter

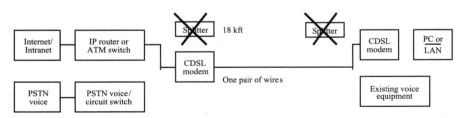

Downstream speed defined to 1.0 Mbps
Upstream speed defined as much less (128 kbps)

for what users currently expect on the Internet and Web. It adds less load and reduces the cost of the CDSL modems.

2. **Easy installation at the premises.** Perhaps this should be #1. With CDSL, there is no need to arrange an appointment for installation, worry about wiring, supply a remote device, or be concerned about future tech support. CDSL service providers are concerned as much about the premises as telephone companies are concerned about user's modems, which is to say minimally.

3. **Simultaneous voice and Internet and Web access.** In spite of the absence of the remote splitter, it is important to point out that CDSL still allows simultaneous voice telephone calls and Internet and Web access over the same local loop. There is no requirement for "alternating" use, as with present modem versus telephone arrangements on a single local loop.

4. **Low cost to customers.** Without the need for added splitter electronics, CDSL modems should be less expensive than ADSL or RADSL remote modem devices, whatever the packaging. Costs should be no more than current prices for high-end modem packages running at 56 kbps. It should never be overlooked that customers often mentally add the price of required equipment to the cost of a service (they are technically separate) when deciding whether a service is too expensive. CDSL minimizes this impact.

5. **Low provisioning costs.** Even with the use of CDSL, a service provider must still install equipment in the local exchange (central office). However, the cost of the equipment can be offset by economies of scale, and the fact that the installers and management are local. The advantage of CDSL is that there is no need to "roll the trucks" to the premises or concerns about managing remote devices scattered all over a service area.

6. **Standards initiative.** Rockwell has had a high degree of success attracting service providers and equipment vendors to the CDSL architecture. The ITU is entertaining CDSL as "G.adsl lite," and, in the United States, Rockwell is working with the ANSI T1E1 committee to create an open standard for CDSL in the near future.

In view of the interest in CDSL, we should take a final summary look at this promising xDSL technology. CDSL modems will not replace 56 kbps or other modems, but they will offer a higher-speed evolutionary path with fewer service provider concerns than any other xDSL solution. However, CDSL is still xDSL and, as such, is not just a seamless upgrade from 56 kbps modem operation. CDSL is still dependent on line conditions for maximum speeds (1 Mbps downstream), but the more modest maximum should allow the maximum to be reached in more varied circumstances than ADSL or RADSL. More importantly, CDSL is not intended as a *replacement* for ADSL or RADSL at all. CDSL's intention is to *complement* service providers' ADSL and RADSL deployments into places where higher speeds are not particularly mandated or feasible, and where there are concerns about remote splitter and wiring installation. ADSL and RADSL are still expected to be popular services in and of themselves.

CDSL still requires the same equipment arrangements as ADSL or RADSL in the local exchange or central office. Eventually, CDSL could be put directly into the voice switch port itself, but this option is available to ADSL/RADSL as well. Because CDSL is a variation of ADSL/RADSL, there should be minimal issues associated with tariffs or contracts for CDSL service. CDSL will not see ADSL or RADSL as much of a "rival" as other technologies altogether, such as cable modems or even 56 kbps modem products.

Very High-speed DSL: Newer and Better?

VDSL is an effort to address at least three issues with regard to broadband services delivered over the local loop. First, the telephone companies are employing more and more fiber feeders, as pairgain and digital loop carrier (DLC) systems. It is not unusual for a neighborhood, especially new housing developments or condominium complexes, to be served by fiber optic feeders, with copper local loops only handling the last few thousand feet to a home. Second, the services that the telephone companies and others want to provide seem to require more and more bandwidth, almost year to year. Third, as xDSL systems evolve to include more mixtures of voice/video/data traffic all in one, some accommodation for *asynchronous transfer mode* (ATM) cell transport may be desirable because ATM easily combines voice/video/data services on the same physical network. While it is true that ADSL is also aimed at ATM networks, but accommodates IP as well, VDSL will virtually demand ATM networks.

Figure 6-7 shows how the basic VDSL concept addresses these three issues. First, VDSL includes an optical network unit to convert and concentrate VDSL signals onto a fiber feeder system, which might be part of

Figure 6-7
The VDSL
Architecture

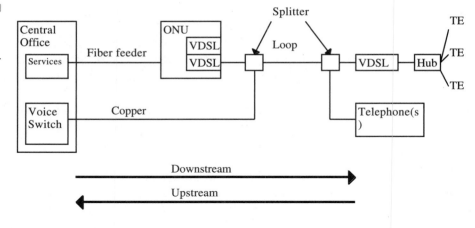

Downstream speeds: 12.96 - 13.8 Mbps   4.5 Kft
25.92 - 27.6 Mbps   3 Kft
51.84 - 55.2 Mbps   1 Kft
Upstream speeds: From 1.5 Mbps to equal to downstream

the "next generation" of DLC systems (called NGDLC). These NGDLC systems are distinguished by a great deal of distributed intelligence, links to Sonet fiber-optic ring configurations instead of RT to COT point-to-point digital links, and easy access to broadband services at the central office.

Second, the bandwidths downstream are far above those defined for ADSL. Speeds of around 13 Mbps will be achievable for 4,500 feet, and 1,000 feet of copper supports a whopping 50 Mbps or so. Upstream, VDSL offers a minimum of 1.5 Mbps and could even be deployed in symmetric configurations.

Third, VDSL is intended to carry ATM cells; that is, VDSL is intended to form a Physical Layer for a full service ATM network. In this case, the home VDSL hub is actually an ATM switch, and the services are on a variety of ATM-attached servers. Other modes of operation are allowed, however. The combination of ATM cells transported on VDSL links is intended to be used to support *switched digital video* (SDV) systems. Of course, a wide range of additional services could be offered as well, including very high speed Internet and Web access.

VDSL is sometimes portrayed as an evolutionary step up from ADSL. For now, VDSL viability seems tied to fiber availability and ATM popularity.

Note that VDSL still offers backward compatibility with existing analog phones via a splitter, as does ADSL. Also, VDSL fully expects there to be multiple devices on the customer premises, in many variations. These are all *Terminal Equipment* (TE) to VDSL and can include PCs, LANs, television sets, and even refrigerators or air conditioners.

So VDSL is more of a "full service network" strategy for service providers. ADSL is more of a "data overlay network" strategy. This is not to criticize either VDSL as a pie-in-sky dream or ADSL as a half-way measure. This just reflects what each is really intended for.

IDSL and the Integrated Services Digital Network (ISDN)

What does the future hold for the Integrated Services Digital Network (ISDN) in an xDSL world? ISDN was intended to digitize the analog local loop many years ago and formed the very first DSL. Is there still a place for ISDN?

It is worth remembering that the whole movement toward xDSL began with the bit-rate limits that channel structures and Basic Rate Interfaces (BRI) imposed on digital loops. The need for higher bandwidths, especially for video services, led first to HDSL and then to ADSL and VDSL.

Now, ISDN channels could be grouped, or bonded, together to provide higher bit rates, and so 768 kbps is definitely possible on the Primary Rate Interface (PRI) (but not the BRI). It was the search for a better (more cost-effective) way to provide PRI bandwidths on analog local loops that led to HDSL in the first place.

The whole problem with ISDN is not BRI or PRI speed limits. It is the fact that ISDN still runs through the voice switch. Here is where trunking and switching tie-ups become serious. After all, these are not just voice services anymore; and when we consider that the upgrade to every central office switch costs about $500,000 to convert to ISDN, ISDN looks more and more like a technology people could live without.

The end result of many years of ISDN conversions and service offerings have been spotty availability (sometimes literally by street address number), price worries (will it go up? down? stay the same?), and continued trunk and switch blocking woes.

The history of ISDN deployment in the United States has not been a happy one. However, its future is not necessarily bleak. ISDN consists of two things: a *digital network* (DN) and *integrated services* (IS). In most residential or SOHO applications, the DN is supposed to be a *digital subscriber line* (DSL) running at 144 kbps (BRI) using 2B1Q (two binary, one quaternary) coding. In addition, the ISs are supposed to be switch-based. The real attraction of ISDN was to be in the integrated voice, video, and data services.

However, in some places, ISDN DSLs for BRI service were often just used as 144 kbps "bit pipes." That is, the two B-channels were permanently bonded and used like a 128 kbps "leased line" running to the switching office. The 16 kbps signaling channel was still present, but messages went into the "bit bucket." At the central office, the line ended up on a cross-connect that led to other DSLs at other sites. Therefore, there were no real ISDN services involved or available.

Why bother with the ISDN DSL then? Well, it turns out that this was a common scheme in many areas of the world outside of the United States for carriers to "bootstrap" themselves into the digital leased line marketplace without running new cable. (Leased lines were scarce and expensive outside of the U.S., mostly for economic and policy reasons.) This plan allowed the use of standard ISDN DSL gear to digitize the loop, and that was all that was needed to supply a *Digital Data Service* (DDS). The plan required no ISDN switch software upgrade and was very popular in South America. How could these ISDN DSLs be converted to xDSL service architectures?

Perhaps xDSL standards could be redefined to use ISDN DSL, and use the newly planned xDSL services (such as video and Internet access) as the "IS" part, optionally of course. This idea makes a lot of sense and is the plan behind what is sometimes known as IDSL (ISDN DSL over xDSL). Such a process would require a modest standards rewrite, but this rewriting goes on all the time anyway.

Figure 6-8 shows two IDSL units operating at 160 kbps over one pair of wires. The line is organized into normal BRI channels as 2B+D service at 144 kbps aggregate. At the premise's end, the user can keep or purchase any ISDN-compliant TAs or TE equipment for excellent backward compatibility. At the local exchange, all D channel signaling messages are ignored because IDSL is still essentially a private line service. This should not be a limitation because most residential users currently employ ISDN to access the Internet or a corporate Intranet. Either one is reachable with ISDL.

Another advantage is that the new xDSL services available through IDSL would no longer require the conversion of a central office switch to ISDN (at about $500,000 per switch).

IDSL seems like an idea worth pursuing further.

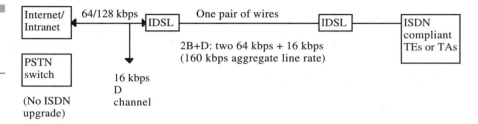

Figure 6-8
Two IDSL units operating over one pair of wires.

xDSL Advantages

Now is the time to examine the advantages that xDSL has with respect to the other access technologies explored in the last chapter. The attraction that xDSL technologies have for service providers is summarized below.

First of all, xDSL goes in only when a customer requests service. A service provider need not spend millions of dollars and then wait for customers to sign up. Initial costs are expected, of course, but they are generally much lower than other competing technologies.

Any xDSL requires no change to central office switch software. In most cases, a splitter carriers normal analog voice into the switch, but all other services are handled through separate servers and routers.

Also, xDSL can be used for residential users, SOHO users, and large organizations alike. The xDSL technology may be different (HDSL, for example), but the service should be essentially the same, with the possible exception of streaming video services.

Another nice thing about xDSL is that some versions, especially ADSL/RADSL and VDSL, can interface with a number of different premises arrangements. Individual set-top boxes and PCs are supported, as well as entire home LANs, such as Ethernet. Even newer electrical wiring schemes such as CEBus (Consumer Electronics Bus) are allowed at the home end of an ADSL/RADSL or VDSL line.

xDSL will even provide an infrastructure for asynchronous transfer mode (ATM) cell transport (especially Very High-bit-rate DSL, but Asymmetric DSL also). This is important because ATM, in turn, forms the basis for the international standard set of broadband services known as *Broadband ISDN* (B-ISDN). It is hard to think of many other technologies accommodating ATM as well as xDSL is able to, especially ADSL/RADSL and VDSL.

Finally, xDSL is not a future technology—it is available here and now!

Cost Comparison of Various Broadband and Other Technologies

Before examining the more important members of the xDSL family in more detail—especially ADSL—we will take a brief look at comparative prices. Some preliminary pricing information is given in Table 6-3. This is a general attempt to allow prospective service providers and customers to get an idea of how competitive the pricing will be for the newer broadband access services being developed today.

This table was compiled from prices appearing in industry sources, such as *Network World, Communication Week*, and so on. These are only generalizations; particulars vary widely. Some details on speeds have been sacrificed for brevity. Prices may drop rapidly for new technologies; these are from late 1997.

The table starts by looking at the price of a 64 kbps DS-0 link, like that commonly used to link routers or other network devices in a purely pri-

Table 6-3

Broadband access cost comparison

Technology	Speed(s)	Monthly cost	Other costs	Comment
DS-0	64 kbps	$ 150 average	Minimal	Router connections
BRI (2B+D)	128 kbps bonded	$185 average	$350 average for TA	$500,000 switch upgrade to ISDN
DS-1 without HDSL/HDSL2	1.5 Mbps	$1000-2000	$500 average	Router connections
PRI (23B+D)	1.5 Mbps	$1300 average	$350 average for TA	$500,000 switch upgrade to ISDN
HDSL (HDSL2)	1.5 Mbps	$175 ($60-120 from RBOCs)	$1200 for equipping line	New DS-1?
ADSL	1.5-6 Mbps	$200 average	$500 average ATU	Cost going down
CDSL	1 Mbps	$40-45 average	$200 average	No remote splitter
IDSL	128 kbps	$200 average	$350 average for TA	Preserves ISDN CPE
Cable modems	1-10 Mbps	$25 - 100	$1000 average "TA"	Only HFC 2-way
LMDS	50 Mbps	$50	$1000 CPE target	2-way? Video also?

vate line network. The cost is modest, but so is the speed. The other costs involved as minimal, but the price rises with distance.

Above the DS-0 speed in many parts of the world, it is possible to use ISDN links configured according to the Basic Rate Interface (BRI). This provides two B channels running at 64 kbps each and a 16 kbps D channel, which can be used for signaling and other purposes. Usually, the B channels can be bonded together to form a single channel running at 128 kbps, but above and beyond the monthly recurring charge of about $185, the user must purchase an ISDN Terminal Adapter (TA) for about $350. Also, the service provider must upgrade the switch software and other hardware, which can cost up to $500,000.

A DS-1 runs at an impressive 1.5 Mbps but can cost anywhere from $1000 to $2000 per month, depending on distance between end points. The equipment needed at each site will also cost the user about $500. These links are also commonly used to link routers or other network devices in a private line network. It is important to note that this entry does not reflect the impact of HDSL on the pricing of the DS-1. That is, this entry is basically a "traditional" T1 running on two pairs of wires using bipolar AMI line coding equipment with repeaters.

Running at the same speed as a DS-1, the ISDN Primary Rate Interface (PRI) consists of twenty-three 64 kbps channels and one 64 kbps D channel for signaling and other purposes. The PRI is just a better BRI in some sense, although it is usually harder to bond the B channels together for higher speeds. The $1300 monthly cost is high, but the $350 for the ISDN Terminal Adapter (TA) is not too bad. However, many TAs are usually needed. Of course, the same ISDN switch upgrade is still necessary.

HDSL runs at the same speed as the DS-1 and ISDN PRI. The cost is typically much less than a "pure" DS-1, however—usually about $175 a month. Some former Bell system companies even offer HDSL at $60 to $120. It does cost up to $1200 to equip a line for HDSL, however, mainly due to tighter physical requirements. HDSL may turn out to be the "new" DS-1 because it looks exactly the same to users. Of course, the temptation is to continue charging DS-1 prices for HDSL provisioned links. This situation is discussed in more detail in the next chapter.

ADSL can run anywhere from 1 to 6 Mbps downstream to a home, but slower upstream. The $200 monthly cost is more of a target, but the relatively expensive $500 cost for the *ADSL Termination Unit* (ATU) that each user must purchase should come down rapidly. Many early ADSL service providers bundle the ATU cost with the service charge itself. Some ADSL service providers have astonished the industry by offering ADSL service for as low as $40 a month.

CDSL shares many of the properties of ADSL, but runs at a lower speed (1 Mbps). These "1 Mbps modems" require no splitter, and costs should be lower. Target prices are $40 to $45 per month for the CDSL service, and about $200 for the CDSL modem itself.

IDSL is essentially "ISDN without the switch." It provides the same 128 kbps bonded B channel speed as the ISDN BRI, but does not require any expensive local exchange switch upgrade or changes to the local loop. The monthly charge is targeted at $200, and metered service might add a lot to the total cost of IDSL. Because IDSL appears to the user as a type of ISDN link, the same ISDN TA equipment can be used, which costs about $350. However, some service providers require a special TA.

Cable modems are a popular alternative to DSL techniques, even among the telephone companies for some strange reason. Unfortunately, in most implementations, cable modems are a sort of half-way solution. That is, cable modems often still need a normal modem telephone connection to function in the upstream direction to the service provider. The 1 to 10 Mbps is usually only on the cable TV system in the downstream direction. However, the faster speed is typically seen as just another premium cable TV service and is priced from only $25 to $100 per month. The $1000 for the "terminal adapter" or cable modem is a drawback, but prices will fall rapidly, as well. Cable TV systems that are based on a *hybrid fiber-coax* (HFC) cable scheme can operate two-way, upstream and downstream, but these are few and far between.

Finally, the local *multipoint distribution system* (LMDS) often described as "cellular cable TV" is the latest entry in the broadband access sweepstakes. LMDS offers an astounding 50 Mbps, but typically only downstream to the home. The targeted monthly price of $50 is unsurpassed, but again, it is only a target. Even the customer premises equipment price is targeted in the $1000 range, there being only one functional LMDS system as yet in the United States. LMDS may yet turn out to be two-way and be as popular as cable TV for general video services. However, the auctioning of bandwidth has been slow, and the long-term future of LMDS as a broadband access technique is not a given.

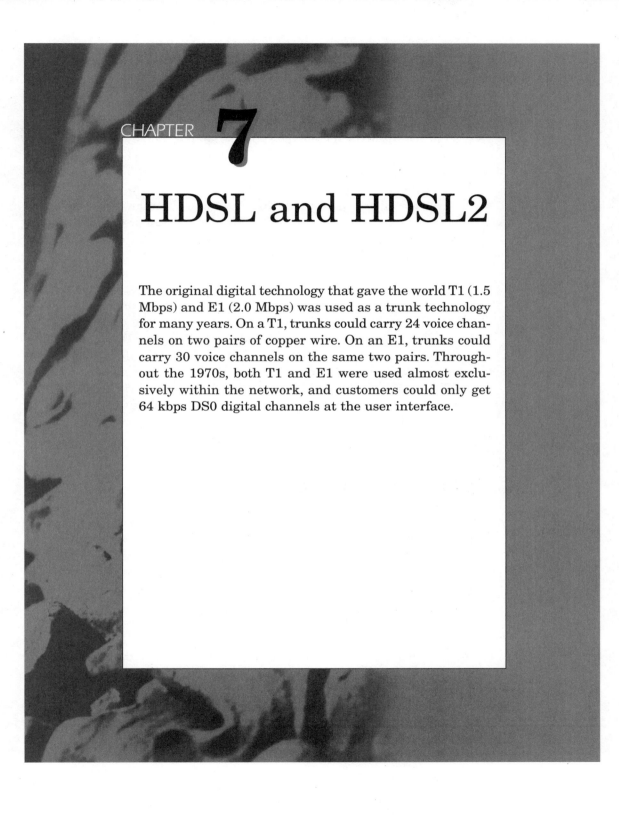

CHAPTER **7**

HDSL and HDSL2

The original digital technology that gave the world T1 (1.5 Mbps) and E1 (2.0 Mbps) was used as a trunk technology for many years. On a T1, trunks could carry 24 voice channels on two pairs of copper wire. On an E1, trunks could carry 30 voice channels on the same two pairs. Throughout the 1970s, both T1 and E1 were used almost exclusively within the network, and customers could only get 64 kbps DS0 digital channels at the user interface.

However, both T1 and E1 have been offered for "last mile" (12 kft, 2.3 miles, 3.6 km) access line applications since the mid-1980s. The divestiture process spurred offerings in the United States, and user pressures around the world for higher speed access spurred deployment elsewhere. Usually, the T1 or E1 offered users 24 or 30 channels at 64 kbps each, but unchannelized versions ran at a full 1.5 Mbps (1.544 Mbps to purists) or 2.0 Mbps (2.048 Mbps).

However, fully 80 percent of existing T1 and E1 circuits are still provisioned on two pairs of copper wire. This is in spite of alternate technologies such as wireless, coaxial cable, or, lately, fiber optics based on SONET or the *Synchronous Digital Hierarchy* (SDH).

The persistence of two-pair copper T1 and E1 is mostly due to few premises fiber interfaces, but lots of copper interfaces. Computers and PBXs expect some form of copper wire to interface with their serial or parallel ports, and it has been expensive to use anything else, even today. Even when the T1 or E1 is provisioned on a SONET or SDH or some other form of fiber ring such as FDDI, the last 1 km or half mile or so is still almost always two copper wire pairs. It makes no real difference whether the T1 or E1 is intended for a customer's private networks or used for an access interface to a public network service like ISDN; the link is still a T1 or an E1.

The Trouble with T1/E1

In spite of its popularity with service providers and customers, both T1 and E1 have some real drawbacks that are mostly the result of the age of the technology upon which both are founded. Although improvements have been made to both over the years, in many respects T1 and E1 remain very much 1980s technologies with features that date back to the 1960s in some cases.

Many of these limitations revolve around the repeaters. In most T1 and E1 installations, repeaters are used every 1 km or 4 to 6 kft to regenerate the signal. This had to be done at the time because the electronics affordable then were not as sophisticated as they are today when it comes to recovering weak signals.

It was mainly the repeaters that made the provisioning of a T1 or E1 so labor-intensive to design, install, and condition the line. Loading coils had to go, the repeaters went in, and bridged taps were taken out as well. Mixed wire gauges needed to be avoided if at all possible for maximum performance, and so on.

Without careful attention to the line characteristics and electrical parameters, the T1 or E1 would not work, so the entire process of provisioning a T1 or E1 could take weeks (or months in some cases) to complete. It was not uncommon for projects to be delayed for weeks at a time due to the unavailability of the links between sites.

Repeaters were a major headache to the service providers because they were numerous (most lines needed at least one pair of repeaters: one in each direction on each wire pair), unsophisticated (they had to be affordable), hard to troubleshoot (none ran network management software), and difficult to maintain (many were in conduits or buried embankments).

It took a while, but the bright idea of the late 1980s was, "Let's try to get rid of the repeaters!" Advances in electronics made this not only feasible, but actually more efficient.

HDSL is Born

Naturally, nobody meant for T1 and E1 to have any drawbacks at all. Both were considered state-of-the-art technology. The trouble is, the time for the core T1 and E1 technology was the early 1960s! In fact, only the familiar EIA-232 serial interface is as old as T1 and E1 when it comes to common standards in the telecommunications field.

But the end electronics have come a long way since the 1960s. Consider the changes in the desktop PC since 1990 alone. End telecommunications electronics today can take advantage of increased processing power, the low cost and availability of a lot of memory, and overall advances in *Digital Signal Processing* (DSP) chipsets, which make it easy to massage bits almost any way one wishes to.

The telecommunications philosophy beginning in the 1980s was more along the lines of "don't adapt line conditions to end electronics; adapt end electronics to line conditions." After all, analog modems have used this concept for years, especially since around 1982. Modern modems do their own equalization across the frequency range (called *self-equalization*), instead of relying on a technician to do it for the whole line (difficult on a dial-up!). Echo cancellation to minimize self-crosstalk (near-end crosstalk or NEXT) circuits were added, and so on.

Now apply this intelligent end device approach to T1 and E1. The result is *High bit rate DSL* (HDSL). HDSL needs no repeaters (usually) or special line conditioning. Even some bridged taps are okay on HDSL links in most cases, as long as there are no more than two and their lengths are limited.

The nice thing about HDSL is that in addition to lowering provisioning costs, HDSL makes copper "look like fiber" in terms of performance, which means that the reliability of the link and bit error rates are much better than copper with T1 or E1. This reliability is a real plus when there is more fiber than ever within the network, but little directly to the customer premises.

HDSL for T1

When HDSL is used to provision T1s for customers, what does the customer see? A T1, or, technically, a DS1 running at 1.5 Mbps, of course. It may be HDSL inside, but the label still says "T1." Figure 7-1, based on the original Bellcore HDSL documentation, shows why this is so.

Note that T1 leased private lines do not go through the service provider's circuit switch. Private lines are routed through a *Digital Cross Connect* (DCS or "DACS") out to a trunking network. In its simplest form, the DS1 input to the DCS at the local exchange does not come directly from a T1 link, but rather from a special card called an *HDSL Termination Unit - Central office* (HTU-C). The HTU-Cs may be individual units or gathered into a rack or unit. The advantage is that several HTU-Cs (numbers vary from vendor to vendor) can be supported with common power supplies, battery backups, management methods and so on.

The opposite end of the HDSL link is the *HDSL Termination Unit - Remote* (HTU-R). The HDSL really takes place between the HTU-C and the HTU-R. In its simplest and original form, the interface between the HTU-

Figure 7-1
A T1 provisioned
with HDSL

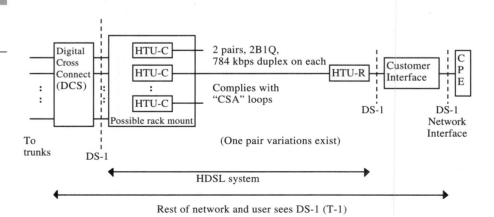

C and HTU-R is two pairs of twisted copper wire. The runs may have some mixed gauges or bridged taps, but must comply with international standards for *Carrier Serving Area* (CSA) local loops.

Each of the two pairs runs at 784 kbps full-duplex upstream and downstream. The 2B1Q line code from ISDN, a type of *Pulse Amplitude Modulation* (PAM) line code, is used. Note that this differs from traditional T1 where each pair only carried bits in one direction. In the original T1, this unidirectional operation was done mainly for simplicity of design of transmitters, receivers, and repeaters.

Note that one-pair variations on HDSL exist, some of which are known by different names, but the point is that the network and customer still see T1s. All user equipment and service provider operations function just as before.

HDSL for E1

When used to provision an E1, HDSL seems to customers to be exactly that-an E1 running at 2.0 Mbps. The same as with T1; it may be HDSL inside, but the link it still an "E1."

Figure 7-2, based on the original HDSL European Telecommunications Standards Institute (ETSI) documentation, shows why this is the case. The figure shows an E1 entering the HDSL equipment at the service provider's

Figure 7-2
An E1 provisioned on HDSL

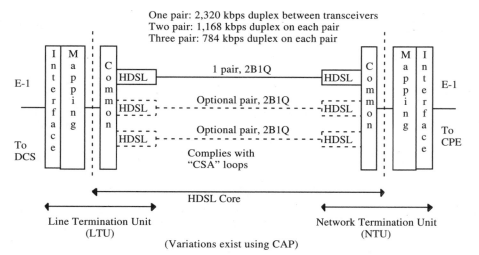

One pair: 2,320 kbps duplex between transceivers
Two pair: 1,168 kbps duplex on each pair
Three pair: 784 kbps duplex on each pair

(Variations exist using CAP)

office on the left of the figure, and an E1 emerging to connect the user's CPE on the premises. In between, the HDSL carries the E1. It should be noted that three pair E1 versions are virtually all phased out today.

HDSL components are gathered into either *Line Termination Units* (LTUs) at the service provider or *Network Termination Units* (NTUs) at the customer premises. Each termination unit consists of four major components.

- First is the HDSL transceiver itself.
- Second is some common circuitry used in all versions of HDSL: one-, two-, or three-pair systems.
- Third is a mapping module to map the E1 frame bits into the HDSL frame structure and back.
- Finally, there is an interface module to accept a standard E1 connector.

The components, from common circuitry to HDSL transceiver at each end, form the HDSL core of the whole system.

Keep in mind that E1 leased private lines, as with T1 leased private lines, do not go through the service provider's circuit switch. Private lines are routed through a *Digital Cross Connect* (DCS, or *DACS*) out on to a trunking network. This is not shown in the figure, nor is the common practice of gathering the LTUs into common racks or equipment.

In its simplest form, the HDSL link between the LTU and NTU is one pair of twisted copper wire. The runs may have some mixed gauges or bridged taps, but they must comply with international standards for CSA local loops. This pair uses the 2B1Q line code from ISDN, exactly the same as HDSL supported the T1 line speed, and runs at a total of 2.320 Mbps between HDSL transceivers. The "extra" bits above the E1 2.048 Mbps speed are used for overhead and compatibility with SDH signal formats.

It is more common to see multiple pairs on HDSL E1 systems. With two pairs in place, each of the two pairs runs at 1.168 Mbps full-duplex upstream and downstream. Note that the total bit rate is the slightly higher (2×1.168 Mbps = 2.336 Mbps) due to increased overhead. Also note that this differs from traditional E1 where each pair only carries bits in one direction, as with T1. And also as with T1, this was done mainly for simplicity of design of transmitters, receivers, and repeaters. When three pairs are used, each runs at 784 kbps full-duplex upstream and downstream. There is more overhead (3×784 kbps = 2.352 Mbps) but each pair runs at a lower bit rate and so might reach farther distances. All of the initial HDSL links running at E1 speeds used three pairs because it was felt within the industry that using existing chipsets running at 784 kbps would offset the cost of needing a third pair of wires. As it turned out, faster chipsets were easy to fabricate, so three-pair E1 speed HDSL systems more or less disappeared in favor of two wire pair systems based on customized chipsets.

Note that variations of E1 HDSL exist using CAP as a line coding technique, but these remain more or less curiosities. The point is that whatever the coding between compatible LTUs and NTUs, all user equipment and service provider operations function just as before.

The HDSL Frame for T1

HDSL, just like almost every other transport link used in high-speed telecommunications (including ADSL), is a *framed transport*-that is, the link sends a series of frames, one after the other, with no pause in between. If there is no "live" used information to send, special idle bit patterns are sent inside the frames. Keep in mind that these transmission frames are distinct from the data link level frames that are used in LANs and some other wide-area protocols.

Figure 7-3 shows the content of an HSDL frame when used for T1 transport on two pairs of wire, based on Bellcore documentation. An HDSL frame is sent once every 6 milliseconds, or about 167 frames per second. A frame consists of a special synchronization symbol in 2B1Q that is 14 bits long. Technically, it is not proper to speak of bits in 2B1Q but rather *quats*, which are 2 bits long. For the sake of simplicity, however, this "bit" terminology will be used throughout this discussion. There is some HDSL overhead (HOH), two to ten bits in length, spread throughout the frame.

The rest of the HDSL frame between the synchronization symbols is divided into 48 *HDSL payload blocks*, which are divided by the HOH bits into four units, each containing 12 HDSL payload blocks. These blocks are numbered B01 through B48, as shown in the figure. The figure also shows the content of any payload block. One pair (Loop #1) carries what are called Group 1 DS-0s (64 kbps channels), and the other pair (Loop #2) carries the Group 2 DS-0s. The Group 1 DS-0s are numbered 1 through 12, and the Group 2 DS-0s are numbered 13 through 24. Together, they yield the T1 transport capacity of 24 DS-0s. A framing bit (not the same as the T1 framing bit) precedes the payload block, making each payload block 97 bits long ($8 \times 12 + 1 = 97$).

Figure 7-3
The HDSL Frame for
T1

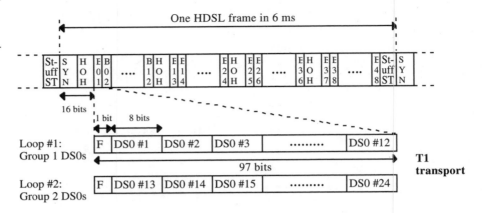

Groups transported 48 times per frame, 12 in each direction:
12 x 64 kbps (DS0) = 768 kbps on each pair

HOH: HDSL overhead (2 bits)   Bxy: HDSL payload block
Stuff: Stuff bits if needed   F: HDSL Framing bit
SYN: Synch symbol (14 bits)

The groups are transported 48 times per HDSL frame, one per payload block, or 48 times in 6 milliseconds. Each pair carries bidirectional traffic, and because 12 times 64 kbps equals 768 kbps, the line rate of 784 kbps on each pair represents the 12 DS-0 channels plus 16 kbps of HDSL overhead.

Recall that one-pair variations on the basic two-pair HDSL exist, but are not part of the original Bellcore HDSL standard.

The HDSL Frame for E1

One of the things that makes HDSL suitable for both T1 and E1 is the fact that both use exactly the same HDSL frame format. This format is a brilliant thing to do, but there are some differences between an HDSL frame carrying a T1, and an HDSL frame carrying an E1.

When used for an E1 transport, just as with a T1, the HDSL link is a framed transport sending a series of frames, one after the other, with no pause in between. Special idle bit patterns are sent if there is no "live" used information to send.

Figure 7-4

The HDSL frame for E1

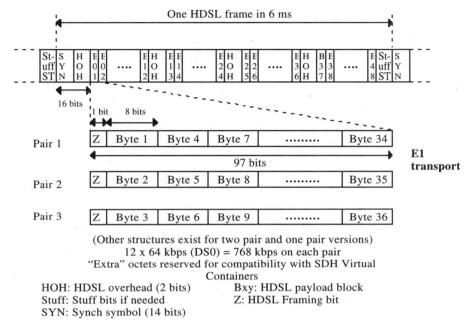

(Other structures exist for two pair and one pair versions)
12 x 64 kbps (DS0) = 768 kbps on each pair
"Extra" octets reserved for compatibility with SDH Virtual
Containers

HOH: HDSL overhead (2 bits)   Bxy: HDSL payload block
Stuff: Stuff bits if needed   Z: HDSL Framing bit
SYN: Synch symbol (14 bits)

Figure 7-4 shows the content of an HSDL frame when used for E1 transport on three pairs of wire, based on ETSI documentation. An HDSL frame is sent once every 6 milliseconds, or about 167 frames per second. A frame consists of a special synchronization symbol in 2B1Q that is 14 bits long (although, as with T1, it is not technically proper to speak of bits in 2B1Q, but rather quats, which are 2 bits long). There also is some HDSL overhead (HOH), two to ten bits in length, spread throughout the frame.

The rest of the HDSL frame is divided into 48 *HDSL payload blocks*, which are divided by the HOH bits into four units, each containing 12 HDSL payload blocks. These blocks are numbered B01 through B48, and each is 97 bits long. So far, this frame structure is exactly the same as the HDSL frame structure used for T1.

The content of any payload block also is shown in the figure. The major differences between T1 and E1 frame structures and terminology are most obvious here. The F bit of the T1 transport is now the Z bit in the ETSI E1 version. The "bytes" (oddly enough, the ETSI documentation calls them bytes rather than octets) are loaded into the HDSL payload blocks on each of the three pairs in a simple pattern. Bytes 1, 4, 7..34 are in the first pair payload block sequence, bytes 2, 5, 8..35 are in the second pair payload block sequence, and bytes 3, 6, 9..36 are in the third pair payload block sequence. Now there are only 32 bytes in the E1 frame structure. The "extra" bytes are used for compatibility with the SDH E1 Virtual Container structure.

The groups are transported 48 times per HDSL frame, one per payload block, or 48 times in 6 milliseconds. Each pair carries bidirectional traffic, and because 12 times 64 kbps equals 768 kbps, the line rate of 784 kbps on each pair represents the 12 DS-0 channels plus HDSL overhead.

Two pair and one pair variations exist and are part of the ETSI HDSL standard, but these are not shown in the fgiure. Basically, the payload block sizes and the line rates increase to keep the frame time at 6 milliseconds. For the sake of completeness, these frame structure variations are shown in Figures 7-5 and 7-6. In each case, 16 kbps must be added to the derived transfer rate for frame overhead, and so the two pair system runs at 1152 kbps + 16 kbps = 1168 kbps on each pair. The one pair system runs at 2304 kbps + 16 kbps = 2320 kbps = 2.304 Mbps.

Figure 7-5

HDSL at E1 speed on two pairs

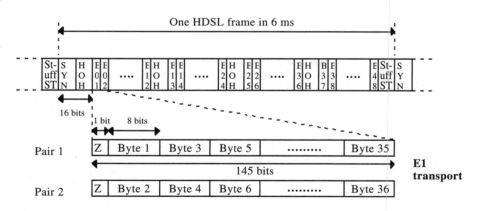

(Other structures exist for three pair and one pair versions)
18 x 64 kbps (DS0) = 1152 kbps on each pair
"Extra" octets reserved for compatibility with SDH Virtual Containers

HOH: HDSL overhead (2 bits)   Bxy: HDSL payload block
Stuff: Stuff bits if needed   Z: HDSL Framing bit
SYN: Synch symbol (14 bits)

Figure 7-6
HDSL at E1 speeds
on one pair

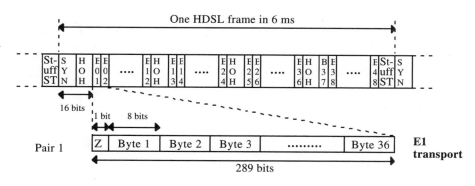

(Other structures exist for three pair and two pair versions)
36 x 64 kbps (DS0) = 2304 kbps on the pair
"Extra" octets reserved for compatibility with SDH Virtual
Containers

HOH: HDSL overhead (2 bits)   Bxy: HDSL payload block
Stuff: Stuff bits if needed   Z: HDSL Framing bit
SYN: Synch symbol (14 bits)

Benefits of HDSL

Using HDSL to deliver either T1 or E1 services benefits service providers and customers. Most of these benefits are on the service provider side of the ledger, but customers have indirect benefits, as well.

First and foremost, only the local exchange card (the HTU-C for T1 or the LTU for E1) and customer card (the HTU-R for T1 or the NTU for E1) are needed for the simplest form of HDSL service.

The loops do not need repeaters every few thousand feet or fraction of a kilometer. Instead, HSDL allows 24 AWG (0.5 mm) copper pairs to provide services for up to 12 kft (2.3 miles or 3.6 km). On 26 AWG (0.4 mm) copper, services extend up to 9 kft (1.7 miles or 2.7 km). Usually, two bridged taps are allowed if the length of each is less than 5,000 feet (1.525 km).

HDSL can be extended to reach about 26 kft (5.0 miles or 7.93 km) by using heavier wire gauges (22 AWG or 0.63 mm) or HDSL repeaters, which are more often called *doublers* in HDSL, since they "double" the length achievable. It sounds odd that HDSL repeaters even exist, but these newer versions are much more efficient and powerful than their early T1/E1 cousins.

Remote HDSL equipment, HDSL repeaters or customer premises cards, can be powered from the local exchange itself. This is a common practice known as *wet HDSL* from the way technicians tested for the low voltage current on the pairs (wet your fingers by licking them and grab the wire . . .).

Another important benefit is that HDSL can be monitored with the same *Operations System Support* (OSS) equipment and software as before. After all, outside of the HDSL core system, the link is still T1 or E1.

Finally, HDSL can be used almost anywhere. Some 80 to 90 percent of copper cable plant is appropriate for HDSL, whether based on CAP or 2B1Q (PAM) line codes.

Uses of HDSL

It may seem obvious just what an HDSL link can be used for. Basically, anywhere a T1 or E1 makes sense, HDSL makes even more sense. However, this might be a good time to list the main uses of HDSL to provision T1 and E1 services from the customer perspective. Because much of this book deals with the uses of ADSL and its related technologies, keeping symmetrical applications in mind is a good idea. The main uses of HDSL are here:

- Internet access to servers, not just from clients
- Private campus networks with installed copper cable plant
- Extend central PBX to other office park locations
- LAN extensions and connections to fiber rings
- Video conferencing and distance learning applications
- Wireless system base station connections
- Primary Rate Access (PRA) for ISDN

HDSL is widely used to provide companies and individuals with fast Internet access to their *servers*, not just from clients. Asymmetrical DSL variations limit upstream traffic to a fraction of the downstream rate. The placement of servers in homes or small offices becomes difficult because, in this case, the downstream traffic from remote clients will be a fraction of the upstream traffic from the local server. Yet the bandwidth is just the opposite. In fact, one of the biggest selling points for HDSL versus ADSL is found in the simple question, "Do you have a server on site?"

HDSL is also used for private campus networks with installed copper cable plant. Many companies, colleges, and universities have many copper pairs between buildings. Until HDSL came along, it was expensive and difficult to squeeze more than 64 kbps out of each pair.

Organizations with a central PBX can extend 24 or 30 voice channels to other office park locations easily and simply with HDSL. The same applies to LAN extensions and connections to fiber rings, although, in this case, the HDSL (really the T1 or E1) is probably unchannelized into its raw bit rate of 1.5 Mbps or 2.0 Mbps.

Video conferencing and distance learning applications are natural for HDSL. Again, the link in this case would probably be unchannelized. The position of the client and server are unimportant, due to the symmetrical nature of the HDSL link.

Moving back to channelized applications, wireless systems have lots of copper wire in use. Every base station transceiver tower must be connected to a central operations center and to switches and trunks, as well. HDSL makes a lot of sense when used for wireless system base station connections.

Finally, HDSL can be used to provide more cost effective *Primary Rate Access* (PRA) for ISDN services. The cost of provisioning the high-speed digital line for ISDN has long been a reason cited for higher-than-anticipated ISDN pricing structures. HDSL makes this process of PRA DSL provisioning simpler and cost effective. Of course, one of the key barriers to the more widespread use of HDSL for T1 service to SOHO and residential users is the fear among service providers that this move would dramatically erode today's high tariff rates for traditional T1 service. Indeed, this has already happened in a number of areas.

HDSL for ISDN

Mention has just been made The use of HDSL to supply ISDN Primary Rate Access was just discussed. The link in this case would operate at the *Primary Rate Interface* (PRI) speed in the country it was deployed in. In the United States, the PRI is 1.5 Mbps, and in most other places it is 2.0 Mbps.

Figure 7-7 shows how HDSL might be used to provision ISDN PRA to a customer premises from an ISDN local exchange (LE). The figure shows HDSL on two pairs of wire, each running at 1168 kbps, giving the E1 transport rate of 2.336 Mbps when HDSL overhead is included. In cases

Figure 7-7
HDSL for ISDN PRA

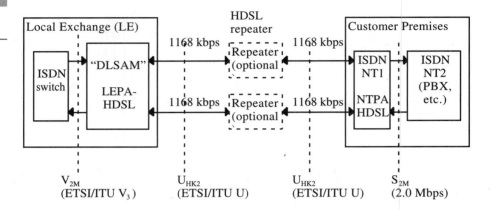

If two S_{2M} interfaces employed on each end, this is just a E1 private line

LEPA: Local exchange primary access
NTPA: Network termination primary access

where the premises exceed a certain distance, HDSL repeaters can be added to each pair. Note the bidirectional functioning of each pair.

In this configuration, the LE side of the link is the Local Exchange Primary Access - HDSL (LEPA-HDSL) equipment. This equipment can be gathered into racks, of course, but the term DSLAM is not really appropriate here. In any case, an E1 input and output port can be found on the ISDN switch.

On the customer premises, the Network Termination Primary Access - HDSL (NTPA-HDSL) equipment basically forms the ISDN Network Termination Type 1 (NT1) unit (technically, a functional grouping). This may in turn be attached to an ISDN Network Termination Type 1 (NT2) unit, such as a PBX or other ISDN-compliant piece of equipment. Note the unidirectional operation of these interfaces.

The interfaces themselves are also shown in the figure. At the LE, the LEPA-HDSL to switch interfaces conforms to the ETSI/ITU V_{2M} interface specification at 2.0 Mbps (E1). At the customer premises, the NT1 to NT2 interfaces conform to the ETSI/ITU S_{2M} interface specification, also at the E1 rate.

The HDSL links themselves conform to the ETSI/ITU U_{HK2} interface specification. HK2 essentially means "HDSL on two pairs" in this case. The point is that ETSI and the ITU have made HDSL an official part of the ISDN PRA interface.

Note that if two S_{2M} interfaces are employed on each end of the link or the V_{2M} interface leads to a digital cross-connect and not an ISDN switch, this is now just an E1 private line. Of course, this private line would still conform to all ETSI and ITU ISDN standards.

The Downward Spiral

In North America, some 60 percent of all new T1s and E1s are now provisioned with HDSL. The customer is still buying a T1 or E1, but delivered with HDSL. This is not to say, however, that the customer does not benefit from HDSL at all. Figure 7-8 shows the impact that HDSL has had on customers. The main benefit to and effect on customers is reduced T1 and E1 pricing.

The figure illustrates the typical pricing of only two United States service providers, and big ones at that, but the effect is the same all around the world. In 1994, Pacific Bell and US West priced T1s monthly prices at about $348 and $441, respectively. That year saw little HDSL. In 1996, when some 60 percent of T1s provisioned in the United States were on HDSL, the prices had fallen to $250 and $340, respectively.

This is a 28-percent drop in the case of Pacific Bell ($98/$348) and a 23-percent drop in the case of US West ($101/$441). In the future, prices should continue to drop as HDSL becomes even more popular.

Figure 7-8
Declining T1 prices

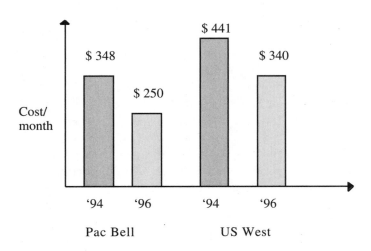

HDSL Limitations

In spite of HDSL's enormous appeal and wide-ranging benefits to service providers and customers, it still labors under limitations that make HDSL less than ideal for all situations.

One of the biggest concerns is that only the barest essentials of HDSL are covered in Bellcore and ETSI specifications, which has meant that there are a number of proprietary implementations of HDSL with wide variations in features which do not allow vendor interoperability at all. This concerns service providers, who would like more consistency between product offerings, and who always like more vendor independence.

Another aspect of HDSL is that the more technical benefits are almost invisible to customers. There might be some better performance in terms of bit rates and some reduced costs, but to the customer it is still a T1 or E1. Customers expect high-performance in any case, and always expect prices to fall. In terms of HDSL CPE, perhaps customers need even more direct benefits,.

Also, HDSL on loops greater than 12 kft (2.3 miles or 3.6 km) still need repeaters. There are exceptions with odd gauge changes and sizes, and some HDSL variations stretch to a full 18 kft (3.4 miles or 5.49 km), but Bellcore expects repeaters to be used on these longer loops. Although HDSL repeaters are much more sophisticated than early T1/E1 models, they are still repeaters and something to avoid.

Whether configured for T1 or E1, HDSL is still supposed to use the 2B1Q line code. The problem is that 2B1Q has real limitations itself of bandwidth efficiency and distances.

In addition, requiring the use of multiple pairs at least halves, and in some cases reduces by two-thirds, the availability of T1 and E1 service in a given area. If HDSL could be standardized on only one pair, as some early variations have done on a proprietary basis, this would maximize the use of the existing cable plant.

The fact is that HDSL equipment has dropped in price so much that the cost of the two pairs of copper wires are significant cost factors.

Finally, it spite of its efficiencies, HDSL can still be slow to deploy. After two or three pairs are located, the rest of the process can take only hours. Some regions have a shortage of copper pairs. However, one still needs to find two or three pairs running between the proper locations. One pair, in contrast, always runs between the same end points.

Beyond HDSL: HDSL2

All of the limitations of current HDSL products are intended to be addressed in the "next generation" of HDSL, commonly known as HDSL2. The "2" refers to "HDSL the second generation" and the name has stuck.

The HDSL2 specification is scheduled to be completed in order for products to appear in 1998. This availability is the result of a lot of hard work by a number of organizations, with the primary work done by the ANSI T1E1.4 committee. However, ANSI's work on HDSL2 is expected to be endorsed by ETSI.

There are three primary goals for HDSL2:

1. 12 kft (2.3 miles or 3.6 km) coverage (which ANSI calls a "full CSA").

2. It must work even with other services on adjacent cable pairs (ANSI calls this *spectral compatibility*).

3. Vendor interoperability (HTU/LTU/NTUs may be mixed and matched among vendors).

In addition, just as with HDSL, HDSL2 must have less than 500 microsecond latency end to end. In other words, the combination of bandwidth and delay effects (from propagation delay on the wire and processing delay building the HDSL frames) must be less than .5 milliseconds. This is done to minimize the effects of far end echoes when HDSL2 is used for voice traffic.

Developing technologies, especially new technologies, is a study of applied engineering. Engineering in turn applies science to the real world. So much of HDSL2 concerns taking electrical formulas and making compromises in order to embody them in equipment that functions reliably in the everyday world under less-than-ideal conditions. This is easier said than done. HDSL2 has already faced its share of compromising situations.

For instance, higher speeds mean smaller bit intervals, which means more errors if all else stays the same. But T1 and E1 specifications limit bit errors, and regulators or customers may demand or be entitled to rebates for elevated error rates. If there is in "plain" HDSL a need to do full T1 or E1 speeds on only one wire pair, one way to make this possible is to add some *forward error control* (FEC) to the HDSL frame. These extra bits can both detect and correct some errors in the HDSL frame, perhaps just enough.

However, adding a FEC function to the HDSL2 device adds to the latency end to end.

Another case is the need in HDSL2 to add support for other signals (maybe analog voice as in ADSL?) on the same wire pair used for HDSL2. This means that HDSL2 should not use 2B1Q (also called "4 level PAM") as the line code of choice because ISDN 2B1Q allows no room for passband voice at all.

However, using other line codes in HDSL2 might mean that the other line codes might be more susceptible to crosstalk from "real" 2B1Q T1s or E1s, or even HDSL T1s and E1s, in the same cable bundle. And so it goes. . . .

CAP or PAM for HDSL2?

Much debate in HDSL2 circles involves the choice of line code for HDSL2. If used for E1 speeds, a PAM code such as 2B1Q needs three pairs of wires, but no passband filters. Passband filters do just what their name implies: exclude frequencies from certain ranges on the wire. Passband filters could be attractive if other services, such as analog voice, need to share the wire, but they add expense and complications to the circuitry.

Initially with HDSL, PAM codes such as 2B1Q were thought to give greater resistance to crosstalk, and implementing 2B1Q was simple and familiar from ISDN. Also, no passband filters were needed.

Even DMT (from ADSL) has been kicked around as a line code for HDSL2. However, this only seems possible in the future. For now, there is just too much latency in current DMT circuitry to meet the HDSL2 design goal of 500 milliseconds delay.

Figure 7-9 shows the basic differences between CAP and PAM (2B1Q), which were the the two leading contenders for HDSL2 line code standard. The figure is similar to the one presented in the previous chapter, but gives more details. Keep in mind that 2B1Q is also called "four level PAM" in some documents.

In the figure, the PAM or 2B1Q signals use all of the bandwidth available. The useful bandwidth tails off at about 400 kHz, but the signal is still present beyond that point (there are no passband filters to exclude these components). Also, the CAP signals go "all the way down to zero Hz," although any electrical engineer would cringe at that terminology. It is more proper to say that 2B1Q signals are coupled directly onto the line, or that 2B1Q can have a dc component, but it all means that same thing: 2B1Q signals go all the way down to zero.

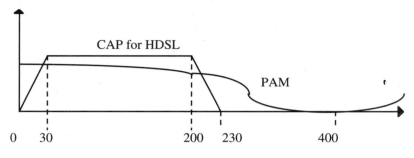

Figure 7-9
CAP and PAM (2B1Q)

CAP for HDSL

PAM

0   30   200 230   400

Frequency in kHz

Early HDSL2 Characteristics

Although the HDSL2 specification and standard is not yet complete, it is not too early to give a breakdown of the major characteristics of HDSL2, according to the design group.

Front and foremost, no analog voice circuit support is planned for HDSL2. Although many requested support for an analog voice circuit in the 0 to 30 kHZ range, like ADSL offers, nothing in HDSL2 addresses this support.

The main point for excluding voice support was that HDSL2 is targeted for the T1/E1 leased line business market and not the residential market.

Next, the initial HDSL2 specification will be PAM-based. Oddly, CAP is firmly expected to out-perform 2B1Q in future HDSL products, which means that an HDSL2 system that runs at 9 kft (1.7 miles/2.7 km) under PAM could now function out to 12 kft (2.3 miles/3.6 km) under CAP at the same signal level. This is CAP's biggest plus, but CAP will most likely have to wait in favor of the simplicity and efficiency of PAM coding.

Finally, echo cancellation is expected to be favored over *Frequency Division Multiplexing* (FDM), which is a somewhat technical issue. The debate revolves around duplex operation on one pair. Sharing the same frequency range for upstream and downstream signals requires echo cancellation to eliminate "self-crosstalk" effects in the circuitry. This can be done, but only with added expense and complexity.

It would be simpler to use different frequency ranges for upstream and downstream signals. This is FDM applied to HDSL2. Using FDM this way eliminates self-crosstalk problems, but requires the use of passband filters to eliminate signals outside the desired ranges. CAP does this easily, but PAM does not. PAM (or 2B1Q) needs to be "optimized" to function in a passband environment.

The candidate line coding schemes for HDSL2 have evolved far beyond the simple alternatives mentioned here, but the issues are more or less the same. Crosstalk, potential voice support, and so on are always important and come up with every new technique. The debate need not be detailed here, but should certainly be noted.

FDM and Echo Canceling for HDSL2

HDSL2 will involve duplex operation on one pair. That is, signals in both directions are carried on the same two wires. To do this, the frequency ranges upstream and downstream may be shared or separate. If they are shared, echo cancellation must be used to eliminate "self-crosstalk" effects in the circuitry. This can be done with the addition of expense and complexity, as pointed out previously.

It is simpler to use different frequency ranges for upstream and downstream signals. This is FDM for HDSL2, which eliminates self-crosstalk effects, but requires the use of passband filters to eliminate signals outside the desired ranges.

Figure 7-10 illustrates these concepts. The top of the figure shows the line code used for upstream and downstream signals in different ranges. This is FDM. The upstream range is due first to the reduced attenuation effects at lower frequencies. Their identical size reflects the symmetrical nature of HDSL2 and HDSL in general.

The lower portion of the figure shows that the same frequency range is now used for both directions, and this requires echo canceling. However, less bandwidth is needed, and, for this simple reason, HDSL2 will use echo cancelling and not FDM.

Note that CAP easily adapts to both environments. PAM needs to be "optimized" to function in a passband environment, as would be needed with FDM. In other words, pure PAM codes, such as 2B1Q, like echo cancellers, which is another reason that the PAM-based HDSL2 specification will use echo cancelling and not FDM.

HDSL2 Report Card

It is never too early to see how something is performing over a period of time, and so here is an HDSL2 "report card." It is still early enough that things may change, but here is how HDSL2 looks to date.

Figure 7-10
FDM and echo can-
celling for HDSL2

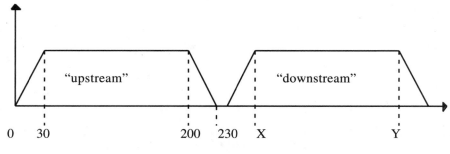

Frequency in kHz

FDM for HDSL2

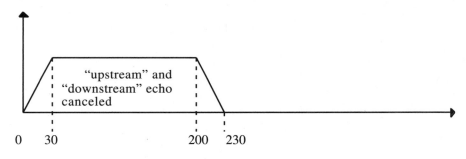

Frequency in kHz

Echo Canceling for HDSL2

First of all, it turns out that lots of new problems arise with the use of a single pair for all HDSL2 situations. HDSL had single pair implementations, but could always use one or two more pairs if necessary. The single pair concerns include performance (squeezing bits through that one pair in all cases), latency (the delays due to processing and sending the bits), and what is known as *spectral compatibility* (the tendency for signals in the same cable bundle to interfere with each other).

The whole "which is better for HDSL2, CAP versus PAM?" debate closely mirrors the whole ADSL CAP versus DMT debate. As a warning to HDSL2 people, even the selection of DMT as the offical ADSL line code did not cause CAP products to disappear or for the CAP proponents to become less vocal. If anything, support rallied to the beleaguered CAP for ADSL.

Also, it is a given that certain vendors will anticipate the HDSL2 standard and introduce products somewhat prematurely. This will likely create the very variants and feature variations that HDSL2 was supposed to avoid. However, those working on HDSL2 intend to provide an easy "migration path" fully compliant HDSL2 equipment.

And finally, a wide array of options and variations in the HDSL2 standard might undermine interoperability intentions. In seeking to satisfy everyone, the specification may end up being so all-inclusive that no one will be satisfied. HDSL2 will strictly limit options and variations to prevent this situation.

Naturally, it is in the best interests of all working on HDSL2 to avoid these pitfalls. And that is the intention all the way along the line.

8

The Asymmetric Digital Subscriber Line (ADSL) Architecture

In spite of the confusion over the relationships between DSLs, xDSLs, and ADSL, one thing is clear: ADSL is the most standardized of all, in terms of available documentation, service trials, and open specifications. It is widely anticipated that many xDSL services will begin with ADSL and may even end with ADSL.

Figure 8-1 shows the basic ADSL system architecture, as documented by the ADSL Forum. Although the figure may seem confusing at first, the overall arrangement of ADSL components is straightforward. The need to standardize makes many of the details necessary. Like most architecture diagrams, the architecture establishes a number of standard interfaces between major components. In between the interfaces, various *functional groupings* are defined that may be gathered by equipment vendors into products that embody the required functions, along with any options or enhancements that the manufacturer feels are necessary. The internal functioning of these components is left to individual vendors.

Where the acronyms mean the following:

ATU-C	ADSL Transmission Unit, CO side
ATU-R	ADSL Transmission Unit, Remote side
B	Auxiliary data input (e.g. set top box)
DSLAM	DSL Access Multiplexer
POTS-C	Interface between PSTN and splitter, CO side
POTS-R	Interface between PSTN and splitter, Remote side
T-SM	T-interface for Service Module
T	May be internal to SM or ATU-R
U-C	U interface, CO side
$U-C_2$	U interface, CO side from splitter to ATU-C
U-R	U interface, Remote side
$U-R_2$	U interface, Remote side from splitter to ATU-R
V_A	V interface, Access Node side from ATU-C to Access Node
V_C	V interface, CO side from Access Node to network service

Figure 8-1
The ADSL
Architecture

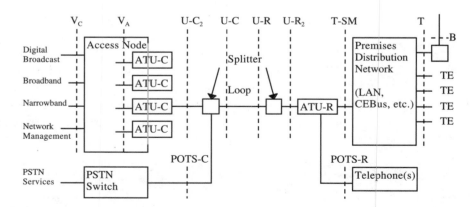

Several features of ADSL are quite important. First of all, note that provision is made to support analog voice service (designated *plain old telephone service*, or POTS). A special splitter device carries the 4 kHz analog channel from switch to premises "under" the digital bandwidth on the ADSL link.

Next, consider that many services are envisioned for ADSL systems, including digital broadcast and broadband (that is, video and Internet access) services, as well as network management. All of these services are accessed outside of the normal *central office* (CO) or *local exchange* (LE) switch, neatly solving the trunking and switch congestion problems. Many ADSL links are serviced by a single ADSL access node in the central office (or local exchange). This access node is sometimes called a DSLAM, or *DSL access module*, but this is somewhat misleading. Although a DSLAM can certainly supply service access to ADSL lines, the full architecture of a DSLAM is much more encompassing than the simple arrangement shown here. More details on the DSLAM architecture are discussed later in this book.

The interfaces listed in the figure might not be self-explanatory. The B interface is straighforward and just indicates a possible auxiliary input, such as a satellite feed into a set top box. The T-SM interface between the ATU-R and Service Module (everything else besides the ATU-R itself) might be the same as the T interface in some cases, especially if the service module is integrated into the ATU-R. If the T-SM interface does exist, it can be more than one per ATU-R and more than one type per ATU-R. For example, an ATU-R might have both 10Base-T Ethernet and V.35 connectors. Likewise, the T interface between premises distribution network and terminal equipment might also be absent if the terminal equipment is integrated with the ATU-R in some fashion.

The various U interfaces might not exist if the splitter is made an integral part of the ATU devices, or if the splitter entirely disappears. This is the latest trend among the equipment vendors at the moment, but this makes support of existing analog telephones on the same line impossible. Also, the V interfaces might be logical rather than physical, which is especially true of the V_A interface if the DSLAM or ADSL access node performs some concentrating or switching tasks. If the V_C interface to the service networks is physical, as it likely will be, this interface is allowed to take a variety of forms appropriate for TCP/IP, ATM, or other service networks.

Finally, the implementation of ADSL on the customer premises can take a variety of forms. Under the architecture, these schemes form the Premises Distribution Network. This could be as simple as individual wire pairs running to devices, such as TV set-top boxes or PCs, or as elaborate as a full local area network (LAN), such as Ethernet in the home. A potential drawback would be the need to run new premises wiring to attached ADSL devices. Newer standards using home electrical wiring for the same purpose (such Consumer Electronics Bus (CEBus)) are supported and very attractive for this reason.

An ADSL Network

Asymmetric Digital Subscriber Line technology is more than just a faster way to dowload Web pages onto a home PC. ADSL is part of a whole networking architecture that has the potential to supply residential and small business users all types of new broadband services. In this context, "broadband" services means services that need network links faster than 1 or 2 Mbps to deliver.

Figure 8-2 shows just what a broadband network based on ADSL would look like. In the simplest version of this architecture, customers would essentially need only a new ADSL modem. This device would have ordinary RJ-11 jacks that would support the existing analog telephone(s) in the home office/small office (SOHO). Other ports, perhaps 10BASE-T Ethernet, would link to PCs or TV set-top boxes for a variety of services, such

Figure 8-2
An ADSL network

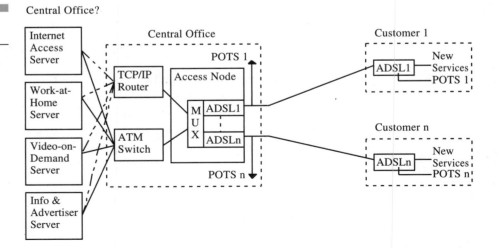

as high-speed Internet access or video on demand. A splitter function separates the POTS service from the digital services. In many cases, additional premises wiring may need to be run, but this is beyond the scope of the ADSL network because inside wire belongs to the owner of the premises, both in the United States and in many other countries.

In the central office (CO), or local exchange, the analog voice service is passed to the CO voice switch with another splitter arrangement. The ADSL local loops now terminate in an ADSL access node instead of leading directly to the CO switch. The access node (which is a type of DSL access multiplexer, or DSLAM) multiplexes many ADSL lines together. On the "back end" of the access node, links to *Transmission Control Protocol/Internet Protocol* (TCP/IP) routers or *asynchronous transfer mode* (ATM) switches can be maintained.

These switches and routers enable users to access the services of their choice. Note that these services may also be housed in the CO, either provided by the *local exchange carrier* (LEC) or a competitor under a collocation agreement. In many cases, the servers may located across the street within a short cable run from the CO.

Typically, services would include Internet access (of course), a work-at-home server (or the corporate Intranet), video-on-demand, and even servers from advertisers of information services (financial or otherwise). Note that access to these services may be either through TCP/IP or ATM —ADSL allows for both.

ADSL Network Essentials

Asymmetric Digital Subscriber Line (ADSL) is a full and complete network architecture. As was mentioned previously, ADSL is more than just a faster way to surf the Web. ADSL has the potential to supply residential and small business users all types of new broadband services. These services can be provided under a competitive environment and can come from a wide range of services, from educational to financial.

Figure 8-3 shows a more detailed look at the possible ways that an ADSL Terminal Unit-Remote (ADSL ATU-R) can be manufactured and configured to support access on the premises. The physical device could be a TV set-top box or a PC; more exotic devices are not precluded. The wiring from the ATU-R to the end device might be as simple as a 10BASE-T LAN, as complex as a private ATM network, or as new as the Consumer Electronics Bus, which uses premises electrical wiring to send information. Whatever form the broadband access takes, the wiring to existing

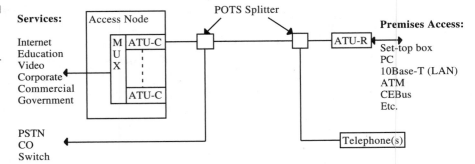

Figure 8-3
ADSL network essentials

analog telephones should not need to be altered because a special splitter separates these analog signals.

In the central office (local exchange), the analog voice service is passed to the PSTN voice switch with another splitter arrangement. The ADSL local loops now terminate in an ADSL access node instead of leading directly to the CO switch. The access node (which is a type of DSL access multiplexer, or DSLAM) multiplexes many ADSL lines together. Naturally, the PSTN switch software need not be upgraded to support these new service offerings (as with ISDN), and ADSL has the added benefit of reducing voice switch and trunk congestion caused by nonvoice services.

The services provided might be CO-based or virtually collocated. They could be offered by the local exchange carriers themselves or by a certified competitor. Services would include Internet access, educational courses, video (on-demand or broadcast), corporate (telecommuting), commercial (bookstores and auto dealers), and even government services (tax information and the like).

Note that the ADSL link could still lead to a *digital cross connect* (DACS) unit before picking up a trunking network leading to the service provider. Of course, if the service provider is also the provider of the ADSL link, then it is possible that all of the services could be directly housed in the CO itself, but it seems more likely that all of the services in the figure could be provided two ways. First, the ADSL links would be aggregated through the multiplexer and put into the DACS equipment. The DACS would lead to a trunk system link, perhaps an unchannelized T-3 running at 45 Mbps, and from there to an Internet service provider. At the ISP, all of the links terminate in an IP router, and the packets are then shuttled onto and off of the Internet itself. Corporate Intranets could be reached this way. This is the simplest method, but of course the sum total bandwidth of all the ADSL links serviced in this way could not exceed 45 Mbps in either direction.

Alternatively, the access node could link directly to an IP router or ATM switch located only a few feet away (or perhaps down the hall) from the access node. The traffic aggregation still takes place in that there is only one physical link from access node to switch/router. However, once the packets are in the IP router or ATM switch network, the route to all destinations is the same as above. This scenario makes sense if the ADSL service provider is also the ISP, but a later chapter details how an independent ISP can still supply ADSL services to their customers even *without* the local telephone company offering or supporting ADSL! This is not only a distinctive feature of ADSL, but one of the major reasons that interest in ADSL runs so high.

The access node has also been a focal point of attention in ADSL standards work. Right now, most ADSL access nodes perform simple traffic aggregation. That is, all of the bits and packets into and out of the access node are carried to the services by simple circuits. In the example above, if there are 10 ADSL customers receiving 2 Mbps downstream and 64 kbps upstream speeds (fairly typical), then the link between access node and service network (the Internet, for example) must be at least 20 Mbps (10×2 Mbps) in both directions to avoid congestion and packet discards. The speeds must be the same "upstream" even though the aggregate bit rate here is only 640 kbps (10×64 kbps) mainly due to the duplex nature of T-carrier links, which are just about all that is currently available to carry this traffic.

So a possible improvement on the basic ADSL system would be to either perform *statistical multiplexing* in the ADSL access node itself or supply some packet switching capabilities in the ADSL multiplexer directly. In the case of statistical multiplexing, based on the bursty nature of the packets themselves, lower speed links are possible because most users will not be active all at the same time. If the ADSL access node has a built-in IP router or ATM switching capability, the benefit is the same and even more direct. Either way, the same 10 customers could be supported on a link running at much less than 20 Mbps, perhaps only 1.5 Mbps. More details on these possible ADSL access node developments, now just beginning to appear, are discussed more fully in a later chapter on DSLAMs.

ADSL and Standards

As with any other technology, ADSL is in need of standards. All technologies go through a phase of exploration and experimentation, and early airplanes and automobiles took on many bizarre shapes and sizes. Before consumers will accept any new technology and pay hard-earned cash for it, however, the technology must become standardized enough to satisfy everyone. People want technology-based products that are consistent in appearance and performance, independent from a particular vendor, and that will work with other devices in the same category.

In the United States, an ADSL standard for physical layer operation was first described in American National Standards Institute (ANSI) T1.413-1995. In other words, this document describes exactly how ADSL equipment communicates over a previously analog local loop. The document does not, nor is it intended to, describe the entire ADSL network architecture and services, or the internal functioning of the ADSL access node, and so forth. It specifies such ADSL fundamentals as line coding (how the bits are sent) and the frame structure (how the bits are organized) on the line.

ADSL products have been manufactured that use *carrierless amplitude/phase* modulation (CAP), *quadrature amplitude modulation* (QAM), and *discrete multitone* (DMT) technology as line coding techniques. Others have been tried in a lab, but these are the most common. Whatever line coding technique is used, whenever the same two wires in a pair are used for full-duplex operation, either the frequency range must be split into upstream and downstream bandwidths (simply frequency division multiplexing) or echo cancellation must be used. (Echo cancellation eliminates the possibility of a signal in one direction being interpreted as a "talker" in the opposite direction, and so "echoed" back to the origin.) In ADSL, both FDM and echo cancellation techniques can be combined and typically are, which means that due to the asymmetric nature of ADSL bandwidths, the frequency ranges may overlap, but do not coincide. So both FDM and echo cancelling are used in a sense at the same time.

In any case, T1.413 specifies that ANSI standard, compliant ADSL must use DMT coding with either FDM or echo cancellation to achieve full-duplex operation. That said, it should be noted that FDM is the simpler method to implement. Echo cancellation is always vulnerable to the effects of *near-end crosstalk*, where a receiver picks up signals that are being transmitted on an adjacent system. The other system may be another pair of wires, or even the transmitter of the same system, since obviously the closest transmitter to a given receiver is its own transmitter in the opposite direction. FDM avoids near-end crosstalk by allowing the receiver to totally ignore the frequency range that the near-end transmitters are sending on. Of course, FDM cuts down on the total amount of bandwidth available in either direction. So echo cancellation makes more efficient use of the bandwidth, but at the price of complexity and sensitivity. Also, echo cancellation enables the lowest possible frequencies to be used, which maximizes performance.

As far as line coding is concerned, ADSL could have used any one of a number of widely used methods. The familiar 2B1Q from ISDN DSLs and HDSL was a possibility, as were other choices, such as CAP or QAM. DMT was chosen for a number of reasons, not the least of which was the inherent rate adaptive nature of DMT devices, which means that DMT devices can easily adjust to changing line conditions, such as moisture or interference. Also affecting the choice were DMT's resistance to noise (mostly am radio) and the presence of digital signals on adjacent wire pairs (crosstalk). However, CAP products have been successful in many ADSL deployment testbed projects. Perhaps CAP will someday form an acceptable line code for ADSL, a possible "Annex" to ANSI T1.413, which has been entertained primarily due to the delay in development of DMT-based systems.

Echo Cancelling and DSLs

Some form of echo control is required whenever the same frequency range is used for sending signals in both directions at the same time on the same physical path.

Echoes commonly arise from impedance mismatches on the signal path. In other words, some of the signal is reflected back to the sender at these points. When the same frequency range is used in both directions, this signal reflection could easily be mistaken for a signal originating at the remote end of the circuit. Echo cancellers electronically subtract the signal sent from the signal received, allowing any signal actually sent from the remote end to be distinguished more easily.

One way to accomplish echo control is to separate the frequency range into upstream and downstream bandwidths (simple frequency division multiplexing, or FDM). Now, no echo control is needed in the end devices.

The top part of Figure 8-4 shows what happens when no echo control is applied to ADSL. A 4 kHz baseband signal range for the analog voice passband is shown, along with a typical ADSL 175 kHz bandwidth for upstream traffic (from the home) and about a 900 kHz bandwidth for downstream traffic (to the home). This is asymmetrical, and this straight FDM approach does away with the need for echo control circuitry in the ADSL end devices.

Figure 8-4
Echo canceling and
DSLs

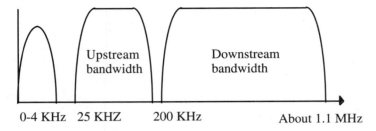

Upstream
bandwidth

Downstream
bandwidth

0-4 KHz   25 KHZ   200 KHz   About 1.1 MHz

FDM for ADSL: no echo cancellation needed

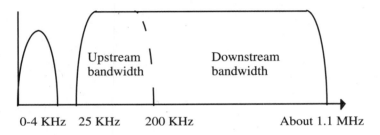

Upstream
bandwidth

Downstream
bandwidth

0-4 KHz   25 KHz   200 KHz   About 1.1 MHz

Echo cancelled ADSL: combined with FDM

However, pure FDM is not the most efficient use of the available bandwidth. The lower part of the figure shows a more effective approach where the upstream and downstream bandwidths actually overlap. Now, even though there is only a partial overlap in frequency range, echo control circuitry is needed in ADSL devices that use this approach.

Note that CAP-based ADSL devices typically use the FDM approach (no echo cancellation), whereas DMT-based ADSL devices typically use the echo cancellation approach, although there are exceptions. This echo cancellation approach is called an echo-FDM "combination" due to the asymmetrical nature of the devices. Basically, there are "FDM ADSL" and "echo cancelled ADSL" systems and equipment.

CAP versus DMT

CAP and DMT are the line codes that are most frequently used in ADSL products (ATU-R and ATU-C). Line codes determine how the zeroes and ones of the digital signal are sent and received. One is carrierless amplitude/phase modulation (CAP) and the other is discrete multitone (DMT).

No other controversy exists in ADSL (or xDSL in general) that is at once so vital to the technology and yet so confusing to non-specialists. Sometimes the debate about the relative merits of each assumes almost religious fervor, but both work just fine. The debate revolves around other issues, such as performance, cost-effectiveness and signal processing delay (that is, how long it takes a bit to make its way through a DMT or CAP device). In other words, which is *better*.

CAP is closely related to QAM, and many treat the two as virtually indistinguishable. DMT is more complex than CAP/QAM and is the open standard chosen by ANSI for ADSL (T1.413-1995). CAP technology is currently only available from a single source (Globespan Semiconductor, formerly a part of AT&T/Paradyne), but may be used by other vendors.

With regard to the CAP/DMT debate, most observers agree that both have strengths in different areas that affect ADSL performance and deployment scenarios. For example, it is generally agreed that DMT is better at rate adaptation (changing speeds due to line conditions), varying loop conditions (bridge taps, mixed gauges), handling noise (digital "crosstalk"), and subcarriers (for voice and other purposes).

CAP is seen to enable more simple echo cancellation (although many CAP products use FDM!), latency (about 25 percent of DMT processing delay is claimed), maturity (based on QAM, which has been around for years), and simplicity.

In spite of the presence of DMT in the T1.413 standard, an active and vocal group within ANSI is pushing for CAP as an acceptable "optional" line code for ADSL. In fact, there is little in the standard that is tied to DMT exclusively; both can be used in ADSL. Some have proposed adding CAP circuits to DMT devices, and vice versa. Even straight QAM has been proposed as an ADSL line code since this would eliminate the need for the CAP rotation function, making devices slightly more cost effective and simpler.

How CAP Works

Several major vendors of ADSL equipment use CAP as the line code. The line code simply determines how the 0s and 1s are sent from the ATU-R to and from the ATU-C, but this is not to downplay the line code's importance. A perfect line code should function well under less-than-perfect line conditions, including the presence of noise, crosstalk, and impairments, such as bridged taps and mixed gauges.

CAP is a close relative of a coding method known as quadrature amplitude modulation (QAM). In fact, they are almost completely mathematically identical, and engineers frequently refer to them in the same breath. The difference is in the implementation. The best description of CAP is probably "carrier-suppressed QAM." Because a carrier relays no information, it is common in a number of coding methods not to bother sending the carrier at all and reconstructing it electronically at the destination. This makes the technique "carrierless," or, more properly, "carrier suppressed."

Carrier suppression requires more circuitry in the end device than QAM does, but circuitry is now much cheaper than even a few years ago (a Hayes 9600 bps modem cost $799 in 1991, for example). So think of CAP as an "improved" QAM.

QAM established "constellations" based on two values of a received signal: amplitude and phase difference. Any point fixed by a given phase difference and amplitude represents a defined sequence of bits (for example, 0001 or 0101, and so on). CAP is essentially a type of QAM in which the QAM "constellation" is free to rotate (because there is no carrier to "fix" it at an absolute value). An element in the CAP circuitry called a *rotation function* can still determine the points of (and so the bits values in) the QAM constellation. So to get CAP from QAM, add a rotation function to the receiver and suppress the carrier at the sender.

CAP uses the entire local loop bandwidth (except for the 4 kHz baseband analog voice "channel") to send bits all at once. In other words, there are no subcarriers or subchannels to worry about. Full-duplex operation is achieved by FDM, echo cancellation, or both, but almost all CAP products to date have used FDM exclusively. CAP, mostly due to its QAM roots, is a mature, stable, and well-understood technology.

CAP/QAM Operation

This section presents a few details on the operation of CAP and QAM.

All analog carrier signals are characterized by amplitude, frequency, and phase. Any one, or even all three, may be used for signaling the 0s and 1s that make up digital information content. As an example, consider phase. Phase may be used in a *differential phase shift keying* (DPSK) scheme to convey digital information. In a differential phase shift keying situation, the phase of the wave is changed by a specific angle to represent each baud (change in line signaling condition) in the transmission. The use of the word "differential" suggests that the phase shift is measured from the current phase of the carrier, not from some absolute or reference phase.

In a simple example, two phase shifts can be employed. If the transmitter wishes to send a 1, it simply changes the phase of the carrier (from its current phase) by 1800. If the transmitter wishes to send a 0, the phase shift introduced is 00 (the current phase of the carrier is not changed). As a technical aside, the spectrum of the signal contains a strong amplitude at the carrier frequency with sideband signals appearing as a consequence of the phase shifting.

A step up from simple DPSK is *quadrature phase shift keying*. QPSK can be understood at two levels. On the surface, it is simply DPSK with two bit per signaling condition, called a *baud*. The phase shifts are measured from the current phase of the wave.

Although this level of understanding may suffice for many, it is misleading when considering more complex situations. In QPSK, the signal that is actually sent is a combination of a *sine wave* and *cosine wave* at F, where F is the carrier frequency. Because a sine and cosine ave function are always 900 out of phase, they are said to be *in quadrature*, which means that two quantities are perpendicular, or 900 different from one another, so the phrase is well used in this situation.

In fact, QPSK really has nothing to do with phase shifting. The phase shift is a consequence of the modulation of the amplitudes of two waves in quadrature. If one understands QPSK at this level, the topic of Quadrature Amplitude Modulation will be trivial. If the reader is not mathematically sophisticated, then a view of QPSK as shifting phases is sufficient. This presents a more detailed (and more correct) view for those who need to understand the concepts of digital communication more deeply.

Quadrature Amplitude Modulation (QAM)

Just as QPSK can be viewed at two levels, so can QAM. On the surface, a QAM modulator creates 16 different signals by introducing both phase and amplitude modulation. By introducing 12 possible phase shifts and two possible amplitudes, it is possible to obtain the 16 different signal types. These 16 signal conditions can represent 4 bits per line condition, or "4 bits per baud" signaling.

But as with DPSK, there is a deeper level of electronics involved. At a deeper level, QAM simply modulates the amplitudes of two waves in quadrature. Instead of using amplitudes of ±1, a simple QAM example uses four different amplitudes for each of the two waves. If the four amplitudes are labeld ±A1 through A4, then the 16 different signal types are simply obtained by using all possible pairs as amplitudes combined with simple sine and cosine wave functions (for example, A1 sin (F_t) + A1 cos (F_t), A1 sin (F_t) + A2 cos (F_t), ...) to create the necessary signal types. This creates the characteristic "constellation" pattern reproduced in all modem manuals based on QAM (and CAP is basically the same). A sample "constellation" is illustrated in Figure 8-5. Although it is much more common to represent these constellations as simple points, arrows have been used to emphasize the phase angles and amplitude differences.

Note that although the example in the figure uses 12 different phase angles, and 4 have two amplitudes of the sine and cosine components (a four-level system) to encode 4 bits per baud (change of signal, in this case phase and amplitude together). Other systems may use greater (or fewer) phase and amplitude level combinations. The greater the number of phase and amplitude levels, the greater the number of constellation points and correspondingly greater numbers of bits per signal. Like all multibit encoding systems, QAM is limited to the number of levels that can be discerned at the receiver in the face of noise.

Consider how QAM/CAP may be used to deliver digital video information. The output bit stream from a digitizer is broken into 4-bit nibbles (one half byte is one *nibble*) because the example uses QAM/CAP with 16 points in the constellation, as in the previous figure. The nibbles are fed (pardon the pun) to the encoder that selects the proper amplitudes for the waves in quadrature.

The output from the encoder is next fed to the actual modulator, which creates the correct output signal. Note that the modulator actually com-

Figure 8-5
CAP (16QAM or 4X4 QAM) modulation.

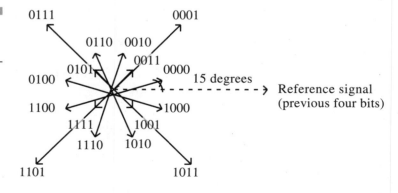

bines the appropriate amplitudes of the sine and cosine of the carrier frequency, thus creating the phase and amplitude shifts associated with the correct constellation point. Finally, the signal is filtered to assure that it does not interfere with other channels. Note that the resulting signal contains substantial energy at the carrier frequency; this should not be surprising.

DMT for ADSL

ADSL devices (the ATU-C and ATU-R) have been built that use QAM, CAP, and DMT technology as line codes. However, the official standard line code for ADSL is DMT, as defined by the American National Standards Institute ANSI T1.413 standard in 1995 for ADSL. Although DMT is often said to be "newer" than CAP or QAM, DMT was actually invented years ago by Bell Labs. DMT was never implemented until recently for many reasons, not the least of which was that CAP and QAM were sufficient for all telecommunications purposes common at the time.

DMT works by first dividing the entire bandwidth range on the formerly analog passband limited local loop into a large number of equally spaced subchannels. Technically, they are called *subcarriers*, but many people still call them subchannels. Above the preserved baseband analog signaling range, this bandwidth usually extends to 1.1 MHz. The entire 1.1 MHZ bandwidth is divided into 256 subchannels, starting at 0 Hz. Each subchannel occupies 4.3125 kHz, giving a total bandwidth of 1.104 MHz on the loop. Some of the subchannels are special, and others are not used at all. For example, channel #64 at 276 kHz is reserved for a pilot signal.

Most DMT systems use only 250 or 249 subchannels for information. The lower subchannels, #1 through #6 in most cases, are reserved for 4 kHz passband analog voice. Because 6 times 4.3125 Hz is 25.875 kHz, it is common to see 25 kHZ as the starting point for ADSL services. Note that a wide *guardband* is used between the analog voice and the DMT signals. In addition, the signal loss at the upper subchannels, such as #250 and above, is so great that it is difficult to use them for information transfer on long loop at all.

There are 32 upstream channels, usually starting at channel #7, and 250 downstream channels, which gives ADSL its distinct asymmetric bandwidth. Each of the subchannels is 4.3125 kHz wide, of course, and only when echo cancellation is used are there actually 250 downstream subchannels. When only FDM is used for echo control, there are typically 32 upstream channels and only 218 or less downstream channels because they no longer overlap. The upstream channels occupy the lower end of the spectrum for two reasons. First, the signal attenuation is less here, and customer transmitters are typcially lower-powered than local exchange transmitters, which is a concern. Second, there is more noise at the local exchange with the possibility of crosstalk, so it only makes sense to use the lower portions of the frequency range for the upstream signals.

When ADSL devices that employ DMT are activated, each of the subchannels is "tested" by the end devices for attenuation. In actual practice, the "testing" is a complex kind of handshaking procedure, and the parameter used is gain (the reciprocal of the attenuation). The noise present in each of the subchannels is measured as well.

Not all of the subchannels are used for information transfer, as mentioned above. Some are reserved for network management and performance measurement functions. For instance, in the downstream direction, only 249 of the 256 subchannels available downstream are typically used for information transfer.

Usually, each of the numerous subchannels employs its own coding technique based on QAM. This may strike some as odd, given the fervor that vendors have when seeking to distinguish CAP/QAM and DMT. Nevertheless, there obviously is at some of QAM in DMT. The real attraction of DMT is not so much that it is different than CAP and QAM, but rather that based on DMT's performance monitoring, some subchannels will carry more bits per baud than others. The total throughput is the sum of all the QAM bits sent on all the active subchannels (some may be completely "turned off").

Moreover, all of the subchannels are constantly monitored for performance and errors. The speed of an individual subchannel or group of subchannels can actually vary, giving DMT a granularity of 32 kbps. In other words, a DMT device might function at 768 kbps or 736 kbps (that is, 32 kbps less), depending on operational and environmental conditions. Just by way of comparison, CAP devices usually offer 340 kbps granularity (768 kbps or 428 kbps), but pure QAM can offer granularity as fine as 1 bps, which means that there is nothing that technically limits CAP/QAM to one level of granularity but not another. In fact, some vendors of CAP-based ADSL equipment have claimed 32 kbps granularity, and even RADSL capabilities, for their latest products. It should be noted that these CAP RADSL products modify their spectrum when the rate changes, and now become a real issue to manage with regard to spectral compatibility.

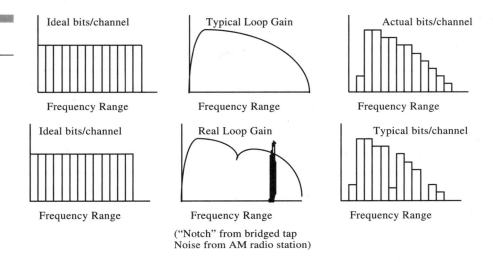

Figure 8-6
DMT in operation

Experts generally concede that finer granularity is a benefit that can maximize user acceptance and deployment situations.

Discrete Multitone (DMT) Operation

Figure 8-6 shows discrete multitone technology in operation in an ADSL device on a typical local loop. The figure actually has two parts. The upper shows a kind of ideal situation, such as that found in a straight run of 24 gauge copper wire less than 18,000 feet without a lot of outside noise (good luck finding one of those). The only real attenuation effects come from the distances and frequencies involved. The lower part of the figure shows a typical local loop in the real world.

Consider the design ideal first. Across the frequency range, on the left, there exists a targeted maximum number of bits per second per subcarrier (channel) that the device would like to send and receive. However, the middle figure shows the situation on a typical loop. The gain (the reciprocal of the attenuation) is better or worse depending on frequency. At higher frequencies, distance effects dominate; at lower frequencies, impulse noise and crosstalk dominate. This leaves a broad middle range (about 25 kHz to 1.1 MHz) for signals, with the gain slowly dropping off with increasing frequency.

DMT devices can measure the gain in each subcarrier and adjust the actual number of bits per second per channel to reflect the actual line gain profile. The upper right part of the figure shows this.

Of course, the real world is not so cooperative. The lower series on the figure shows a real loop gain profile in the middle. Two line impairments are added. The first is a distinctive "notch" caused by the effects of a bridged tap. The bridged tap acts like a long delay circuit as the signal travels out and back. The returning signal interferes with the main signal. The placement of the notch in the frequency range depends on the wavelength of the bridged tap itself; however, one frequency or another will be affected by the standing wave. The second impairment is noise from a nearby am radio station. These stations broadcast in the same high kHz (or "kilocycles" in am radio talk) range as that of ADSL devices trying to listen in. Because local loop wires are long antennas, it comes as no surprise that this signal is picked up (analog voice, operating at 4 kHz baseband, is immune to this am radio interference).

DMT devices can measure the gain in each subcarrier and adjust the actual number of bits per second per channel to reflect the actual line gain profile. The right part of the figure shows this. Note that some channels are even "turned off." Also note that DMT is inherently rate adaptive, as the figure illustrates, as channels increase or decrease their ability to support a given bit rate.

DMT for ADSL Advantages

The choice of discrete multitone technology as the ADSL line coding standard in ANSI T1.413-1995 was not made by flipping a coin. When used for ADSL, DMT has a number of distinct advantages over CAP line coding. It remains to be seen, however, whether newer CAP-based (and even pure QAM-based) devices can make up for their perceived shortcomings with newer or better features. There is an active movement within ANSI to approve an alternative line code standard for ADSL, and perhaps based on CAP or QAM . In fact, some within ANSI have suggested devices that can do both, or even all three, are the only realistic, long-term product strategies to pursue.

In any case, the following lists reasons commonly given why DMT was chosen as the ADSL standard. After going through them all from a DMT perspective, this section also considers all eight from the perspective of a vendor using CAP-based ADSL in their products.

1. Built-in subchannel optimization (RADSL)
2. Ongoing active monitoring
3. Maximum loop variation coverage
4. Highest level of rate flexibility
5. Superior noise immunity for greater throughput
6. Broad industry chipset support
7. Interoperability through standards
8. Virtually future proof

First of all, DMT is capable of built-in subchannel optimization due to the "testing" procedure on each subcarrier. This means that DMT is inherently RADSL, a trait that needs to be added to CAP with great effort and may never be done consistently. Single-carrier techniques such as CAP are not capable of adjusting for different frequency gains. Additionally, DMT features constant, active monitoring of channels for performance. CAP has no similar built-in capability. Both of these combine to give DMT a maximum loop coverage. In other words, DMT might still be deployed in places where CAP just will not work.

DMT gives a higher level of rate flexibility. The granularity in DMT is typically 32 kbps, whereas CAP's is generally 340 kbps (that single carrier again). Because DMT can selectively "turn off" channels due to noise effects (such as am radio stations), greater throughputs under impaired conditions are possible. The fewer errors, the fewer re-transmissions of data, and greater throughput results. CAP would struggle with am radio noise.

In addition, DMT is an open standard (and so has, by definition, only a "fair and reasonable" licensing fees associated with it), so chipsets can be developed by anyone. The industry gives broad support for DMT chipsets, but only a single (but licensable) source for CAP (Globespan Semiconductor). Of course, the Issue 2 standard for DMT is also meant to assure interoperability. Finally, DMT is virtually future proof. CAP/QAM has been around for a while, but the brightest days of DMT are still to come.

For these reasons, DMT became the ADSL standard. This is not to disparage CAP. Both work fine (although there is evidence that DMT works better on many types of real world lines), and CAP has advantages over DMT in some areas—but ANSI decided on DMT.

In all fairness, testing of ADSL devices has shown that they are all quite tolerant of bridged taps, mixed gauges, radio noise, and even adjacent T-carrier circuits using bipolar *Alternate Mark Inversion* (AMI) line code. This appears to be true whether the ADSL devices are based on CAP or DMT, although DMT seems to be marginally better. The granularity issue is likely to be the most critical one right now.

But it may be worthwhile to quickly revisit the DMT list again, but from the CAP perspective. After all, if the whole CAP versus DMT debate were so clear, then one set of products would quickly disappear. This has not been the case.

The CAP response to RADSL has been to add its capabilities to products. "It's only electronics," goes the CAP approach. Active monitoring has to be added? Big deal. Similar functions are added to circuit boards all the time. One is built in and the other is added on. No one can tell the difference.

The loop coverage issue has not been the decisive factor either The rate flexibility issue is another of those "so we'll add it" responses from the CAP camp.

Although DMT maintains interoperability through standards, CAP maintains interoperability through single-sourcing. The CAP people say that the method is not the point, only the result is what is important. , It is arguable that single-sourced implementations provide *better* interoperability features than does a standard with many options and interpretations.

Although it is true that DMT had many years of development preceding it, there is still plenty of life left in CAP and even QAM. It is far too early to write off the CAP method of ADSL.

The controversy will continue for a while. It seems in the long run, however, that vendors will always seek to push the boundaries of their product's performance a little further. There is little doubt that DMT offers a far richer field for this type of exploration than CAP does.

9

The ADSL Interface and System

The ADSL standard technically defines a "metallic interface" between a network (the PSTN) and customer. However, the ADSL Forum has extended this interface into the full architecture introduced previously and even proposed several ways for services to be accessed and delivered over the ADSL interface. This chapter describes the standard ADSL interface in full. This standard is based on ANSI T1.413-1995, the *Network and Customer Installation Interfaces—Asymmetric Digital Subscriber Line (ADSL) Metallic Interface* specification. It should be noted that this specification is now updated as of Issue 2. But many ADSL products are based on the information here. The chapter then details some of the aspects of the overall ADSL architecture and service platforms outlined by the ADSL equipment vendors and service providers.

ADSL Unidirectional Downstream Transport

ADSL interfaces can do more than support a single bit stream to and from the customer premises, although this is certainly an option, but ADSL can do a lot more. ADSL, like most transports, is a *framed transport*; the bit stream inside the ADSL frames can be divided into a maximum of seven bearer channels (just called *bearers* in ADSL) at the same time. The bearers fall into two main classes: There can be up to four totally independent downstream bearers that always function unidirectional ("simplex" in the specification) downstream. That is, these four bearer channels can only carry bits downstream to the customer. The four bearers are designated AS0 through AS3. The "AS" has no real meaning except perhaps to its inventor. In addition to the AS channels, there may be up to three bidirectional ("duplex" in the specification) bearers that can carry traffic both upstream and downstream. These bearers are designated LS0 through LS2. The "LS" is apparently as meaning-free as "AS." Note that these bearer channels are *logical* channels, and that bits from all channels are transmitted at the same time over the ADSL link and do not use dedicated bandwidth.

Any bearer channel can be programmed to carry bits in any multiple of 32 kbps (the "natural" DMT granularity). Bit rates that are not simple multiples of 32 kbps (70 kbps, for example) can be supported, but only by carrying the "extra" bits (in the 70 kbps example, 6 kbps) in the shared overhead area of the ADSL frame.

Now, allowing bearers to run at almost any multiple of 32 kbps might not be the best idea in the world, especially when interoperability is a concern. Accordingly, the ADSL specification has established four *transport classes* for the downstream simplex bearers. These are based on simple multiples of 1.536 Mbps (the user transfer rate of a T1). The transport classes are 1.536 Mbps, 3.072 Mbps, 4.608 Mbps, and 6.144 Mbps. The duplex bearers can carry a control channel and some ISDN channels (the BRI and 384 kbps). Please note that ADSL is not limited to any particular transport classes in and of itself. Future specifications may be made for 1.544 Mbps transport (the full T1 line rate) or 2.048 Mbps (the E1 line rate). No maximum speed is defined for any bearer. The upper limit is dependent on the total carrying capacity of the ADSL link.

ADSL products have established various sub-channel data rates for the default bearer bit rates. The maximum transport class speed of 6.144 Mbps is not allowed on all AS bearers at the same time. The limitations are shown in Table 9-1.

Table 9-1

ADSL Sub-channel
Rate Restrictions for
Default Bearer
Rates

Sub-channel	Sub-channel Data Rate	Allowed Values of n_x
AS0	$n_0 \times 1.536$ Mbps	$n_0 = 0, 1, 2, 3,$ or 4
AS1	$n_1 \times 1.536$ Mbps	$n_1 = 0, 1, 2,$ or 3
AS2	$n_2 \times 1.536$ Mbps	$n_2 = 0, 1,$ or 2
AS3	$n_3 \times 1.536$ Mbps	$n_3 = 0$ or 1

Rates equivalent to multiple DS1s are also supported

Support for at least AS0 is mandatory. The maximum number of sub-channels that can be active at any given time and the maximum number of bearer channels that can be transported at the same time in an ADSL system depends on the transport class. Transport class support depends on the achievable line rate of the specific ADSL loop and the configuration of the sub-channels, which may be configured to maximize the number of sub-channels or the line rate. Switching between configured sub-channel speeds and numbers is intended for future study. For now, whatever structure and rate is put in place on an ADSL link remains fixed.

The transport classes themselves are numbered 1 through 4. Support for transport classes 1 and 4 are mandatory, and support for transport classes 2 and 3 are optional. In addition, a series of optional transport classes prefixed by "2M" are intended for use with 2.048 Mbps E-carrier-based systems in common use outside of the United States.

Transport class 1 is mandatory and is intended for the shortest loops, but it offers the highest downstream capacity of any ADSL configuration. This class carries 6.144 Mbps downstream and can be made of any combination of one to four bearer channels running at simple multiples of 1.536 Mbps. Support for at least one sub-channel running at 6.144 Mbps on AS0 is mandatory. Transport class 1 can carry the following optional configurations, all of which add up to 6.144 Mbps:

- One 4.608 Mbps bearer channel and one 1.536 Mbps bearer channel

- Two 3.072 Mbps bearer channels

- One 3.072 Mbps bearer channel and two 1.536 Mbps bearer channels

- Four 1.536 Mbps bearer channels

Transport class 2 is optional and carries 4.608 Mbps downstream. This class may be comprised of any combination of one to three bearer channels running at multiples of 1.536 Mbps. Systems can provide any or all of the bearer rates because none are mandatory. AS3 is never used in transport class 2. Transport class 2 can carry the following configurations, all of which add up to 4.608 Mbps:

- One 4.608 Mbps bearer channel
- One 3.072 Mbps bearer channels and one 1.536 Mbps bearer channel
- Three 1.536 Mbps bearer channels

Transport class 3 is also optional and carries 3.072 Mbps downstream. This Class may be made up of one or two bearer channels running at simple multiples of 1.536 Mbps. Systems can provide any or all of the bearer rates because none are mandatory. AS2 and AS3 are never used in transport class 3. Transport class 3 can carry the following configurations, all of which add up to 3.072 Mbps:

- One 3.072 Mbps bearer channel
- Two 1.536 Mbps bearer channels

Transport class 4 is mandatory and runs on the longest loops, but offers the minimum downstream carrying capacity. The bearer channel is just 1.536 Mbps running on AS0.

The ADSL specification also takes into account networks based on the E-carrier hierarchy of 2.048 Mbps, which is common outside of the United States. In fact, all of the common local loop structures found outside of the United States are addressed in the specification, specifically in Annex H to T1.413-1995. Only AS0, AS1, and AS2 are supported using this 2M structure, as shown in the Table 9-2.

As with the 1.536 Mbps structure, support for AS0 is the minimum requirement. The maximum number of sub-channels that can be active at

Table 9-2

ADSL sub-channel rate restrictions for 2.048 Mbps (optional)

Sub-channel	Sub-channel Data Rate	Allowed values of n_x
AS0	$n_0 \times 2.048$ Mbps (optional)	$n_0 = 0, 1, 2,$ or 3
AS1	$n_1 \times 2.048$ Mbps (optional)	$n_1 = 0, 1,$ or 2
AS2	$n_2 \times 2.048$ Mbps (optional)	$n_2 = 0$ or 1

any given time and the maximum number of bearer channels that can be transported at the same time in an ADSL system depends on the transport class. Furthermore, transport class support depends on the achievable line rate of the specific ADSL loop and the configuration of the sub-channels, which may be configured to the maximize number of sub-channels or the line rate. Switching between configured sub-channel speeds and numbers is intended for future study. For now, whatever structure and rate is put in place on an ADSL link remains fixed.

For 2M structures, the transport classes themselves are numbered 2M-1 through 2M-3. Support for all 2M transport classes is optional. The configurations of the 2M transport classes closely follow the 1.536 Mbps transport classes. That is, transport class 2M-1 still runs at 6.144 Mbps downstream in total.

Class 2M-1 can be made up of any combination of one to three bearer channels running at simple multiples of 2.048 Mbps. All transport class 2M-1 configurations all optional and can carry the following configurations, all of which add up to 6.144 Mbps:

- One 6.144 Mbps bearer channel
- One 4.096 Mbps bearer channel and one 2.048 Mbps channel
- Three 2.048 Mbps bearer channels

Transport class 2M-2 is optional and carries 4.096 Mbps downstream. 2M-2 may be made up of one or two bearer channels running at simple multiples of 2.048 Mbps. Systems can provide any or all of the bearer rates because none are mandatory. AS2 is never used in transport class 2M-2. Transport class 3 can carry the following configurations, all of which add up to 4.096 Mbps:

- One 4.096 Mbps bearer channel
- Two 2.048 Mbps bearer channels

Transport class 2M-3 is optional and runs on the longest loops, but offers the minimum downstream carrying capacity. The bearer channel is just 2.048 Mbps running on AS0.

Another point must be made about ADSL transport classes. Support is optional for carrying Asynchronous Transfer Mode (ATM) cells downstream. ATM cells are fixed-length 53 octet (8 bits, a byte) data units. ATM cells each have a 5-octet header and a 48-octet payload. Information is carried in the 48-octet payload according to ATM Adaptation Layer 1 (AAL1) rules. The AAL defines how information is formatted within the ATM cell payload area. In AAL1, which offers a *constant bit rate* (CBR) and stable delay through the network with connections between end-points, one payload octet is used for additional overhead, and the other 47 octets carry user data. AAL1 is the easiest and simplest way to make ATM cell streams look and act like traditional circuits.

When ADSL is used for downstream ATM cell transport, only AS0 is used, so there is only one configuration option—AS0—which can run at one of four different rates. These rates are defined as ATM transport classes 1 through 4 and run at 1.760 Mbps, 3.488 Mbps, 5.216 Mbps, and 6.944 Mbps, respectively. The odd-sounding speeds are a result of the desire to preserve compatibility with existing AAL1 and circuit definitions already entrenched in ATM documentation, and they take into account the effects of ATM overhead on user data rates.

It should be noted that the above discussion concerns ANSI T1.413 Issue 1. Issue 2 is now ready and involves several further "tweaks" to accommodate ATM.

ADSL Bidirectional (Duplex) Transport

So much for the unidirectional (simplex) downstream channels. Upstream simplex channels are being studied, but up to three bidirectional (duplex) bearer channels can be transported at the same time on an ADSL interface. One of these is always the mandatory control channel, designated the C channel. The C channel can carry signaling messages for selection of services and call setup. All user-to-network signaling for the simplex downstream channels is carried here, and the C channel can also carry signaling for the duplex channels, if present, as well.

The C channel is always active and runs at 16 kbps in transport classes 4 and 2M-3. In transport classes 4 amd 2M-3, the C channel messages are always carried in a special overhead section of the ADSL frame. All other transport classes use a 64-kbps C channel, and the messages are transported in duplex bearer channel LS0.

In additional to the C channel, an ADSL system can carry two optional bidirectional bearer channels: an LS1 running at 160 kbps and an LS2 running at either 384 kbps or 576 kbps. The exact structure of the bidirectional channels varies by transport class as defined for the simplex channels, so the easiest way to relate them is by using a table. Table 9-3 relates the duplex channel structures to the simplex bearer channel transport classes. The two notes mean the following

Table 9-3

Maximum Optional Duplex Bearer Channels Supported by Transport Class

Transport Class	Optional Duplex Bearers that may be Transported (note 1)	Active ADSL Subchannels
1 or 2M-1 (minimum range)	Configuration 1: 160 kbps + 384 kbps	LS1, LS2
	Configuration 2: 576 kbps only	LS2 only
2, 3, or 2M-2 (mid range)	Configuration 1: 160 kbps only Configuration 2: 384 kbps only (see note 2)	LS1 only LS2 only
4 or 2M-3 (maximum range)	160 kbps only	LS1 only

NOTE: **1.** *When the 160 kbps option al duplex bearer channel is used to transport ISDN BRA, all signaling associated with the ISDN BRA (160 kbps) is carried on the D channel of the 2B+D signal embedded in the 160 kbps. Signaling for the 576 kbos, 384 kbps, and non-ISDN 160 kbps duplex bearers may be included in the C channel, which is shared with the signaling for the downstream simplex bearer channels.*

2. *Whether transport classes 2, 3, or 2M-2 should support the 576 kbps optional duplex bearer is left for further study.*

1. If the channel carries ISDN, then the D channel can still be used for signaling; otherwise the C channel is used.

2. Nobody knows yet whether mid-range ADSL loops can run at 576 kbps upstream.

As might be expected, the bidirectional channels have an option to transport ATM cells. If supported, the ATM cells run on the optional LS2 channel. In this case, in addition to AAL1, the LS2 channel carries ATM cells formatted according to AAL5, which is just another way of mapping user information into a series of ATM cell payloads. The attraction of AAL5 is that it has the minimal overhead of any AAL. However, AAL5 is intended for *variable bit rate* (VBR) applications and will not guarantee a stable delay through the network (therefore, the application must if a stable delay is required). Also, AAL5 usually uses connections through the ATM network, but these are not required. Most ATM switches support both AAL1 and AAL5, although the AAL5 support is usually connection-oriented.

The data rates for the LS2 channel when used for ATM cell transport is 448 kbps or 672 kbps. Again, the odd rates are due to a desire for compatibility with existing ATM documentation and takes into account the effects of ATM overhead on user data rates.

Combining the Options

The establishment of standard bearer channel structures and speeds at least prevents a free-for-all by ADSL equipment designers and vendors. However, the presence of a set of transport classes for simplex downstream operation and multiple bidirectional bearer channel options does not completely achieve this goal. There still exists a bewildering array of options in both directions that technically comply with the specifications. What is needed is a way to combine AS channel structures with LS channel structures in a way that is both meaningful and standardized. Fortunately, the ADSL specification does this as well. Each of the simplex and duplex channels can be configured independently, as shown in Table 9-4. The table forms a useful summary for the major points mentioned previously. The exact configuration is specified by certain parameters carried in the ADSL frame for each of the bearer channels. Naturally, there is a corresponding table for bearer channels based on the 2.048 Mbps structures. This is shown in Table 9-5.

At the risk of providing an overwhelming amount of information, one more attempt must be made to organize the transport classes, in terms of downstream simplex and bidirectional duplex channels. The final table of this type, Table 9-6, presents the channel structures for the optional ATM bearer rates.

Note the relatively restricted structure of an ADSL system employing ATM cells. The upstream direction (really bidirectional) is much slower than in the other transports, amounting to nothing more than the signaling C channel. The lack of structural options in the downstream simplex direction is due mainly to the capability of ATM to establish its own virtual circuits, namely the *virtual path* (VP) and *virtual channel* (VC). All the previous notes have said is that first, if there are not enough cells containing live data to send to the customer (downstream), then the network equipment inserts just enough idle cells to bring the cell rate up to the line rate. This process, called *cell rate decoupling*, is common in ATM networks. Second, the C channel is not even represented as a separate channel in the ADSL frame structure, but is carried in a special overhead section of the ADSL frame.

Table 9-4

Bearer Channel Options by Transport Class for Bearer Rates Based on Downstream Multiples of 1.536 Mbps

Transport Class:	1	2	3	4
Downstream simplex bearers:				
Maximum capacity (in Mbps)	6.144	4.608	3.072	1.536
Bearer channel options	1.536	1.536	1.536	1.536
(in Mbps)	3.072	3.072	3.072	
	4.608	4.608		
	6.144			
Maximum active sub-channels	Four (AS0, AS1, AS2, AS3)	Three (AS0, AS1, AS2)	Two (AS0, AS1)	One (AS0 only)
Duplex bearers:				
Maximum capacity (in kbps)	640	608	608	176
Bearer channel options	576	see Note	see Note	
(in kbps)	384	384	384	
	160	160	160	160
	C (64)	C (64)	C (64)	C (64)
Maximum active sub-channels	Three (LS0, LS1, LS2)	Two (LS0, LS1) or (LS0, LS2)	Two (LS0, LS1) or (LS0, LS2)	Two (LS0, LS1)

NOTE: *Whether transport classes 2 or 3 should support 576 kbps optional duplex bearer is under further study.*

ADSL Overhead

It should come as no surprise that ADSL includes overhead in the channel bit rate figures, as well as the capacity to carry user information. In addition to the simplex (downstream) and duplex (upstream and downstream, but mainly upstream) bearer channels, ADSL includes overhead

Table 9-5

Bearer Channel Options by Transport Class—Optional Bearer Rates Based on Downstream Multiples of 2.048 Mbps

Transport class:	2M-1	2M-2	2M-3
Downstream simplex bearers:			
Maximum capacity (in Mbps)	6.144	4.096	2.048
Bearer channel options	2.048	2.048	2.048
(in Mbps)	4.096	4.096	
	6.144		
Maximum active sub-channels	Three (AS0, AS1, AS2)	Two (AS0, AS1)	One (AS0 only)
Duplex bearers:			
Maximum capacity (in kbps)	640	608	176
Bearer channel options	576	see Note	
(in kbps)	384	384	
	160	160	160
	C (64)	C (64)	C (64)
Maximum active sub-channels	Three (LS0, LS1, LS2)	Two (LS0, LS1) or (LS0, LS2)	Two (LS0, LS1) or (LS0, LS2) Two (LS0, LS1 or (LS0, LS2)

for a variety of functions. One crucial function is *synchronizing* the bearer channels, which means that the devices at each end of the ADSL link know which channels are configured (the ASs and LSs), at what rate they run, and where their bits are located in the stream of ADSL frames.

Other overhead functions in ADSL include an *embedded operations channel* (eoc), an *operations control channel* (occ) used for remote reconfiguration and rate adaptation, error detection by means of a *cyclical redundancy check* (crc), more bits set aside for *operations, administration, and maintenance* (OAM), and bits used for *forward error correction* (FEC, carried in "fe" bits) so that some errors can be corrected without a need for retransmitting information. Oddly, many of the ADSL overhead acronyms are not capitalized. Presumably, the role of the overhead bits in ADSL is not diminished by this apparent loss of alphabetic status.

Table 9-6

Bearer channel options by transport class for optional ATM bearer rates

Transport class:	1	2	3	4
Downstream simplex bearers:				
Maximum capacity (in Mbps)	6.92834	5.196255	3.46417	1.732085
Bearer channel rate (in Mbps) (see Note 1)	6.944	5.216	3.488	1.760
Maximum active sub-channels	One (AS0 only)	One (AS0 only)	One (AS0 only)	One (AS0 only)
Duplex bearers:				
Maximum capacity (in kbps)	64	64	64	64
Bearer channel options (in kbps)	C (64)	C (64)	C (64)	C (64)
Maximum active sub-channels	1 (LS0)	1 (LS0)	1 (LS0)	1 (LS0) see Note 2

NOTE: 1. *The bearer channel rate is equal to the ATM data cell bit rate rounded up to the nearest integer multiple of 32 kbps (an ATM cell processor on the network side of the V-interface performs the rate adjustment by inserting idle cells).*

2. *The 16 kbps C channel is carried entirely within the synchronization overhead; the LS0 sub-channel does note appear as a separate byte within the ADSL frame.*

All of the overhead bits in ADSL are sent in both the upstream and downstream directions. In most cases, the overhead bits are sent as 32-kbps bit streams, but there are exceptions. For higher-speed channel structures, there is a maximum bit rate of 128 kbps and a minimum bit rate of 64 kbps, with a default of 96 kbps downstream and a maximum of 64 kbps and minimum of 32 kbps, with a default of 64 kbps (the maximum) in the upstream direction.

In some cases the overhead bits are embedded in the overall bit rate of the ADSL frames and do not consume additional bandwidth. In other cases, the overhead bits add marginally to the overall bit rate in one direction or the other. For example, transport class 1 running at 6.144 Mbps downstream adds a maximum of 192 kbps and a minimum of 128 kbps to the overall bit rate. When coupled with the maximum overhead rates on the duplex channels, the transport class 1 line rate rises from 6.144 Mbps to either 6.976 Mbps (maximum) or 6.336 Mbps (minimum), with 6.912 Mbps being the most typical rate using the overhead default speeds. Other transport classes are similarly affected.

When overhead considerations are added to ADSL channel capacities, the rate determined becomes known as the ADSL *aggregate bit rate*.

A key concept in ADSL overhead is the need for synchronization among the bearer channels. This concept was touched briefly above as concerning the structure of the bearers on the ADSL link, but ADSL synchronization has another important function. As it turns out, not all bits sent on an ADSL link are considered equal, which makes sense because some of the bits may represent delay-sensitive audio and video services, other bits may represent Web page access, and others may represent simple bulk file transfers of e-mail.

This being the case, the ADSL specification established two major bit categories. The terms chosen to represent these are rather unfortunate, but they are firmly established. All bits transported across an ADSL link come from either a "fast" data buffer of an interleave data buffer. A better term for the fast data buffer would have been *low-latency data buffer*, and this is the definition provided in the documentation. The interleave data buffer transports bits that can function properly given a bounded *interleaving delay*, whereas the fast data buffer transports bits that cannot function properly unless there is no additional interleaving or buffering delay. The term *interleaving* refers to how the bits in this buffer are protected from errors, rather than how the buffer area is serviced by the sending device, as might be expected. Better terms might have been *delay-sensitive buffer* and *delay-tolerant buffer*, but these phrases are never used in ADSL standard documents.

In other words, some of the bits sent on ADSL channels are never buffered at all beyond whatever buffering is needed to format the frame they are to be transported in. Other bits may sit in an interleave buffer until there is room in the ADSL frame to transport them. This interleave buffer may be configured in various sizes, as well. The whole system forms a rather neat, but limited, arrangement of raw priorities independent of priorities assigned to traffic streams by the routers or switches in the networks themselves.

Space is assigned in each ADSL frame to transport bits from the fast data buffer separately from the interleave data buffer bits, so there is never any question as to which bits are being sent and where they are located.

The ADSL Superframe

ADSL devices, specifically the *ADSL Terminal Unit-Central Office* (ATU-C) and the *ADSL Terminal Unit-Remote* (ATU-R), exchange bits by using a line code, usually DMT (the standard) or CAP. Bits are just bits, though. It is what the bits represent that is important. How are IP packets represented as ADSL bits? What about ATM cells? Motion Pictures Expert Group (MPEG) video? Dolby digital audio? How is a sender to tell an ADSL receiver what the bits represent when all types may be transported at the same time to or from a multimedia-capable device? It is all done with ADSL superframe.

All protocols today function in layers, and ADSL is no exception. At the lowest level of any protocol, there are bits, which are represented by the DMT or CAP line codings. The bits are organized into frames and are gathered into what ADSL calls *superframes*. Frames are "first order bit structures" and are the last things that bits are before they are sent, and frames are the first thing that bits become when they are received. Note that the ADSL superframe has much more in common with the T1 frame or superframe than the familiar Ethernet LAN frame. In fact, an Ethernet frame can form the content of an ADSL superframe. The overall structure of the ADSL superframe is shown in Figure 9-1.

In ADSL, the superframe is broken into a sequence of 68 ADSL frames. Some frames have special functions. Frames 0 and 1, for instance, carry error control information (that is, a cyclic redundancy check, or CRC) and indicator bits (ib) that are used to manage the link. Other indicator bits are carried in frames 34 and 35. A special synchronization frame follows the superframe and carries no user information. One ADSL superframe

Figure 9-1
The ADSL superframe

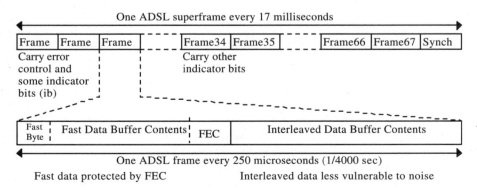

One ADSL superframe every 17 milliseconds

| Frame | Frame | Frame | | | Frame34 | Frame35 | | Frame66 | Frame67 | Synch |

Carry error
control and
some indicator
bits (ib)

Carry other
indicator bits

| Fast Byte | Fast Data Buffer Contents | FEC | Interleaved Data Buffer Contents |

One ADSL frame every 250 microseconds (1/4000 sec)

Fast data protected by FEC   Interleaved data less vulnerable to noise

(Frames are scrambled and size varies based on line bit rate)

is sent every 17 milliseconds. Becaise ADSL links are essentially point-to-point links, no frame addressing or connection identifier is needed at this level of ADSL.

Inside the superframe are the ADSL frames themselves. One ADSL frame is sent every 250 microseconds (1/4000 of a second) and consists of two main parts. The first part is the fast data. *Fast data* is considered to be delay-sensitive, yet noise tolerant (audio and video, for instance), by the equipment vendor, and ADSL tries to keep the latency associated with this to an absolute minimum. The contents of the fast data buffer of the ADSL device are placed here. A special octet called the *fast byte* precedes this section and carries the CRC and indicator bits where necessary. Fast data is protected by a FEC field in an attempt to correct fast data errors (audio frames can hardly be re-sent).

The second part of the frame contains information from the interleaved data buffer. Interleaved data is packaged to be as impervious to noise as possible, at the cost of increased processing and latency. The interleaving of the data bits makes the data less vulnerable to the effects of noise. This part of the frame is mainly intended for pure data applications, such as high-speed Internet access. All frame contents are scrambled before transmission to minimize the possibility of false superframe synchronization. This is common practice on all framed transports.

Note that there are no absolute frame sizes for the ADSL superframe. Because ADSL line rates vary and are asymmetrical, the frames sizes themselves may vary. However, the frame size is fixed in the sense that frames must be sent every 250 microseconds (fast and interleaved every 125 microseconds), and one superframe must be sent every 17 milliseconds. Naturally, the maximum line rates for ADSL establish a maximum frame size. The buffer sizes are determined by the speed and structure of the bearer channels when configuration is first done. There is nothing to prevent re-configuration of buffer sizes during operation of the ADSL link, but no such provision is currently made in ADSL specifications.

Also, there is nothing that determines how or which user bit streams fill the fast and interleaved buffers. This problem is beyond the scope of the ADSL standard. ADSL simply provides the information transfer mechanism.

As was mentioned previously, frames 0 and 1, and 34 and 35, have special roles in the ADSL superframe. These frames carry a cyclical redundancy check for the superframe and also the various indicator bits representing the overhead functions. The other frames, namely frames 2 through 33 and 36 through 67, carry overhead information as well, but overhead representing the embedded operation channel (eoc) and synchronization control (sc). All of this information is carried in the fast data byte position of each ADSL frame in the superframe.

To make things even more confusing, the fast data overhead bits have different structures depending on whether the frame is an even (0,2,4,...) or odd (1,3,5,...) numbered frame. The structure of all these bits in the fast data byte is shown in Figure 9-2.

Four main overhead functions are detailed in the figure. Bits exist for error detection (crc) and indicator bits (ib) for OAM functions in frames 0 and 1. Bits exist for other OAM functions in the ib bits in frame 34 and 35. The other frames carry the configuration bits (eoc) and synchronization control bits (sc) for determining bearer channel structures and the like.

The "r1" bit is reserved for future use and must be coded to a 1 value. Note that four other bits must also be coded as a 0 value or 1 value. These help to identify the eoc or sync control frames. Also note that this 0 or 1 bit will be the first bit of the fast data byte, as shown in the figure.

The largest portion of the overhead is for the indicator bits. These have a variety of functions that are defined for the downstream direction in Table 9-7.

The indicator bits 0 through 7 and 14 through 23 (the first 8 and the last 10) are reserved for future use. Indicator bits 8 through 13 have explicitly defined functions.

It is easy to understand the los or rdi bits (ib12 and ib13). The los bit is for *loss of signal* and is used by an ADSL ATU to indicate whether the pilot signal in the opposite direction disappears or drops below a certain threshold. The los bit is 1 when there is no los condition to report and 0 when there is. The rdi bit is for *remote defect indication* and is used by an

Figure 9-2
The fast data byte overhead bit structure

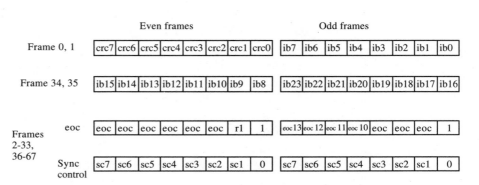

Direction of transmission

Table 9-7

Indicator bit functions in downstream direction

Indicator Bit	Definition
ib0-ib7	reserved for future use
ib8	febe-i
ib9	fecc-i
ib10	febe-ni
ib11	fecc-ni
ib12	los
ib13	rdi
ib14-23	Reserved for future use

ADSL ATU to indicate whether a *severely errored frame* (sef) is received. A sef is defined as occuring when two consecutive ADSL superframes do not have the expected symbols in the synchronization frame following frame 67. This synchronization frame should not be confused with the synchronization control (sc) bits, which are totally separate. The value of the rdi bit is 1 when there is no sef to report and 0 when there is—the same pattern as the los indicator.

The other four indicator bits are ib8 through 11. All four bits represent *far-end* error conditions, which is just the device (ATU-R or ATU-C) on the opposite end of the ADSL link. Near-end errors exist, as well, but these are beyond the scope of the current discussion. Ib8, called the febe-i, stands for *far-end block error on the interleaved data* in the ADSL superframe. This bit is used to indicate when the cyclical redundancy check for the interleaved data (the crc-i) in the received superframe does not match the local calculation result. The bit is 1 when the crc is correct, and 0 otherwise. Ib10, called the febe-ni, performs the same function on the non-interleaved (or "fast") data, with the same values. Ib9, called the fecc-i, stands for *forward-error correction code on the interleaved data* in the ADSL superframe. This bit is used to indicate when the forward error correction (FEC) code is used to correct errors in the received data. The bit is 1 when there is no error to correct and 0 otherwise. Ib11, called the fecc-ni, performs the same function on the non-interleaved (or "fast") data, with the same values.

All this might seem somewhat strange considering that the ADSL specification seems to firmly establish the FEC for the primary method of "fast" data buffer protection and the crc in the superframe as the primary method of interleaved data buffer protection. However, the indicator bits show that both methods are used for fast or interleaved data. Indeed, a crc is calculated as well on the "fast" data buffer, and a special FEC code is generated, not by each individual interleaved data frame, but rather on a sequence of them.

The structure and meaning of the eoc bits for the embedded operations channel and the sc bits for synch control are quite complex and need not be discussed in detail.

The ADSL Frame Structure

Only one other major topic needs to be discussed regarding the ADSL bit stream between ATU-R and ATU-C—the structure of the bits within the individual frame that make up the ADSL superframe. As it turns out, this structure is easy to describe, but difficult to explain in detail. It is simple because each ADSL frame in a superframe has a fixed structure. For each data buffer, fast or interleaved, the frame simply takes a given number of bytes (octets) for the AS0 bearer channel, followed by AS1, and so on up to AS2. These bytes are followed by LS0 bytes, then LS2, and finally LS3. If there are no bytes for a particular AS or LS, these areas are empty. Finally, there are some added overhead bytes shared by the channels.

The structure is complicated by the fact that ADSL has many line rates that are different in each direction, of course. Only the transport classes add any overall organization to this loose arrangement. Note that any AS or LS bits may be transported in either the fast or interleaved data buffer area within an ADSL frame. Each user data stream is assigned to either the fast data buffer or the interleaved data buffer during an initialization process. However, if the AS0 user stream (unidirectional downstream) is assigned to the "fast" data buffer area, it cannot be simultaneously assigned to the interleaved buffer. In other words, if an ADSL frame contains bits for AS0 in the "fast" data buffer area of a frame, the exact same number of bits must exist for AS0 in the interleaved buffer area of the frame.

Configurations for the default number of bytes in the ADSL frame based on transport classes is shown in Table 9-8. These are defaults and can be changed. As well, this information is for ADSL frames transmitted to the ATU-R. Note that if a particular buffer area has a non-zero value, the corresponding value in the other buffer must be zero.

Table 9-8

Default buffer allocations for transport classes based on 1.536 Mbps multiples

Signal	Interleave Data Buffer				"Fast" Data Buffer			
	Trans Class 1	Trans Class 2	Trans Class 3	Trans Class 4	Trans Class 1	Trans Class 2	Trans Class 3	Trans Class 4
AS0	96	96	48	48	0	0	0	0
AS1	96	48	48	0	0	0	0	0
AS2	0	0	0	0	0	0	0	0
AS3	0	0	0	0	0	0	0	0
LS0	2	2	2	255 (Note)	0	0	0	0
LS1	0	0	0	0	5	0	0	5
LS2	0	0	0	0	12	12	12	0

NOTE: *When coded as an all 1s byte (255), the LS0 bearer channel is a 16 kbps C channel carried in the synchronization control overhead.*

To bring about some closure on the discussion of ADSL frames and superframes, note that in transport class 1, the default configuration allocates 96 bytes to AS0 and AS1 in each ADSL frame. Because there are 8 bits in a byte and 4,000 ADSL frames are sent each second, regardless of total frame size, the bit rate on both the AS0 and AS1 bearer channels will be 3.072 Mbps (96 bytes $\times$ 8 bits/byte $\times$ 4000/second). And indeed, two downstream bearer channels running at 3.072 Mbps is an established option for transport class 1. In fact, in this case, based on default buffer size, this is effectively the default configuration. Note also that the LS0 channel runs at 64 kbps in both directions (2 bytes $\times$ 8 bits/byte $\times$ 4000/second).

The structure of an ADSL frame based on transport class 1 as sent by the ATU-C is shown in Figure 9-3. Note that the fast byte must be present, but all of the data bytes come from the interleaved data buffer.

Of course, there is a companion default buffer allocation for the 2M transport classes based on multiples of 2.048 Mbps services. These are shown in Table 9-9.

Table 9-9

Default Buffer Allocations for Transport Classes Based on 2.048 Mbps Multiples

Signal	Interleave Data Buffer			"Fast" Data Buffer		
	Trans Class 2M-1	Trans Class 2M-2	Trans Class 2M-3	Trans Class 2M-1	Trans Class 2M-2	Trans Class 2M-3
AS0	64	64	64	0	0	0
AS1	64	64	0	0	0	0
AS2	64	0	0	0	0	0
LS0	2	2	255 (Note)	0	0	0
LS1	0	0	0	5	0	5
LS2	0	0	0	12	12	0

NOTE: *When coded as an all 1s byte (255), the LS0 bearer channel is a 16 kbps C channel carried in the synchronization control overhead.*

As before, consider transport class 2M-1 with AS0, AS1, and AS2 all sending 64 bytes in each ADSL frame. This means that there are three downstream bearers running at 2.048 Mbps (64 bytes × 8 bits/byte × 4000/second). And three downstream bearer channels running at 2.048 Mbps is an established option for transport class 2M-1. In fact, in this case, based on default buffer size, this is effectively the default configuration. Note also that the LS0 channel runs at 64 kbps in both directions (2 bytes × 8 bits/byte × 4000/second) in this configuration.

Figure 9-3

ADSL Frame Structure Based on Default Buffer Sizes for Transport Class 1

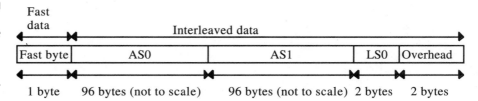

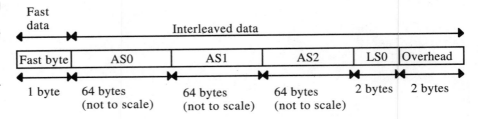

This technically challenging chapter ends with a look at the ADSL frame structure sent from the ATU-C based on the transport class 2M-1 default buffer allocations (see Figure 9-4). As before, the fast byte must be present.

10

Inside the ADSL Frames

The previous chapter spent a lot of time describing the overall architecture of the ADSL physical layer standard. The interfaces were introduced, and emphasis was placed on the flow of information downstream to a user's premises and upstream to the service provider's access node. The ADSL transport classes were introduced, and the structure and overhead of the ADSL superframe was covered. Finally, several ADSL frame structures were described, based on the transport classes.

One rather obvious feature of the ADSL frames was investigated. That is, the flow of bits across the ADSL link as described were just that: bits. An ADSL link is every bit as much a circuit as a 33.6 kbps modem connection. ADSL just runs faster. When ADSL is configured and operating with an AS0 running at 6.144 Mbps, the ADSL frames carry 6.144 Mbps of bits all the time from ATU-C to ATU-R. These bits may represent useful information, or they may possibly represent a bit pattern that essentially forms an idle channel. The point is that bits are just bits. Circuits in ADSL are the same as circuits in ISDN or the analog content PSTN. Circuits use all of the bandwidth all of the time. Circuit switches always switch bandwidth to bandwidth, point to point, and do nothing else. As described to this point, all that ADSL does is remove the bandwidth from the PSTN and place it onto the ADSL Access Node in the service provider's office. This would be enough if the only goal of ADSL were to relieve PSTN congestion, but ADSL can do more, much more.

The time has come to look inside the ADSL frames and look at some of the possibilities that make ADSL not just an adjunct to the PSTN, but a true alternative to the PSTN.

Bits sent across an ADSL link are just an unstructured stream of information. The ADSL link is, loosely speaking, just a "bit pipe" or "bit pump." Bits go in one end and emerge unchanged from the other end of the link, between ATU-C and ATU-R. For ADSL to provide useful services to a customer, however, the bits must mean something, must represent information according to the service.

Only a small degree of organization is provided for these bits by the ADSL superframe structure. An ADSL superframe is sent every 17 milliseconds (almost 59 per second) and consists of a sequence of 68 ADSL frames. ADSL frames contain both fast (audio and video, for example, which is delay sensitive and needs a stable delay) and interleaved (Web data, for example, usually delay insensitive but error sensitive) bits. This is as far as the ANSI T1.413 standard goes.

The question is what is inside the ADSL frames? For this important piece, the ADSL Forum is the indispensable source of information. The ADSL Forum is an association of service providers and equipment vendors and other interested parties who need to go beyond the low-level technical aspects of ADSL if full service networks are to be deployed based on ADSL. Forums are not an uncommon practice in the telecommunications industry. The ATM Forum and Frame Relay Forum, among others, have provided ample precedent in this regard and have been instrumental in bringing their technologies to market and into the public consciousness.

By the end of 1997, the ADSL Forum had some 75 principal members who write and vote on *implementation agreements* regarding ADSL. These implementation agreements are binding on all members of the ADSL Forum, including the 170 or so auditing members who have access to ADSL Forum documentation, but who have a reduced role in formulating and approving the implementation agreements. That said, please note that there is no mechanism in the ADSL Forum (or any other forum of its kind) to punish members for not observing the implementation agreements or taking the drastic step of banishing the member from the ADSL Forum itself. Compliance with implementation agreements is a matter of honor, and most members try scrupulously to comply with ADSL Forum decisions. However, it is always good to bear in mind that there is always room for variations in the name of improvement. The ADSL Forum's web site is **www.adsl.com**.

The ADSL Forum has defined four distinct *distribution modes* for *all x-type Digital Subscriber Line* (xDSL) technologies, ADSL included. The distribution mode determines just what form is taken on by the bits inside the ADSL frames when they are sent. Figure 10-1 shows the main characteristics of the four distribution modes.

Figure 10-1

The four ADSL distribution modes

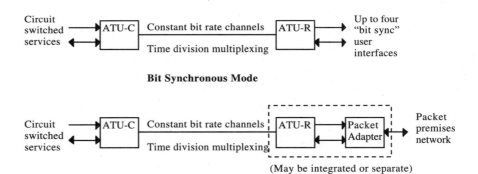

Bit Synchronous Mode

Packet Adapter Mode

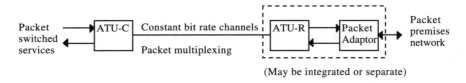

End-to-end Packet Mode

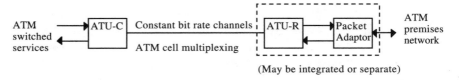

ATM Mode

At first glance, the figure seems complex and confusing. Part of this is due to the fact that four distribution modes are combined into one figure, but this is done to make it easier to compare the four. Although there are many similarities between the four configurations, their significant differences must be appreciated to understand what ADSL can do and what other network components and capabilities ADSL needs in order to provide services to customers. Some time will be spent discussing the four distribution modes in detail.

The first, and simplest (and also least interesting), distribution mode established by the ADSL Forum is known as *bit synchronous mode*. The name is awkward, but it basically means that any bits placed in a buffer (the "fast" data buffer or interleaved buffer) in a device at one end of the link (the ATU-R) will pop out at the buffer in the device at the other end (the ATU-C). In another document, the ADSL Forum suggests that the "fast" data buffer operate about 10 times faster that the interleaved buffer in terms of latency. Latencies of about 2 milliseconds for "fast" data and 20 milliseconds for interleaved data are mentioned. The difference is due mainly to the way the bits in the two buffers are handled with regard to error protection.

At the top of the figure is the bit sync distribution mode. Up to four "bit sync" symmetric user devices can be hooked up to a given ATU-R at the customer premises when functioning in bit sync distribution mode, which makes sense because there are four downstream bit flows established in ADSL (AS0 through AS3). Naturally, if only AS0 is present on the ADSL link, then only one flow is present at the ATU-R, and presumably there would be only one device. This user device may be a TV set top box or a PC, but all of the bits are delivered to the attached device. The upstream and directional channels must consist of at least the control C channel and may also include the LS structure. In its simplest form, the flow would consist of AS0 downstream at 1.536 Mbps in the United States and 64 kbps upstream (also used for bidirectional functions).

In the bit sync mode, the ATU-C in the ADSL Access Node at the service provider's office simply hands off the user's bits arriving on the LS or C channel to some circuit-switched services at the service provider's office. In this mode, the ADSL link is essentially a bit pipe to a fixed end device (just like a leased line). The ADSL link always operates at the line speed, or in what is called a *constant bit rate* (CBR). The ADSL link can be divided into channels, but always with straight *time division multiplexing* (TDM) establishing time slots in the ADSL frames for bits.

That many initial ADSL service offerings will consist of bit sync ADSL was widely anticipated. At present it seems that bit sync ADSL will attract little interest. But if the equipment located at the end of the ADSL channel "behind" the ATU-C is an IP router attached to the Internet, even this simple mode can do many things for users and service providers. First, it enables LEC service providers to move traffic off their PSTN switches onto a network of its own. Second, it enables users to access the Internet at speeds that have been available only in large office environments. Third, it enables Internet service providers to eliminate any bottlenecks resulting from users dialing in to the ISP office over the PSTN because the ADSL bit sync channel is essentially a leased private line. Of course, it remains to be seen whether many user PCs and ISPs can handle traffic at ADSL speeds, and most cannot. This issue is one of a number of related ADSL issues that are discussed in a later chapter in this book.

The second distribution mode in the figure is *packet adapter mode*. The only change occurs at the customer premises. The main difference from bit asynchronous mode is that the devices at the customer premises are now expected to send and receive packets instead of just a stream of bits. The packets are placed in the ADSL frames by some packet adapter function, which may be a separate device or built-in to the ATU-R. Note that this mode expects to interface with some packet-enabled premises network and not just a simple bit-consuming and generating end-user device.

Why bother? Well, bit sync ADSL requires that each user device be plugged directly into the ATU-R where it can only send or receive bits on one of the ADSL channels, such as LS1. It is anticipated, however, that there may be many devices at a customer's premises that might take advantage of ADSL speeds. Many people have more than one PC. Some even have 10BaseT Ethernet-type LANs in their homes (kits for linking two PCs this way start at a couple of hundred dollars). When linked to a SOHO customer, there may be more than just four devices needing to use the ADSL link. Not only traditional LANs may be used as the premises packet distribution network. Newer techniques are allowed as well, such as the use of a building's power distribution wiring to carry information. In fact, this *consumer electronics bus* (CEBus) architecture may actually get a boost from ADSL.

Packet adapter mode allows this to take place. Packets from many sources and destinations at the customer premises can share a single LS1 channel on the ADSL link. Of course, the ATU-R still maps these packets to fixed channels on the ADSL link, but if the other end of the link beyond the ATU-C and Access Node is an Internet router, the arriving and departing packets can be dealt with effectively.

Also, note that in packet adapter distribution mode the packets are still sent as a stream of bits on TDM channels to wherever the end device is beyond the ATU-C because the endpoints are still reached only by circuits. That is, each packet flow needs its own TDM channel on the ADSL link, and each ADSL channel is still a CBR transport.

The third distribution mode in the figure is the *end-to-end packet mode*. Actually, this is not really a distinct distribution mode at all from packet adapter mode. In fact, both packet adapter and end-to-end packet modes must be used to produce an overall packet service. However, the two tpyes of packet distribution modes are typically presented separately, as they are here. The major difference between this end-to-end packet mode and the packet adapter mode is that packets are now multiplexed on the ADSL channels. In other words, the packets to and from a variety of user devices are not mapped to a sequence of ADSL frames representing ASs or LSs, but are all sent on an "unchannelized" ADSL link running upstream and downstream at a given rate. Note that the big change still 20 occurs at the customer premises, but the user choices are now more constrained; user packets must be the same as those used by the service provider's portion of the link. User devices send packets to and from the packet adapter device. At the service end of the link at the ATU-C, the packets are not all just shuttled to the endpoint represented by the LS channel. In packet mode, the packets are sent to their proper "servers" based on packet address. This packet switched service network can be based on X.25 or, more likely, TCP/IP.

In this mode, IP packets can be multiplexed and switched (that is, routed) onto the Internet at the service side of the ADSL link. Note that using ADSL to send and receive packets can also be accomplished with bit sync or packet adapter mode. The difference is that the packets traveling on the ADSL system in these two modes are *invisible* to the ADSL system. In packet distribution mode, the packet switching is part of the ADSL network. The bits in packet mode *must* be organized as packets, not *possibly* organized as packets as in bit sync and packet adapter modes. In the case of packet mode ADSL, the ADSL link becomes much like a link between an IS (intermediate system) router and a small office router in an Internet access arrangement. Of course, other types of packets also may be used in this mode. The transport of video packets is just as possible as IP packets, as long as both the client and server understand the type of packet being transported.

There has been a lot of interest in ADSL products and services based on packet modes.

The last distribution mode in the figure is *asynchronous transfer mode* (ATM), or more properly end-to-end ATM mode. ATM multiplexes and sends ATM cells from an ATM adapter (at the ATU-R) instead of IP (or some other) packets. At the ADSL service provider side, the ATU-C passes the cells to an ATM network. Remember that the contents of these ATM cells may still be IP packets, and the ADSL Forum recently decided to adopt the IP point-to-point protocol (PPP) over ATM for this mode. But the ADSL network deals with the ATM cells, which must form the contents of the ADSL frames. Of all the distribution modes, ATM mode has evoked the greatest amount of interest from ADSL vendors and service providers.

As a final word about the figure, remember that the ATU-C devices may be (and usually are) part of a Digital Link Subscriber access multiplexer (DLSAM) assembly.

ADSL Possibilities

ADSL devices (essentially the ATU-C and ATU-R) send bits back and forth inside ADSL frames and superframes. The four ADSL distribution modes establish the contents of the frames as unstructured bits to the network, packets as the source of the bits at the ATU-R (Remote), packets across the ADSL link, or ATM cells across the ADSL link, respectively. However, there is more to the ADSL network architecture than the ADSL link from ATU-C (Central Office) to ATU-R (Remote). The four ADSL distribution modes combine with the formats of information on other sections of the architecture to form at least six different ways (according to the ADSL Forum) that users at home or in small offices could access broadband services delivered over ADSL links. These six distinct possible network arrangements are shown in Figure 10-2.

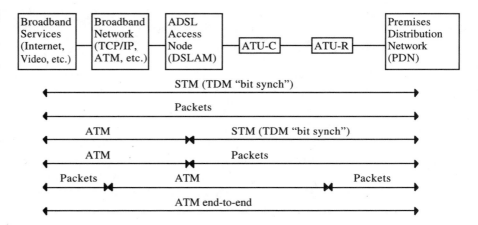

Figure 10-2
Six possible ADSL networks

The figure shows a simplified picture of a full ADSL network at the top, but all the essential elements are present. Note that ADSL itself is just a part of the whole architecture, albeit a crucial part. At the user end, the ATU-R hooks up to one of a variety of *premises distribution networks* (PDN), such as a 10BASE-T local area network or a CEBus power distribution network. At the service provider office, the ATU-C hooks up to (or forms part of) a DSL access multiplexer (DSLAM). Technically, the DSLAM for ADSL is just an ADSL Access Node, but a DSLAM is the usual choice for ATU-C packaging in ADSL.

The DSLAM might have a number of, but at least one, ports that access a variety of networks to reach the information servers needed by the user. This is the "back end" of the ADSL network and forms the source of the bits or packets or cells sent to the ATU-R. These possible service network types include the regular circuit switched PSTN, a router attached to the Internet, or an ATM switch. In its most effective form, this network has enough bandwidth and low enough delays to be called a "broadband" network, but this is not always the case.

Finally, any one of the network possibilities would have access to a number of information servers that deliver broadband services. Again, the intention is to provide broadband services or even *Broadband Integrated Services Digital Network* (B-ISDN) services if the network is ATM, but this may not be the case in all instances. Typical services would include Internet or Intranet Web site access, video-on-demand, and so forth.

The six possible ADSL scenarios in the figure are derived by combining the four ADSL link distribution modes with the typical types of traffic on the rest of the broadband network. In ADSL Forum documentation, the time division multiplexing (TDM) "bit synch" mode is usually called *synchronous transfer mode* (STM) to distinguish it from asynchronous transfer mode. All four possible ADSL distribution modes are present in the figure: bit synch (STM), packet adapter, end-to-end packet, and ATM.

The main point is that "beyond" the DSLAM on the service provider side of the ADSL network, there are really only three modes: STM (bits on TDM circuits), packets (usually, but not limited to, IP packets), and ATM (cells). Even the case where IP servers are accessed over an ATM network, and those cells are then delivered as IP packets again, is covered by this architecture.

Most ADSL service providers and equipment vendors consider ATM end-to-end to be the most viable long-term scenario, due to ATM's graceful handling of merged traffic types. Although IP is being heavily modified to provide adequate voice and video support, ATM has had such support since its inception.

The first possible ADSL network shown in the figure uses bit sync mode from end to end. Whatever the format of the information taken from a server and delivered to the user device, the ADSL network is totally indifferent to this structure. The ADSL network is a passive bit pipe (or "bit pump" to the ADSL Forum specification) and provides only time division multiplexing and a constant bit rate on the established ADSL channels, such as AS0 or LS1.

The next possible ADSL network shows packets flowing end to end through the network. The packets are in all likelihood IP packets, but this is not always true. The packets may represent other protocols or video services, as long as both ends understand the packet format. The advantage of packets end to end is that the network can combine flows of packets, switch packets to and from various end points, and in general become a more active partner in the services provided by the ADSL network.

The third ADSL network in the figure combines bit sync mode and ATM cells. The ADSL Access Node (probably a DSLAM) still handles bit pipe and TDM circuits, but the network "behind" the DSLAM now handles ATM cells right through to the servers with the information content provided to the users. Note that this approach requires at least minimal ATM capability in the DSLAM itself and most likely a lot of ATM capability. The advantage is that service providers who have deployed extensive ATM backbones (and many have) can take advantage of the traffic type merging that ATM was intended for while not requiring users to adapt ATM in any shape, manner, or form.

The fourth ADSL network in the figure still uses ATM on the "back end" but now employs packets (IP or others) on the ADSL link itself. This scenario is very attractive because it does away with passive bit pipes and at the same time preserves the potential of ATM backbone network traffic aggregation.

The next to the last ADSL network in the figure is an odd mixture of ATM and (IP) packets. The whole idea here is that although many service providers have been busily deploying ATM backbones, both on the server side and on client PC side, ATM will remain rare for the foreseeable future to the rest of the world. This being the case, this scenario allows a service provider to furnish ADSL services based on ATM while still maintaining packet interfaces at the server and client end. All in all, this approach seems hardly worth the effort, but it is allowed.

Finally, the last ADSL network in the figure employs ATM end to end. The only difference between this network and the second, using packets end to end, is that there is a flow of cells through the network instead of packets, IP or otherwise. A substantial percentage of ADSL network trials have used this approach, especially for corporate customers or those who have ATM equipment on the premises. However, this might remain an evolutionary goal for most ADSL service providers or perhaps will not happen at all. The attraction of ATM in giving the lowest latency and highest bandwidth even with a mixture of traffic types should soon be matched by IP itself. And given the prominence of entrenched IP hardware and software on the Web and in corporate Intranets and Extranets, the lure of ATM for users may fade.

ADSL for TCP/IP: Adapter Mode

In spite of all the possible ADSL networks outlined above, it very much remains a TCP/IP world. If ADSL is primarily deployed to get long holding time Internet and Web users off of the PSTN and onto a separate packet network of their own, the packet network will most likely be based on IP.

It is all well and good to say that ADSL allows for a packet mode of distribution, where packets may be sent and received by filling the ADSL frames and superframes with packets. In the real world, however, this really translates to this: Is the transport of packets in ADSL to be based on the TCP/IP protocol suite (more properly, the Internet protocol suite) and not another packet protocol, such as X.25 or those based on the *Open System Interconnection Reference Model* (OSI-RM)? The continued growth and popularity of the Internet and Web, and such variations as Intranets and Extranets, make other packet protocols besides the Internet Protocol increasingly uncommon.

This being the case, the ADSL Forum allows two different ways for an ADSL network to support TCP/IP. The first method has two variations. Both of these variations should be familiar to you from the previous discussion of possible ADSL networks. The only real element added is that the packets are now emphatically based on the Internet Protocol Suite (formerly TCP/IP, the packet format of which is IP). This first method is shown in Figure 10-3.

The figure shows two basic scenarios in which ADSL links can be used to support the transport of TCP/IP packets. The network consists of a *premises distribution network* (PDN), such as a 10BASE-T LAN or CEBus network, as before. At the service provider switching office, the ATU-C hooks up to (or forms part of) a DSL Access Multiplexer (DSLAM) as a form of ADSL Access Node. The DSLAM, in turn, has access to a "broad-

Figure 10-3
ADSL for TCP/IP:
Adapter Mode

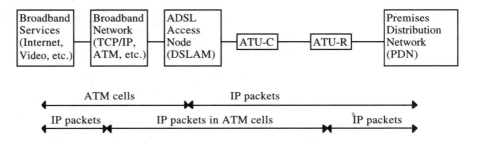

band" (the bandwidth and delay determine this) network and access to various servers delivering "broadband" services.

The ADSL Forum allows two different types of traffic flows through the network in adapter mode for TCP/IP. In the first, the network "behind" the DSLAM is an ATM network, and the DSLAM interfaces directly to an ATM switch (or the DSLAM actually contains some ATM switching capabilities). However, on the ADSL link side of the DSLAM, the ATM cell contents are translated ("adapted") to a stream of packets based on TCP/IP. The intent is to deliver ATM-based broadband services to users who may not necessarily have, want, or be able to afford ATM equipment at their end. This scheme takes advantage of the universality of TCP/IP software on PCs and most other platforms today.

However, not all servers that provide audio and video services, or other forms of broadband services, are ATM enabled. There is a lot of video, audio, graphics, and the like on TCP/IP-based servers, especially when the Web is considered. So a second TCP/IP adapter mode variant allows TCP/IP-based servers to be accessed by the ATM network. In this case, the transport to the DSLAM is still by way of ATM cells. The TCP/IP packet content, however, is not translated to ATM cells, but rather carried in a stream of ATM cells to the ATU-R (Remote). All of the possible ATM features are not present, just the capability of ATM cells to carry IP packets by using a method of ATM adaptation known as *ATM Adaptation Layer 5* (AAL5). Once at the ATU-R, the TCP/IP packets are extracted from the ATM cells, again using AAL5, and sent over the PDN to the end user's device.

This variation enables a service provider to take advantage of a true broadband network such as ATM, and at the same time preserve information available on TCP/IP-based servers (which is actually just about everything). In this case, any TCP/IP-to-ATM server migration undertaken by the service provider at any point in time can proceed more leisurely.

ADSL for TCP/IP: End-to-End Mode

So the ADSL Forum defines a TCP/IP adapter mode for an ADSL network with both ATM and TCP/IP protocol elements. If the information is available only on an ATM network and server, adapter mode will allow a premise's TCP/IP to get this information. Even if the information is available on a TCP/IP server reachable through the ATM network, an adapter mode variation allows for the transport of TCP/IP packets inside of ATM cells.

However, it has not escaped the notice of the ADSL Forum that TCP/IP is just about everywhere, and ATM is still rarely seen outside of specialized application and network environments, such as the service providers' backbones. The ADSL Forum has therefore established support for a TCP/IP end-to-end mode that allows all traffic inside both ADSL superframes and frames to be TCP/IP packets from server to *DSL access multiplexer* (DSLAM) to ATU-R, all the way to the user and back. This method is shown in Figure 10-4.

The figure illustrates the TCP/IP end-to-end mode. All "broadband" services (it depends on the bandwidth and delays of the network) are Internet-based in this mode (not a bad initial ADSL network assumption). Therefore, the services are all accessed through an Internet router running TCP/IP protocols. Now the access to the ATU-C (technically, through the DSLAM or ADSL Access Node) is by way of the more common (and less expensive) 10BASE-T or 100BASE-T, and perhaps soon Gigabit Ethernet running at 1,000 Mbps, instead of ATM.

The figure also shows how TCP or *User Datagram Protocol* (UDP) messages inside IP packets would be placed in *Point-to-Point Protocol* (PPP) frames and sent inside of Ethernet (technically, IEEE 802.3) frames on

Figure 10-4
ADSL for TCP/IP: End-To-End Mode

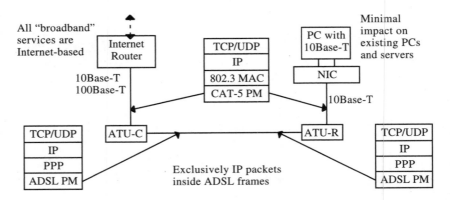

Category 5 (Cat-5) unshielded twisted-pair copper wire. The ATU-C is inside the ADSL Access Node or DSLAM, which now has a regular Ethernet 10Base-T or 100Base-T interface port on the "back." The DSLAM or ATU-C extracts the IP packet and then sends it out onto an ADSL Physical Media Layer. This amounts to placing the PPP frame (which is used because it has its own link control and error checking) inside the ADSL frame and superframe.

It is important to realize what this flow of information would look like on the ADSL link, not the least of which because this will probably be the most common ADSL method of operation for TCP/IP compatibility. Consider the downstream flow from IP server to user PC. The ADSL link consists of a stream of ADSL superframes. The ADSL superframe is made up of a series of ADSL frames. At its very least, there must be an AS0 and LS0 or C channel configured on the ADSL link. Frames consist of a fixed amount of bits from AS0, fast or interleaved, but not both, followed by a fixed amount of bits from the LS0 or the C channel buffer, fast or interleaved, but not both, giving the constant bit rate on the line.

But packets are bursty—sometimes extremely so. What bits flow on the ADSL link inside the ADSL frames when there is no live user information to take from a buffer and place inside the frame? There is no problem at all, at least not in this scenario. The Internet PPP is a close relative of all standard wide area network protocols such as those used in IBM's *Systems Network Architecture* (SNA) and ISDN. As such, PPP can constantly generate an idle bit pattern when not actively sending user information in the form of IP packets inside of PPP frames. This idle bit pattern is the repeated eight bit sequence 0111 1110, or 7E in hexadecimal notation, pronounced "seven easy" by those who work on IBM networks. This is known as *synchronous* operation, and is yet another variation on the term *synchronous* that has been applied in this book to multiplexing. Here it applies to low-level line operation. In any case, a PPP frame by definition occurs between two 7E bit patterns (a special rule prevents the 7E pattern from appearing inside a PPP frame).

Look at the flow across an ADSL link. The superframes are made up of a sequence of frames. Inside the frames are fixed amounts of bits from the configured channels, and always the same number of bits, but, inside, the channels are just streaming 7E bit patterns a large part of the time (bursty). Once in a while the 7E pattern in interrupted. When it is, this interval of bits represents a PPP frame. And yes, PPP frames are inside ADSL frames: one is a transport frame (ADSL) and the other is a *data link frame* (PPP).

What's inside the PPP frames? Why, IP packets, of course! And inside those are the TCP segments that form part of the entire message being sent, from voice to video to e-mail. Data communications is not particularly difficult, just acronym- and terminology-rich.

At the ATU-R, the IP packets are removed from the PPP frames, which are inside the ADSL frames when they occur. So the line rate of ADSL may be 1.536 Mbps all the time, but only when packets are actually being sent do the bits represent anything useful. The IP packets are placed inside Ethernet frames and are sent on yet another 10Base-T LAN on Cat-5 wire to the PC. The message and packet's journey is complete.

In is important to realize that only IP packets are ever sent across the ADSL link in TCP/IP end-to-end mode. At the ATU-R, another 10BASE-T LAN leads to a PC (the device need not be a PC, but few TVs and other devices run TCP/IP today). The PC, with a low cost ($30) 10BASE-T *network interface card* (NIC) and proper TCP/IP software (usually free), can handle the transfer on Cat-5 unshielded twisted-pair (UTP) house wiring carrying what are now not PPP frames, but IEEE 802.3 Media Access Control (MAC) LAN frames, with the IP packets and TCP/UDP messages inside.

This arrangement has a minimal impact on existing PCs and servers (and the Internet), and so is likely to be a very popular ADSL arrangement. This point alone justifies the time spent detailing the arrangement here.

ADSL for ATM

Internet service providers are excited about ADSL for two main reasons. First, it offers users high-speed Internet access; second, it offers users access to broadband services that, for the past several years, have only been a dream. However, the Internet is not yet a broadband network, and it may be many years before it becomes one. It would be nice if the same networks and servers accessed by the *Digital Subscriber Line access multiplexer* (DSLAM) could deliver both high-speed Internet access and broadband services. However, broadband services require a substantial amount of bandwidth, as well as low and stable network delays (on the order of 10s of milliseconds). The present structure of the Internet is characterized by limited bandwidth in many places, and anything but low and stable delays.

Therefore, using ADSL for Internet access based on the TCP/IP is a good idea. People will pay a lot for faster Internet access. However, not all of the services, especially broadband services, are or will be available through TCP/IP servers or networks. Now, people have used the Internet for voice, but TCP/IP voice, although rapidly getting better, is somewhat primitive ("hel . . . lo"), and there has been little interest in delivering broadcast-quality video over the Internet (although there were experiments with things such as MBONE—Multicast Backbone—a portion of the Internet that uses special routers to broadcast radio and video programming). Some major national ISPs have begun experiments in delivering broadcast video using heavy compression across the Internet. However, most observers agree that many changes must be made to the structure of the Internet before it is anywhere as good as ATM at merging traffic types easily and transparently.

For the time being, it may be better if the more sophisticated broadband services are based on ATM servers. This does not means that ATM is an operating system, but rather that the server sends and receives only ATM cells, usually at about 155 Mbps. Why ATM? One big reason is that the large telephone carriers have built large backbone networks based on ATM, so the infrastructure is available and the interface clean. But ATM also excels at delivering combined traffic streams (voice/audio/video/data, for example) on the same network. This is what ATM was designed for, as the network portion of the *Broadband Integrated Services Digital Network* (B-ISDN) protocol stack. Using ATM for broadband services is not like adding voice capability to data-based frame relay or video "flows" to TCP/IP. Using ATM for broadband networks and services is to use ATM for what it was invented to do.

When used for B-ISDN services, ATM transport is to take place over *Synchronous Optical Network* (SONET) fiber networks. Given the prevalence of SONET fiber in the public switched telephone network today, this makes using ATM for as much as possible within ADSL a natural thing to do.

Full Service Network with ADSL and ATM

You have a *full service network* when all forms of services, such as voice, video, and data, are accessed over the same physical network. This has been the dream of many service providers, be they LECs, IXCs, or ISPs. The international standard for such broadband full service networks is

Broadband ISDN. B-ISDN calls for ATM cells and ATM switches as the transport network, and SONET fiber as the physical links between major components of the network. However, if the requirement for SONET fiber could be relaxed in a few places, it is possible to build a full-service network by using ADSL. Figure 10-5 shows how this can be done.

On the customer premises in the figure, all access to devices, whether TV set-top box, PC, or some other device (stereo, refrigerator, and so on), are connected to a premise's ATM switch with 25-Mbps Cat-5 UTP wiring. This premise's ATM switch can even be integrated with the ATU-R device. Naturally, ATM cells are sent inside ADSL frames across the ADSL link. In the case of ATM cells inside ADSL frames, the bit stream is a constant flow of fixed-length, 53-byte (octet) ATM cells packed head-to-tail. There is no 7E pattern as with PPP or other data link protocols. Instead, special ATM *idle cells* are inserted when no live information is present in a buffer to be sent. This practice is known as *ATM cell rate decoupling*. The effect of cell rate decoupling is the same as with 7E patterns. When there is no useful information, idle cells fill the line to the constant bit rate that the ADSL link is running at.

At the DSL access multiplexer, access to services is provided by means of 155 Mbps ATM running on SONET fiber (usually configured as SONET rings). This part of the network complies in every way with the B-ISDN specifications and standards. Many service providers, especially telephone companies, are extremely conscious of B-ISDN compliance, mostly because it comes from the ITU, which has overseen telephone service in terms of international standards compliance for years. Although vendors and service providers may sometimes seem to wink at the various Forum's

Figure 10-5
Full Service Networks with ATM

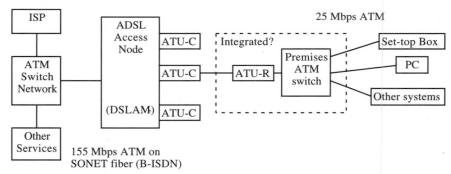

implementation agreements, not many are willing to risk non-compliance with ITU specifications, which basically have the force of law.

The figure also shows an ATM switching network. This network in turn provides access to all services, from an ISP offering Internet access to other services, such as video or graphics applications.

The whole point in this section is that a full service network based in large part on B-ISDN and ATM could be built without the need to run costly SONET fiber to every home and office. Full-service ADSL networks, also based on TCP/IP, have been proposed, but these are in no way compliant with B-ISDN, so those concerned with B-ISDN and ITU compliance have little choice but to at least contemplate the use of ATM with ADSL. Because many of these service providers already employ ATM internally on their backbone, this use of ATM should not be an excessive burden on anyone in the long run.

ADSL and ATM: PPP Using PVCs

It is very attractive for service providers to be able to deliver ATM-based services over an ADSL link to users. ATM networks are designed for multimedia applications, and many services are defined by the B-ISDN standards that ATM networks support. It makes even more sense to transport the ATM cells over the ADSL link to the premises because *Quality of Service* (QOS) support, in terms of stable delays and guaranteed bandwidth, is therefore guaranteed by the ATM portion of the network, and no "ATM to something else" conversion is necessary. Once at the premises, the ATM cells may become IP packets inside PPP frames, but only due to the utter absence of ATM products in subscriber client PCs. Note that the PPP frame structure is kept even as the IP packet is placed inside a string of ATM cells. The attraction of this method of keeping the PPP frame is the capability to retain key existing PPP services, such as authentication on the link.

How might ATM cells carry PPP frames all the way to the user device such as a PC? As it turns out, there are basically two ways outlined by the ADSL Forum. The ADSL Forum recently selected PPP over ATM for ADSL. Figure 10-6 shows one way to achieve IP transport by using PPP frames over an ADSL link with ATM cells. The task is to connect subscriber client PCs at the premises with an ATM access switch, which in turn mediates access to a larger network (ATM or not) where the servers are located.

Figure 10-6
ADSL and ATM: PPP
using PVCs

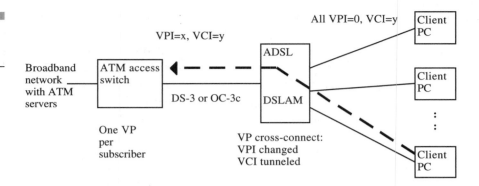

As shown in Figure 10-6, the simplest way to support PPP over ATM is to provision a series of *permanent virtual circuits* (PVCs) over the ADSL links and through the ATM-enabled DSLAM. This can be done at service subscription time when traffic profiles and QOS parameters are established for a particular user. As mentioned previously, ATM carries IP packets and PPP frames inside a series of ATM cells with ATM Adaptation Layer 5 (AAL5). The client end employs PPP over ATM with *null encapsulation* using AAL5 (also called "PPP directly on AAL5"). Note that the client PC still must be able to send and receive ATM cells. The attraction of this method is the capability to retain key existing PPP services, such as authentication. Multiple PVCs can be defined to allow users to reach different networks.

The DSLAM performs a crucial task in this scenario. All ATM connections consist of two identifiers in the ATM cell header. A *virtual path identifier* (VPI) is used mainly for site-to-site connectivity, and a *virtual channel identifier* (VCI) for device-to-device connectivity. As an example of VPI and VCI usage in ATM, consider a home with an ATM-enabled PC that wants to access an ISP for Web access and a telephone company to make a telephone call, all through the PC (obviously, this example is set somewhat in the future). All ATM cells sent to the network from the home itself can be delivered by an ATM switch simply by looking at the VPI field in the cell header. It is a quirk of ATM that this header value can actually change as the cell is sent from switch to switch, but that is not the important point. What is important is that the VCI field never needs to be examined by an ATM switch to deliver a cell to the service provider in question. But obviously cells going to the Internet service provider must be distinguished if two different servers are to be accessed there. That is the role of the VCI. Each device will have the same VPI in cells sent to the ISP, but different VCIs. This allows the two streams of cells to be delivered properly.

What if there is only one destination for all ATM cells sent from the home, such as a single ISP server? In that case, only the VPI needs ever be examined by an ATM network device to send the cell on its way. The VPI may be changed, but the VCI is said to be *tunneled*, in this case. For ADSL services, the VPI assigned for all users will be VPI=0, with the VCI value to be determined. This works because all ATM cells will have their VPIs changed once they are flowing between the DSLAM and ATM switch. However, note that all cells sent on VPI=0 must end up at the ISP. If there is only one VPI defined for all cell streams, they all must end up at the same end point. This is okay, however, because the PPP protocol is a protocol that can only connect two end points. On a PVC, as in the figure, the end points are always as fixed as the ends of a private line.

In Figure 10-6, the DSLAM provides a simple *virtual path cross-connect* function and does not act on or even look at the VCI values in any arriving cell stream. Each subscriber's PVCs are sent through the DSLAM to the ATM access switch over a DS-3 or SONET OC-3 link. Once at the ATM access switch, user traffic may be sent to various other ISPs and local content service providers on an equal access basis, but not based on the single VPI. Several concurrent connections may run to more than one ISP and corporate Intranets—all through multiple PVCs and table configurations using values other than VPI=0, however.

The key point is that the PPP frames inside the ATM cells all end up at the same device, in this case the ATM access switch. There must be at least one virtual path for each subscriber, which makes this method rather awkward to administer and wasteful if access is intermittent.

ADSL and ATM: PPP Using SVCs

It is more attractive to use *switched virtual circuits* (SVCs) rather than *permanent virtual circuits* (PVCs) to link subscriber clients to network servers in an ATM scenario using ADSL. PVCs are there all the time and need to be reconfigured or added by hand before any communication can take place over the ATM network. On the other hand, SVCs are set up and come and go depending on signaling messages to the network from the clients. This keeps configuration chores to a minimum and allows for demand connections to be set up to locations where, previously, connectivity was unforeseen. Unfortunately, there is little to no ATM SVC support (through ATM UNI signaling 3.1 or 4.0) in public ATM networks. However, this may change.

Figure 10-7
ADSL and ATM: PPP
using SVCs

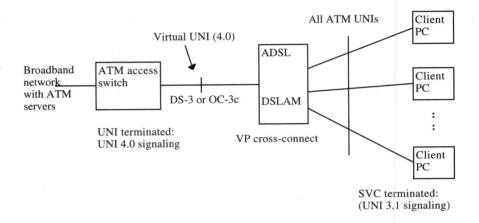

Figure 10-7 shows how to achieve IP transport by using PPP frames over an ADSL link with ATM cells in an SVC environment. As with the PVCs, the IP packets are still placed inside PPP frames using null encapsulation over AAL5 to preserve PPP features. This architecture uses the ATM UNI 4.0 concept of a virtual UNI. In this architecture, there is still a VPI for each user, which is cross-connected in the DSLAM, as in the PVC scenario. These are gathered into a single virtual UNI with one VPI to the ATM access switch where the UNI signaling protocol is processed. In other words, the signaling from client to ATM access switch is transparent to the DSLAM, which is still cross-connecting virtual paths.

The ATM access switch still has to support ATM UNI signaling 4.0 to allow multiple signaling channels (one from each client), but the client PC only needs to support ATM UNI signaling 3.1. The virtual UNI, in turn, is transparent to the client PCs.

In this case, only one VPI is ever needed to each subscriber PC, which is the whole point. The DLSAM handles all of the virtual UNI chores, such as network management (called the *Interim Local Management Interface* (ILMI) in ATM) protocols on behalf of the users. Equal access connections to ISPs and concurrent connections to intranets can all be handled with SVCs, although this configuration does not rule out multiple PVCs in any way.

Resource allocation at the DSLAM can be a little trickier in this case because it must be done on the fly as SVCs are established and released. Also, the virtual UNI requires the support of many signaling channels (one per user) at the DSLAM. It might be better to terminate the UNI signaling at the DSLAM, which would reduce signaling requirements on the

ATM access switch. This is strictly a DSLAM/ATM access switch design tradeoff and does not affect the users in any way.

There seems to be three approaches to the problem of supporting user devices with ATM on the ADSL network. After all, ATM end devices will be few and far between for some time to come. In the meantime, ADSL equipment can:

1.   Put the ADSL ATU-R inside the PC. The ATU-R can perform required AAL and segmentation and reassembly functions. This could also be done in software (slow for ATM) or on a separate board altogether.

2.   The ATU-R can funnel the ATM inside of ATM cells from the ATU-R out onto a 10Base-T LAN, a process called *cells in frames*. The ATM cells are interpreted and processed by software in the end device.

3.   The ATU-R can send the cells to the end device on a universal serial bus (USB) interface. The USB ports are appearing on some PC boards now.

This chapter has ended with a rather elaborate and detailed analysis of ATM and ADSL for a number of reasons. First and foremost, the position of ATM in ADSL is by no means assured. Perhaps it would be much easier to put IP packets inside PPP frames and carry them inside ADSL superframes. Nevertheless, the point here is that a lot of thought has gone into addressing these issues of IP on ADSL with ATM. Also, even those unfamiliar with ATM terms and concepts and who are so struggling to follow the whole ATM PVC/SVC cross-connect discussion can appreciate the versatility of ADSL. As a low-level transport, ADSL is equally at home with IP, ATM, any combination of the two, and even almost anything else for that matter.

ADSL Packet and ATM Service

This might be a good place to summarize the relationship of ADSL Forum work and the ANSI T1.413 specification. The current state of both packet and ATM service work in progress by the ADSL Forum will be outlined as well.

Figure 10-8 shows both the relationship of the ANSI ADSL work and the work of the ADSL Forum, as well as the state of the implementation agreement work on various components.

Looking at packet services first, note that ANSI T1.413 just specifies the physical transfer of bits from place to place. In other words, this is a physical layer specification. There is nothing wrong with this, every network needs one, but there is much more to networking than shuttling bits, as this chapter has shown and the next will make clear.

So one top of the ADSL physical layer, and inside the ADSL frames, sits the packet layer. This is the work of the ADSL Forum and is essentially done, but had no channelization or mapping included. This will be added by ADSL Forum work on the higher layers still.

The same ANSI T1.413 comments about physical transfer apply to the ATM services stack as well. The inclusion of an ATM transport function on top of the physical layer is also done. In fact, the whole PPP over ATM architecture is coming soon.

Figure 10-8
ANSI and ADSL Forum ADSL work

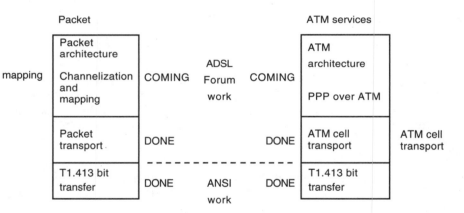

11

ADSL in Action

Thus far this book has emphasized the networking aspects of the xDSL family in general and ADSL in particular. The time has come however to shift the focus back onto the customer. That is, what does ADSL mean to a prospective user on a service provided by ADSL? After all, ADSL itself does nothing except shuttle bits back and forth. What can ADSL do for a user? What might a user have to do to prepare their home or business for ADSL?

Suppose a local service provider, be it a cable TV company or the local telephone service provider, announces the availability of ADSL in a given area. Interested users contact the service provider to find out what they can do with it. Many only want faster Internet access. These customers would be extremely happy because almost all ADSL pilots, trials, and active services start with basic Internet access at 64 kbps upstream and 1.5 Mbps downstream.

But what else can be done with this bandwidth? It seems silly to expend all this time and effort on ADSL just to have it form a bit pipe through the service provider onto the Internet, and all the faster access line might do is clog up the Internet even more. Not that Internet access will not happen plenty of times anyway, but perhaps the local ADSL service provider can add value to the ADSL link by providing local servers for local users. These servers can still be based on IP, of course, and may still be attached to the rest of the Internet, but the key here is that messy Internet bottlenecks can be avoided because the server is right at the end of the big ADSL bit pipe.

With this in mind, the ADSL Forum has outlined a series of speeds at which ADSL should run upstream and downstream to provide a variety of services. Remember that all of these services, with few exceptions, could still be provided directly from the Internet, as well; it may be simply more efficient to get them from the local service provider.

ADSL Target Service Speeds: Video

As was mentioned previously, many discussions of ADSL network services revolve around references to high-speed Internet access and vague assurances of unspecified broadband services gathered under the "video" umbrella. As important as Internet access is, it might be more important that ADSL perform more than one neat trick to assure itself a place in the hearts and minds of potential customers. Therefore, video services might assume a more important role in ADSL as time goes on. All are characterized by a need to transfer full motion video to the end-user.

ADSL promises enough bandwidth (and perhaps sufficiently stable delays and low bit error rates) to deliver many types of video services to the home. In an effort to be more helpful and specific, the ADSL Forum has documented a number of possible video services that can be provided with ADSL and has even included the suggested downstream and upstream bandwidths that these services would need. It should be noted that this particular document is quickly being replaced by newer specifications. However, the emphasis here is on the characteristic speeds associated with the services, and not so much on how the services will actually be delivered on an ADSL network. These are listed in Table 11-1.

These video services were originally considered bit sync mode services in ADSL-the traffic both ways forms an unstructured stream of bits deciphered only by the end devices linked by the ADSL network. This maxi-

Table 11-1

ADSL Target Service
Speeds for Video

Application	Downstream	Upstream
Broadcast TV	6 to 8 Mbps	64 kbps
Movies on Demand	1.5 to 3 Mbps	64 kbps
Near Video on Demand	1.5 to 3 Mbps	64 kbps
Distance learning	1.5 to 3 Mbps	64 to 384 kbps
Home Shopping	1.5 Mbps	64 kbps
Information Service	1.5 Mbps	64 kbps
Computer Gaming	1.5 Mbps	64 kbps
Video Conferencing	384 Kbps to 1.5 Mbps	384 Kbps to 1.5 Mbps
Video Games	64 Kbps to 2.8 Mbps	64 kbps

Source: ADSL Forum

mizes the flexibility of the system as a whole (the only requirement is that the end devices understand the format(s) of the bits transferred), but reduces the role of the ADSL network to be, effectively, a passive bit pipe connecting the end devices with a fat path (or paths) using straight *time division multiplexing* (TDM). However, newer implementations of ADSL will use packet or ATM transports to deliver these video services.

Video services such as the following will function well with 1.5 Mbps downstream and a modest 64 kbps upstream:

- Home shopping
- Information services, such as a person announcing library hours
- Computer gaming-usually used to mean a type of game system played over the network, such as Monopoly or Doom, but sometimes used as a polite term for lotteries and gambling).

Others, such as videoconferencing and traditional video games, may require a range of speeds, depending on the video quality and detail required. Still others, such as distance learning, near voice-on-demand (movies start every 15 minutes), and movies-on-demand will require downstream bandwidths as high as 3 Mbps. Broadcast-quality TV images, given state-of-the-art digital compression techniques, will still require 6 to 8 Mbps. Bit rates above 1.5 Mbps might be hard to come by in early ADSL systems.

Only a couple of the services, distance learning and videoconferencing, would require an upstream bandwidth of more than 64 kbps according to the ADSL Forum. In many implementations, videoconferencing would require symmetric bandwidth, up to a full 1.5 Mbps in both directions. Such video applications enter the realm of *High-bit-rate DSL* (HDSL) or *Symmetric DSL* (SDSL), rather than pure *Asymmetric DSL* (ADSL).

How could an ADSL service provider use the information in Table 11-1 to provide services to prospective customers? Well, at the very least, there could be a LAN located right next to the DSLAM. The LAN could have routers and servers attached to it, of course-perhaps several of each. Local libraries, government offices, and other local organizations could have pre-recorded video clips of information about hours, meetings, or almost anything else on this server.

Retail stores at the local mall could have a separate server for a home shopping service, all based on video. The customer could "walk" through the "virtual mall," enter a store, browse around, even "talk" to a video sales associate. People tend to sneer at such lazy and easily pleased shoppers, but this type of service has real attraction for retailers. Most retailers do almost half of their annual business between Thanksgiving and Christmas, regardless of religious mix in the area. In fact, some only stay in business from one holiday season to the next from the profits of that brief period. How damaging to their revenues is a November ice storm or December blizzard? When people do not come out to shop, retailers suffer a disaster. Now, customers-no matter how much they prefer to socialize at the mall with real humans-do not have to brave 12 inches of snow to pick out a hideous tie for a relative. The presence of bookstores on the Internet has already revolutionized the retail book industry. It is not unusual for people to browse the bookstore as before, but order the books on the Internet. Stores can charge a shipping fee, but people seem to prefer paying that over state sales tax.

The same thinking applies to local schools. School closings due to weather are normally thought to be a symptom of life in the Northeastern United States. However, when a freak snow fall strikes Georgia, the resulting paralysis is much worse than in the snow-savvy areas of the country. After all, how many snow plows are there in the Sun Belt? Distance learning with the help of video can keep things going when everything else has stopped.

A corporate video conferencing service could help with telecommuters. The effectiveness of work-at-home employees, or *telecommuters*, is often limited because they are in contact with the office through e-mail and faxes, but are cut off from human contact and often vital social interactions. Video conferencing services help telecommuters to participate more fully, even from home.

Movies could be provided by a local video server also, but few plan to offer or even deem feasible offering cable TV-like services with ADSL. Even if half of the cable TV customers in a given area switched to ADSL TV there would probably not be enough of a revenue stream to justify the expense of bandwidth- and gigabyte-gobbling digital video.

This list could easily be extended, but the point is made: ADSL needs services to be useful. The service could simply be Internet access, and, in fact, most if not all of these services detailed could easily be provider on the Internet itself. The attraction of the services to the local ADSL provider and customer is that the information they represent could easily be localized for more immediate relevance. This approach has already been tried with smaller, local ISPs, with some success. All of the services could still be IP-based, and Internet access through a router would still be available.

ADSL Target Service Speeds: Miscellaneous

The ADSL Forum series of target speeds supports a wide range of video services, but the target speeds do not exclude other types of services, and miscellaneous services, such as imaging and "legacy" services, are included. Legacy services provide what the customer is used to, and the others add to service capabilities. These types of service are shown in Table 11-2.

Data communications covers Internet access, remote LAN access (from a home office to a business, perhaps), and distance learning (non-video-based). These should function well downstream with anywhere from 64 kbps to a full 1.5 Mbps available. Internet and LAN access should have 10 percent or more of the downstream available bandwidth (for example, more than 150 kbps upstream at 1.5 Mbps downstream). Distance learning should have marginally more, topping out at 384 kbps upstream.

Image-based services, such as home shopping from an online catalog or information services in the form of a graphical page showing business hours, should have anywhere from 64 kbps to 1.5 Mbps available downstream and a modest 64 kbps upstream because user commands should be terse in this environment.

Legacy services include plain old telephone service (POTS) and ISDN (not Broadband ISDN, but narrowband ISDN). ISDN needs 160 kbps in

Table 11-2

ADSL Target
Speeds for
Miscellaneous
Services

Service	Application	Downstream	Upstream
Data Communications	Internet Access	64 kbps to 1.5 Mbps	>10% of downstream
	Remote LAN Access	64 Kbps to 1.5 Mbps	>10% of downstream
	Distance Learning	64 Kbps–1.5 Mbps	64–384 kbps
Image-based	Home Shopping	64 Kbps to 1.5 Mbps	64 kbps
	Information Service	64 Kbps to 1.5 Mbps	64 kbps
Legacy Services	POTS	4 kHz	4 kHz
	ISDN	160 Kbps	160 kbps

Source: ADSL Forum

both directions (144 kbps for Basic Rate Interface (BRI), plus some overhead), and POTS requires only 4 kHz analog for support on an ADSL link.

Some of the services, such as distance learning or home shopping, are the same in concept as the video equivalents, just without the video component. The others, such as remote LAN access, are familiar to anyone working in an organization of almost any size. The nice thing about these services is that people do not have to go to the office to get them.

ADSL Target Speeds and Distances

It is a frustrating fact of life that ADSL is evolving so rapidly that hard and fast speeds and distances are difficult to translate into absolutes. ADSL equipment that provided 1.5 Mbps downstream at 12,000 feet now runs at 3.0 Mbps, and it stretches to 15,000 feet and beyond in its new product packaging.

Nevertheless, the ADSL forum has put together a number of documented speeds and distances that makers of ADSL equipment should "target" in their products. Keep in mind that even these speeds and distance are a moving target (pun intended), but they are still helpful.

Table 11-3 shows the target speeds and the distances at which they should be achieved, for both 24 AWG (American wire gauge) and 26 AWG copper loops (the most common gauges used, although they are frequently mixed).

Note that with the exception of a few landmarks that stand out, many of the distances are not yet well-known enough, in terms of equipment tri-

Table 11-3

ADSL Target
Speeds and
Distances

Downstream Speed:	24 AWG	26 AWG
1.544 Mbps (T1)	18,000 feet	15,000 feet
2.048 Mbps (E1)	16,000 feet	12,000 feet
3.088 Mbps (2 × T1)	?	?
4.096 Mbps (2 × E1)	?	?
4.632 Mbps (3 × T1)	14,000 feet	12,000 feet
6.312 Mbps (T-2)	12,000 feet	9,000 feet
8.448 Mbps (Upper limit)	9,000 feet	?

Source: ADSL Forum

als and real world deployments, so a "?" marks the unknown. However, a rough estimate involving a form of interpolation (3 Mbps should be achievable roughly halfway between 14,000 and 18,000 feet) might be a start for fixing these distances.

The 18,000 feet on 24 AWG loops (15,000 on 26 AWG) for 1.5 Mbps seems to be a given. In other words, all ADSL equipment vendors seem to be comfortable with offering this speed at this distance. Beyond 18,000 feet, loading coils and line extenders enter the equation. The T-2 speed of 6.312 Mbps is another benchmark speed as well, pegged at 12,000 feet and 9,000 feet for 24 and 26 AWG loops, respectively. Note that dropping the speed to 4.632 Mbps (three times the T1 rate of 1.5 Mbps) only marginally increases the distance achievable (from 12,000 to 14,000 feet, and 9,000 to 12,000 feet).

The upper limit of 8.448 Mbps defined for ADSL is based on the current technology line code speed limitations and a wish to clearly define speed limits. Aside from this, however, nothing prevents an equipment provider from trying to exceed these target speeds and distances. A lively competition will likely result from this.

Note that the figure mentions nothing about impairments due to the possible presence of bridged taps or mixed wiring gauges.

ADSL in the House!

The hypothetical ADSL service provider has a DSLAM and perhaps even a group of servers and routers on a LAN to provide all or some of the services shown in the tables above. All that is needed is a pool of customers eager to hook their local loops up to the DSLAM instead of the PSTN switch (of course, the splitter still handles the analog voice). But wait a minute. What about the ATU-R? What about the unit with the remote splitter and ADSL interface at the customer side of the link? Where does that come from? How is it packaged? Is there anything else that needs to be done to prepare a home for ADSL?

First, some background to the issues raised here. Under the 1984 Divestiture agreement, all telephones and premises wiring in the United States are owned by and under the control of the customer (technically, the owner of the premises, but this is usually the customer anyway). Customers are free to buy and use any FCC-approved telephony device and may hire anyone they want (even the telephone company) to run new wiring or maintain old wiring. Customers can install their own wiring and often do, but few are able to comply with all relevant electrical or power standards, or are even aware of them.

The boundary between what is controlled by the customer (the devices and premises wiring) and what is controlled by the network (the telephone company) is established by a demarcation point, or just *demarc*. This demarc is typically a small box mounted in the basement or garage in residential environments. In newer housing developments, the demarc is located outside of the premises. In ADSL networks, once the local loop is digitized, this becomes the *network interface device* (NID).

In many other countries around the world, the wiring and telephony devices continue to be owned and controlled by the service provider. Deregulation movements in many countries will change this eventually, but this section emphasizes the current situation in the United States and few other places.

After a firm NID or demarc has been established, the location and arrangement of the ADSL POTS splitter to provide continued analog telephone support, as well as the *ADSL Terminal Unit-Remote* (ATU-R), become an issue. What should be the relationship between the splitter and the ATU-R, and how should the devices that connect to the ATU-R be wired? Fortunately, the ADSL Forum has addressed this issue. There can be no universal or "correct" way to do this for the simple reason that the ADSL devices, the existing telephone(s), and the premises wiring are beyond the demarc and beyond the control of the network-an ADSL network or otherwise. ADSL service providers have as little control over the ATU-R and splitter as telephone companies have over modem or telephone handset appearance, location, and function.

Nevertheless, the ATU-R equipment vendors have a lot of control over the situation. If a particular configuration is not offered by any vendor, it will not matter whether customers want it or not. This does not mean that all ATU-R packages will be identical. There is always variety for the sake of product distinction. However, ATU-R and splitters can be arranged in products in several fairly standard ways, each with its own implications for the customer, especially where wiring is concerned. This section examines the issues involved with ADSL installations at the customer premises. Please note that Consumer DSL (CDSL) does away with the issues regarding the position of the splitter on the premises because there is no remote splitter in CDSL. Also, CDSL requires no new wiring except to a location that has no previous modem connectivity (but CDSL will not support analog voice either). In other words, CDSL makes most if not all of these premises issues inoperative.

It should also be noted that several of the architectural scenarios discussed below, particularly those siting the splitter in a television set top box or PC and using low pass filters on all telephones, are still being intensively studied. These methods have not yet proven to be reliable and have technical risks and problems, most of which go far beyond the scope of this discussion. But this should be kept in mind.

Installing ADSL: Splitter at the Demarcation?

The simplest way to accomplish an ADSL installation is shown in Figure 11-1. The customer acquires an ATU-R in different ways (purchase, lease, or straight give-away). In any case, it is the customer's decision. In this scenario, the splitter is built into the ATU-R, which is mounted as close to the demarc as possible. At this time, the demarc becomes a NID. Existing premises wiring for existing telephone(s) is then simply unplugged from the NID (the demarc) and plugged into the ATU-R splitter, RJ-11 connector to RJ-11 connector. (RJ-11 is the standard analog connector that is built into the splitter at the ATU-R). The ATU-R is plugged into the NID-so much the better for analog telephone support.

However, because the ADSL devices such as PCs or cable TV set-top boxes are likely to be far from the NID, new wiring must be run to support the PC or set-top box that wants to access the new ADSL services. While not an insurmountable problem, the lack of control that the service provider can exercise in this situation could be troublesome. If the customer wants to install their own wiring, the job might not be done properly. It seems reasonable, however, that most customers will defer to the advice of the service provider ("let us do it-for a small price").

Figure 11-1
Installing ADSL:
Splitter at the
Demarc

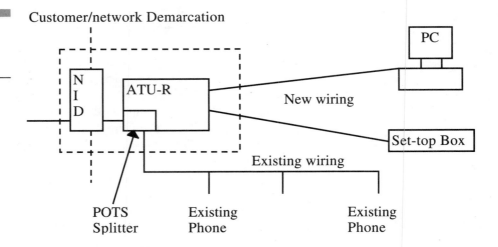

Customer/network Demarcation

PC

NID

ATU-R

New wiring

Set-top Box

Existing wiring

POTS
Splitter

Existing
Phone

Existing
Phone

Benefits of this arrangement include minimal risk of appliance or POTS coupled noise; plus, it is the best solution if the service provider "owns" the ATU-R. Drawbacks include lack of the ready alternating current (AC) power needed for the ATU-R at most demarc locations (garages, basements, and the outside of homes), as well as the presence of environmental extremes (heat and cold, wet and dry) in most of these locations. Small business environments should be better in this regard.

Another potential drawback is the presence of the splitter and ATU-R in the same device. This circumstance maximizes the possibility of interference from POTS ringing and other high-voltage signals on the ADSL bit stream. In fact, putting the splitter under the same covers is possibly the worse place to put the splitter for these reasons. Nevertheless, most ATU-R equipment vendors have packaged the splitter and ATU-R for convenience and simplicity of design. Interference between POTS and ADSL has so far not been a problem or concern.

Installing ADSL: Splitter in the PC/Set-Top?

Again and again, regardless of where the ADSL splitter is located, the overwhelming concern is the amount of new premises wiring required. As mentioned previously, the control that a service provider may exercise over the type and workmanship of new premises wiring installed is exactly none whatsoever. Although standards exist for electrical and power requirements covering premises telecommunications wiring, inspections are rare and violations common. With any new ADSL installations, it might be a good idea to try to minimize the new wiring required.

It is anticipated that most initial ADSL service offerings will feature high-speed Internet access quite prominently (some would say exclusively). This being the case, it might be a good idea to package the ATU-R to closely resemble a modem. Most PC users are familiar and comfortable with modems and may already have a telephone line at the desktop location. Alternatively, the ATU-R and splitter could be integrated as a new type of TV set-top box, with which customers are also familiar. This ADSL arrangement is shown in Figure 11-2.

When the ATU-R and splitter are packaged as an internal modem board or external modem device, the wiring task might be easier. Only one new cable run-from the NID to the ATU-R-might be needed to connect the splitter in the ATU-R to the existing POTS. If a second line has been used for Internet access along with a line for a telephone next to the telephone handset before the ADSL install, this wiring can be used to link the ATU-R splitter to the existing analog wiring without any new wiring at all. Even so, the wiring at the NID would have to be rearranged, as shown in the figure.

Naturally, other new wiring runs to other ADSL devices, PCs or set-tops, would still need to be installed. The runs should be much shorter, however, as the figure illustrates. Additionally, no wiring is needed to connect to the device that holds the integrated ATU-R. Many early implementations of ADSL devices will be based on this modem model. Because few ADSL service providers or equipment vendors will support more than a single upstream and downstream ADSL channel, this approach makes a lot of sense.

Figure 11-2

Installing ADSL:
Splitter in the PC or
Set-top

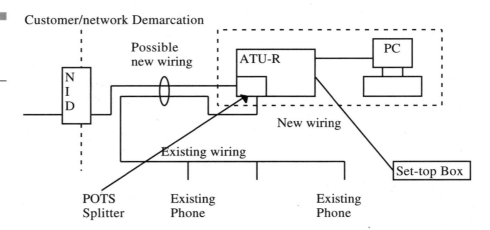

The benefits of this arrangement include the minimum requirement for new wiring and the need only for new POTS wire (very tolerant of "non-standard" installations). Drawbacks include the fact that continued POTS service now depends on the presence of the ATU-R, which is similar to a modem. In other words, unplugging the POTS wiring from the "modem" would interrupt analog voice service. A special wall connector (known as RJ-31X in the United States) might help solve the problem, which is not really an ADSL problem at all. A more serious drawback is that the long parallel run of POTS and ADSL signals from NID to splitter, as seen in the figure, cross-couples all of the POTS noise onto the ADSL link. POTS noises include ringing, pulse dialing, hook flashes, and so on. The ADSL Forum has reported that just a few feet in parallel have been shown to cause ADSL errors. Interleaving within the ADSL frame and higher layer error control in the transfer protocols such as IP or ATM will help, but this arrangement could impact perceived service quality.

All in all, there are serious technical problems with this scenario. Most critical is the coupling problem on the parallel wiring run.

Installing ADSL: Low Pass/High Pass Filters?

In any ADSL installation, there are two problems with combining the POTS splitter to provide continuing support for existing analog telephones inside the ATU-R (Remote) housing itself. The first is that this arrangement is sure to maximize the risk of coupling POTS noise onto the ADSL wiring, to the detriment of ADSL signal quality. The second is that this arrangement forms a single point-of-failure for both new ADSL services and analog voice services.

This being the case, it might be better to separate the POTS splitter from the ATU-R altogether. Unfortunately, the combined splitter and ATU-R configuration is practically universal in current products. Therefore, this separation forms a radical break in ATU-R product evolution.

Figure 11-3 shows how analog and digital signals would both exit the NID. A small *low pass filter* (LPF) is mounted near the NID. The low pass filter is needed because POTS support is provided by the continued presence of analog voice baseband signals at the low end of the frequency range (4 kHz is needed for analog voice support). The filter, which needs no AC power, passes these voice signals to existing telephones over existing wiring.

The ATU-R now contains a *high pass filter* (HPF) that filters out the low frequency analog voice signals, allowing only ADSL signals to enter

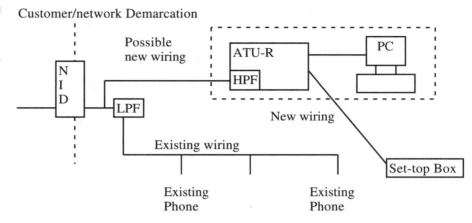

Figure 11-3
Installing ADSL: Low
Pass/High Pass Filters

the ATU-R-this is basically just another form of splitter. Now the ATU-R can be packaged exactly like an analog modem and can be located at the most convenient place on the premises. Wiring to the ATU-R might be new or existing if the PC previously used a second line for Internet access.

The benefits of this arrangement include short new wiring distances and elimination of parallel POTS-ADSL wiring that could cause signal coupling and interference. Of course, if an existing telephone is adjacent to the PC, this benefit could disappear. Also, the LPF can actually be installed before the ATU-R, perhaps months earlier in preparation of an ADSL service offering, and the splitter is now independent of the functioning or presence of the ATU-R (that is, no inadvertent disconnects of voice service). The ATU-R can now be purchased, installed, and employed in precisely the same way as a modem.

The drawback is that all splitters are intended to function with a specific ATU-R from the same vendor. Another potential problem is in countries where lines need an active POTS splitter. In this case, the ATU-R (or some other network component) would have to provide power to the splitter over the ADSL wiring.

Installing ADSL: Low Pass Filters for all Phones?

Figure 11-4 shows a distributed split POTS splitter arrangement. This variation is sometimes known as the *bridged configuration*. Each tele-

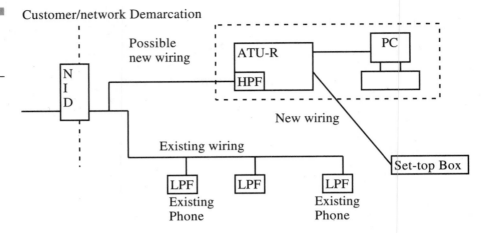

Figure 11-4
Installing ADSL: Low Pass Filters for all Telephones

phone and jack now has a low pass filter installed, possibly as a faceplate-mounted device. The low pass filter is needed because POTS support is provided by the continued presence of analog voice baseband signals at the low end of the frequency range (4 kHz). The filters, which need no AC power, passes these voice signals to existing telephone over existing wiring.

The drawback here is the same as before-all splitters are intended to function with a specific ATU-R from the same vendor. Another potential problem is that removing any low pass filter, even from unterminated (phoneless) jacks, might affect the performance of the ADSL link.

This is another technically problematic scenario that has yet been shown to work effectively and is currently under intensive study.

The only other point about ADSL on the premises that should be made involves Consumer DSL (CDSL) proposals, often packaged as "ADSL Lite" or "G.lite." The big attraction of CDSL is that there is no remote splitter on the premises. However, any line used for CDSL services will not support existing analog telephone equipment. The tradeoff in cost and simplicity is the main attraction of this method.

ADSL Premises Issues

It should be clear from the preceding discussions that the whole issue of the best customer premises arrangement is still up in the air. The best or even most common configuration could take a while to figure out. However, some general guidelines regarding the customer premises should be followed when installing ADSL devices, including ATU-Rs (or ADSL terminal Unit-Remote) and POTS splitters, no matter what the details. Most of these were established by the ADSL Forum.

First of all, the POTS splitter should be isolated from the ATU-R. This eliminates a single-point-of-failure scenario and minimizes the chance for POTS-to-ADSL signal coupling. It is important to note that little to no ATU-R equipment available today does this.

Also, it is important that premises inside wiring essentially belongs to and is controlled by the customer in the United States and some other countries. Therefore, it can be quite hard to control and maintain the quality of the wiring, especially in cases where the wiring was installed by nonprofessionals (perhaps even the customer) with little regard or respect for standards. Indeed, they might not even be aware of the relevant standards.

This being the case, installations that minimize the requirement for new inside wiring are always the preferred approach. Additionally, this will also minimize the cost to the customer if they want to have the service provider install the new wiring.

Keep in mind that the full ADSL architecture supports a variety of *premises distribution networks* (PDN), not just point-to-point wiring. A 10BASE-T LAN or Consumer Electronics Bus (CEBus, which sends digital signals over electrical wiring) is possible, along with other more exotic arrangements. However, for most ADSL installs, a single new Category 5 (Cat-5) wire that runs to the family PC should be all that is necessary.

All things considered, two or three basic forms of ATU-R will be available initially. In the first basic architecture, the ATU-R and splitter are housed in a device resembling an external modem. Placement of the device will not be mandated by the vendor, but it makes sense to place it as close to the NID as is feasible for environmental and power reasons. The customer interface will consist of three forms: a basic serial connector for carrying serial bits to a device over Category 5 wire; a 10Base-T LAN-compliant jack (physically identical to the RJ-type) for linking the ATU-R to a 10Base-T network interface card (NIC) in the PC using the ADSL service; or both. In the future, the ATU-R will include a *universal serial bus* (USB) connector, a standard T1 connector, or something even more exotic. A slightly different packaging allows the ATU-R to act like a 10Base-T LAN hub or even be attached to a 10Base-T LAN hub (in which case the ATU-R is sometimes called an "ADSL bridge").

Another basic form would be an internal modem board. This ATU-R would fit into a spare PC board location on the PC bus just like any other

PC or modem card and might use a USB connector. The splitter function could be a full splitter or high pass filter. If an HPF is used, the location of the LPFs can vary according to one of the scenarios above.

Finally, although not a true variation, many people will be encouraged to employ the ATU-R on their premises, at least initially, in the following way. The ATU-R is packaged as a modem, internal or external, and designed to be in or by the PC using the ADSL service for Internet access. There is no splitter function in the ATU-R device; if there is, it is not used. A second line-either newly installed by the service provider (perhaps at no charge to the customer) or existing already for Internet access (in which case the ADSL service provider should check the wiring for compliance)-is used for ADSL services. The first line into the home remains in use for regular telephony services. Note that this arrangement minimizes the ADSL installation task. However, it also minimizes the features of ADSL with regard to POTS support and maybe even other ADSL services.

Packaging can vary more widely than even this survey suggests, and combinations of all three are possible before one form or another emerges as the most common.

Again, please note that CDSL makes most of these premises issues regarding ADSL more or less inoperative. However, packaging CDSL modems could still be accomplished in a variety of ways, as well.

12

The Other Side of ADSL: The DSLAM

As important as the ATU-R (Remote) and ATU-C (Central Office) are for the operation of an ADSL link, the Digital Subscriber Line Access Multiplexer is just as important. Without the DSLAM, the bits have nowhere to go once they reach the ATU-C. The DSLAM links the customer side of the network (the DSLs) with the service side of the network (the Internet, video servers, and so forth).

This chapter focuses on the DSLAM and related issues. The emphasis is on the network side of the ADSL link in all cases. Other xDSL technologies are included because the use and function of a DSLAM is in no way limited to ADSL.

On that note, it is important to realize that DSLAMs are used with more than just ADSL. In the ADSL network architecture, the device that interfaces with the ATU-Cs in the serving office technically is the *ADSL access node*. Now, an ADSL access node is a very good example of a DSLAM, but the DSLAM is a more generic device and is not just tied to ADSL. In other words, all ADSL access nodes are forms of DSLAMs, but not all DSLAMs are ADSL access nodes.

DSLAMs have their own architectures apart from ADSL. Additionally, DSLAMs support a variety of DSLs, not just ADSL. A DSLAM might support one or more of the following DSLs: *High-bit-rate DSL* (HDSL), *Symmetric DSL* (SDSL), *Asymmetric DSL* (ADSL), and so on—this is just on the DSL side. Usually, the ATU-C splitter is built into the DSLAM itself, just as the ATU-R splitter is usually housed in the ATU-R. Of course, the DSLAM side splitter can be implemented in a variety of ways, just as the ATU-R splitter was in the previous chapter, but for the rest of this chapter, an integrated splitter arrangement is assumed for the DSLAM.

On the service side, the DSLAMs can interface with ATM switches, TCP/IP routers, *switched digital video* (SDV) servers, LANs, and many other possibilities. If this seems like a lot of functionality and, thus, room for a wide range of competitors and product capabilities, it is.

In fact, the battle for the DLSAM and services side of the ADSL network architecture is just heating up. Premises ADSL equipment must be relatively inexpensive, and so profit margins from this equipment might be low. Perhaps added revenue from active DSLAM markets could help profits.

DSLAM Architecture

Figure 12-1 shows the basic architecture of a DSL access multiplexer. Note that the architecture is not tied to any type of DSL, nor is it tied to any particular service. It all depends on the support that the vendor builds into a particular product. In one of its simplest forms, the DSLAM might be an ADSL access node and support ADSL links and Internet access through a TCP/IP router exclusively.

Figure 12-1
The DSLAM
architecture

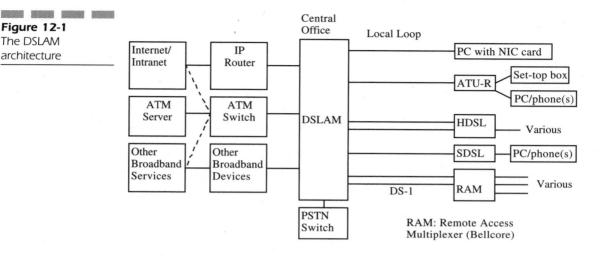

The figure shows both the central location of the DSLAM within the ADSL network, but also the versatility of the DSLAM when it comes to service and line configuration support. Some of the roles of the DSLAM are not obvious.

The DSLAM usually sits in the central office (CO). Other interconnection possibilities exist with virtual collocation arrangements, some of which will be discussed later in this chapter, but the CO seems most likely. In some competitive environments, the DLSAM could be owned and operated by a service provider other than the incumbent *local exchange carrier* (LEC). In any case, the DSLAM interfaces with the PSTN switch if it must continue to provide analog voice services, which it typically will.

On the customer side, the DSLAM can support any one of a number of DSL techniques and loop arrangements. The local loop can terminate directly at a PC ADSL *network interface card* (NIC, in which case the ATU-R is internal), or at a separately packaged ATU-R for a number of ADSL devices (but probably initially only one). High-bit-rate DSL (HDSL) with two pairs of wires can support a variety of devices. However, this would require a T1- or E1-compatible interface on the premises—not always a given in residential environments. Symmetric DSL (SDSL) can also be supported, even with POTS in some cases, in the same fashion as ADSL. Although not shown in the figure, ISDL is a distinct possibility for DSLAM support also. Bellcore has defined a *remote access multiplexer* (RAM) configuration with ordinary DS-1 support on two pairs of wire for small businesses. One or all of these xDSL technologies may be supported by the DSLAM. However, it should be kept in mind that T1/E1 units in a DSLAM will be extremely rare.

On the service side in the figure, the DSLAM might support IP routers, ATM switches, or even other broadband devices to access services on things such as SDV servers (which could also be on the ATM network). Again, support may be for one or all. Usually, the IP router would be used for Internet or corporate intranet access, and the ATM switch for ATM-based servers. Alternatively, the DSLAM architecture allows for all servers and services to be accessed through an ATM switching network. Typically, on this side, the interface to the network service side of the DSLAM would be a 10Base-T LAN interface.

Please realize that the basic DSLAM is not a switch or router. It is in more ways than one just a type of multiplexer. That is, the DLSAM combines streams of bits in the channels coming upstream from homes and small offices and splits up a large bit stream coming downstream from the IP or ATM network. The DLSAM splits this bit stream based totally on channels, as a multiplexer always does. So the link from the DSLAM to some other piece of equipment, whatever it may be, must be able to carry the sum total of the traffic arriving from customers all at once.

This aggregation of traffic based on the sum total of the input bit rates is commonly known as *tine division multiplexing* (TDM). However, other types of multiplexing are used in networks that send and receive digital information. Often considered a step above straight TDM is a technique called *statistical time division multiplexing*, or statistical TDM, or just "stat muxing." Statistical multiplexing takes advantage of the fact that many applications that generate these bits are bursty applications, as has been mentioned many times. Lulls often exist in the stream of bits, during which other applications can share the same bandwidth. This is the basis of the whole concept of packet switching in the first place. In some sense, packet switching in statistical TDM applied not on a link-by-link basis, but on a network-wide basis; this is an oversimplification of the situation, but it is a useful analogy.

What has all this to do with the DSLAM? A DSLAM today will rarely use simple TDM. The DSLAM usually includes features like tagging of priority traffic, traffic shaping to smooth out packet or cell bursts, and even cross-connect features. This makes the DSLAM more expensive, but more effective. Here is why.

Suppose a DSLAM services 64 ADSL links in a communications rack, 8 per shelf, and there are 8 shelves in the rack. This is by no means an uncommon number for a small DSLAM. Naturally, multiple racks can be used to service more ADSL customers. Now, suppose that each ADSL link operates at a modest 1.5 Mbps downstream and 64 kbps upstream. Again, these numbers are quite representative, although ADSL is capable of much more. How fast would the links coming out of and going into the back of the DSLAM have to be to handle this traffic load?

This is not such an easy question to answer. There are several ways that an ADSL link can operate. For instance, one of the simplest forms of ADSL support and configuration is the "bridging" mode, which carries packets over the ADSL links onto some service network behind the DSLAM and back again. Perhaps the DSLAM just funnels these packets into a router located in the local exchange office. It is up to the router to figure out just what the bits in the packet represent, as was detailed in the previous chapter.

Consider upstream first in this example. If there are 64 inputs to the DSLAM (the upstream user traffic) running at 64 kbps each, the link from the DSLAM to anything else must be able to handle 64 x 64 kbps = 4,096 kbps or 4.096 Mbps. The easiest way to do this is with a series of T1 links to separate serial ports on the router (three will do), but consider the situation downstream. Each ADSL link operates at 1.5 Mbps downstream. The same 64 outputs must operate at 64 x 1.5 Mbps = about 96 Mbps into the back of the DSLAM from the router. It is not feasible to put 64 T1 links onto the router, and even two T3 at 45 Mbps falls short of 64 T1s (a T3 can only carry 28 T1s, or 56 total for two T3 links). Now, as it turns out, given the asymmetrical nature of ADSL, the upstream traffic aggregate of 4 Mbps can be carried on the same T3s (however many) used for downstream transport. This is quite wasteful, however, because a lot of upstream bandwidth is being unused.

However, if the DSLAM is just a passive TDM device, little else can be done to carry the bits through the DSLAM to and from the ADSL links. Perhaps some amount of intelligence could be added to the DSLAM to allow the DSLAM to stat mux the traffic from and especially to the ADSL links themselves. Recall that the AS and LS channels in bit synch ADSL contain idle bits, as well as information bits. Perhaps the DSLAM could filter out the idle bits on each ADSL link and only send live data bits to the router. And indeed, there are a variety of DLSAM products that perform such stat muxing for each ADSL link. In fact the Massachusetts Institute of Technology has called this concept not a DSLAM, but a Dynamic Access Multiplexer (DAM). Of course, this stat muxing need not be done right at the DSLAM. In terms of bandwidth, it would even more efficient to perform this muxing task as close to the customer as possible. This is whole idea behind Bellcore's Remote Access Multiplexer shown in Figure 12-1. The RAM stat muxes ADSL traffic from many users onto a series of DS-1s (T1s) far from the DSLAM itself. The DLSAM could even perform further stat muxing, all in the name of efficiency.

Such stat muxing of ADSL traffic is not part of the ADSL specification, however. And, of course, this would no longer be entirely ADSL bridging in the DSLAM, but maybe that's okay.

However, there are other ways that a DSLAM can make better use of the bandwidth between the back of the DSLAM and whatever other switches or routers used to access the ADSL services provided. Suppose that the DSLAM and the router communicate not over a T-carrier serial link at all. Suppose both the router and DSLAM are linked by an Ethernet LAN running at 10 Mbps or even 100 Mbps. Technically, this is either a 10Base-T or 100Base-T LAN, but the point is the speed, not the architecture. However, the DSLAM cannot simply dump bits onto the Ethernet LAN. The DSLAM network interface card in the DSLAM must send and receive Ethernet frames, just like any other LAN-attached device, and so bits coming in from the ADSL links must be packaged as a series of Ethernet frames and sent to the router. The router must respond in kind, but this is not a difficult situation at all.

What is difficult is determining the content of the frames. By definition, the content of a frame is a *packet*. The DSLAM must be able to send packets inside frames across the Ethernet LAN. No one should be surprised that one of the allowable ADSL distribution modes is "packet bridging the ADSL links, and ATM on the DSLAM back end." Although this ADSL distribution mode emphasizes ATM cells beyond the DSLAM, in the actual world of DSLAM products, an IP packets interface for Internet and router communication is much more common.

On the return path to the DSLAM, the router sends packets (most likely IP packets), as well. It is important to realize that if the packets sent to and from the DSLAM are also sent to and from the customer on the ADSL links, this is nothing more that ADSL packet distribution mode in action. Note also that this implementation of a DSLAM has already moved far beyond the simple TDM device at the beginning of this section. Why bother? The whole point is to make the DSLAM more versatile and efficient. Why stop with simple stat muxing, this philosophy goes. Why not put some packet switching or routing intelligence directly in the DSLAM? This idea is explored more fully later on in this chapter.

The "Typical" DSLAM

This chapter has mentioned DSLAM architectures and capabilities several times. However, the information might have seemed vague and unsatisfactory; there is a reason for this. The basic DLSAM form and function is not covered by any set of ADSL or other xDSL standards documents. The fundamental concept of the DLSAM as a housing for multiple ATU-Cs, HTU-Cs, or something else is a given in all cases. Beyond that, what the DLSAM does and how it does it is completely up to the DSLAM equipment vendor.

This does not mean that the entire field of DSLAM operation is a free-for-all. Most DSLAM products all support some common features quite consistently, but DSLAM products vary with other features, in terms of numbers and speeds. This section attempts to sort out just what features would be found on a "typical" DSLAM, if there is such a thing in these early days of ADSL.

Recall that the DSLAM occupies the key position in the entire ADSL architecture. All traffic to and from the users passes through the DSLAM. All traffic to and from the servers on the network behind the DSLAM must pass through the DSLAM also. The DSLAM is typically located in the wire center of the local exchange because of the need to have access to the local loops. This location is common when the service provider is a LEC or CLEC, but other types of ADSL service providers are possible. All that is needed is access to some local loops leading to customers, a DSLAM, and some services to offer.

It might be useful to divide the entire ADSL network architecture into three parts, especially from the DSLAM perspective. This model has been common in many DSLAM vendors documentation. In this model, the ATU-Rs, or other remote xDSL devices such as an HTU-R, form the *service user* (SU) portion of the network. The ATU-C, HTU-C, or other interfaces in the DSLAM form the *network access provider* (NAP) portion of the network. The access network itself and the network or networks on which the services may be found are part of the *network service provider* (NSP) portion of the network. The central role of the DSLAM as NAP is to form the connection between service user and service provider. This model of the DSLAM's role is shown in Figure 12-2.

In the figure, the DSLAM does more than just house the ATU-Cs. In its most general form, the DSLAM also includes what is called in the figure the *access network* to the services. This may be as simple as a single ATM switch or IP router, or as complex as a network all its own. The type of access network supported depends on the type of connectivity offered by the DSLAM vendor, of course.

With this DSLAM-centric model in mind, the common DSLAM features shown in Table 12-1 takes on special meaning. This table does not identify any specific vendors. The whole point is not to try to evaluate products but to make readers familiar with the range of DSLAM vendors products. In an ADSL world, the DSLAM is the ADSL Access Node, but the

Figure 12-2
A DSLAM-centric
View of ADSL

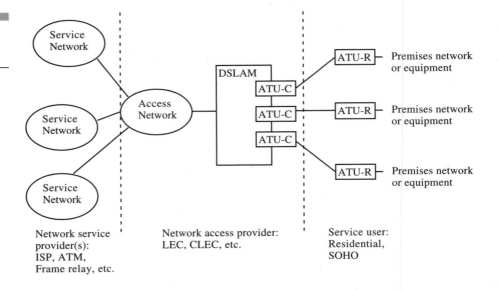

Network service
provider(s):
ISP, ATM,
Frame relay, etc.

Network access provider:
LEC, CLEC, etc.

Service user:
Residential,
SOHO

Table 12-1

The Range of
DSLAM Vendor
Products

DSLAM Feature	Almost All Do	Some Also Do
xDSLs supported	ADSL/RADSL	HDSL, IDSL, SDSL, VDSL
Line Code	CAP, DMT, or both	QAM, 2B1Q
Access network connection	ATM Sonet, 10BaseT, 100BaseT	HSSI, T-1, T-3
Bridging/routing protocols	ATM, IP	IPX, PPP, FR, NetBIOS, SDLC
Network management	SNMP	CMIP, proxy agent
	Minimum	**Maximum**
Total loops served in rack	48 ADSL or RADSL	1440 SDSL
Price	$4,500	$300,000

DSLAM is a more general-purpose device. DSLAM vendors make products that support many types of xDSL, not just ADSL. This is obvious from the table.

The table shows one of the reasons why a fair and impartial evaluation of DSLAM packages is so difficult. There are big DSLAMs and little

DLSAMs intended for different markets and situations. A few words about the table entries are still in order.

A DSLAM services local loops in a *card* that contains the software and hardware for the proper type of xDSL supported and the proper line code. The cards are gathered into *shelves*, which usually share a power supply and may have their own network management capability. The shelves are housed in *rack*, which is the basic unit of the DSLAM. That is, the size of the rack determines the total capacity of the DSLAM. Almost all DSLAMs will do ADSL, ADSL, or both, either with CAP, DMT, or both. In addition, many other DSLAMs will support cards for HDSL, IDSL, SDSL, VDSL, and many variations on HDSL especially. When using these other xDSLs, the DSLAM will generally employ QAM or 2B1Q line code.

The access network connection type will most likely be an ATM connection or 10 or 100 Mbps Ethernet (10Base-T, 100Base-T). However, some DSLAMs will support the *high-speed serial interface* (HSSI) connections common in frame relay networks or some form of T-carrier (useful for pure bit synch xDSL modes), most likely T1, T3, or both. The DSLAM itself might also perform some bridging and routing functions. The DSLAM might be able to send and receive ATM cells or IP packets in most cases. Sometimes, other protocols might be supported, such as Novell's *proprietary packet protocol* (IPX), the Internet's *point-to-point protocol* (PPP), *frame relay* (FR), an older LAN protocol called NetBIOS, and IBM's *synchronous data link control* (SDLC) protocol for SNA networks. More details on this bridging and routing with ATM and FR are explored in the next section.

As far as network management goes, almost all DSLAMs support the widely supported *Simple Network Management Protocol* (SNMP), first used on the Internet but now seen everywhere. In addition, some support the international standard *Common Management Interface Protocol* (CMIP). Also, some DSLAMs do not employ SNMP directly, but allow the DSLAM to be indirectly managed by means of a proxy agent, usually a PC that understands SNMP and whatever network management protocol the DSLAM uses. Proxy agents are common when a product's network management software is proprietary and not standard, such as SNMP or CMIP. Proxy agents are generally frowned upon except when absolutely necessary because of their perceived inefficiencies and added complexity.

The number of loops serviced by a single rack of DSLAM shelves vary from less than 100 to more than 1,000. The pricing varies accordingly, depending on more than just numbers of local loops supported, such as network management capabilities.

In summary, it seems fair to say that a "typical" DSLAM will support ADSL or RADSL with either DMT or CAP. The ATU-C splitter is integrated into the DSLAM itself. The access network connection is most likely to be an ATM interface (generally, SONET speeds of 155 Mbps) or an Ethernet-type LAN connection at 10 or 100 Mbps. The typical DSLAM is not a passive TDM device, but is capable of some bridging and routing based on ATM cells or IP packets. Also, the Internet- and industry-standard SNMP protocol is used for network management of the DSLAM. Finally, there is a wide range of loops served and the price of the configuration.

The Expanding Role of the DSLAM

The simple time division multiplexing DSLAM has already given way to newer units that not only aggregate the traffic from many xDSL links, but also concentrate the traffic with simple techniques, such as filtering out idle bit patterns. Obviously, there is lot more functionality that could be built into a DSLAM to expand its already central role in the ADSL.

Many DSLAM products come with 10Base-T ports. These DSLAMs can be linked by means of a 10Base-T LAN hub to a router. The router links to the Internet or some other IP-based network, but this scenario requires the service provider to purchase not only the DSLAM, but also the 10Base-T hub and router. This increases the complexity of the configuration, not to mention the total cost before a single ADSL user can be offered Internet connectivity.

It is therefore attractive to service providers to consider buying a DSLAM with IP routing already built in inside the DSLAM. This cuts down on the total cost of the configuration, eliminates the hub entirely, and is a more compact device. In either case, the ADSL links are presumably running as packet mode ADSL because the packets have to go to and come from somewhere outside of the DSLAM. The differences between these two configurations is shown in Figure 12-3.

This is not a perfect solution. Router purchasers have historically been more comfortable with separate routers than with routers built into other products. For example, early 10Base-T hub vendors put a board in their hub that could do everything a separate router box could do. Potential buyers did not like the idea. They preferred routers separate from everything else. Apparently, they felt that a separate router reduced the complexity of the hub and made troubleshooting the entire configuration

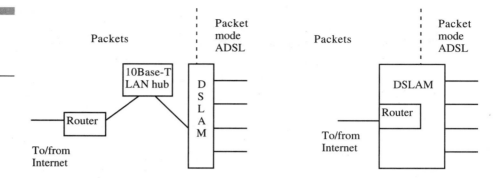

Figure 12-3
DSLAM with and
without Routing
Capability

easier because the router could be isolated from the hub. Perhaps the same thing will happen with DSLAMs, and the router will always remain separate from the DSLAM.

Several of the ADSL modes also include ATM switches. Several DSLAM vendors support interfaces capable of linking to an ATM switch port. From there traffic would make its way onto a broadband ATM network where the servers are located. The evolutionary path for the ATM switching could be the same as for the router. That is, the ATM switching function could be incorporated into the DSLAM itself. All of the same advantages and disadvantages of the DSLAM-based router apply to the ATM switch scenario.

As ADSL—and, in truth, all of the DSLs—matures, the switching and routing capabilities of the DSLAM will continue to expand. Even if the potential buyers of DSLAMs reject the integrated switch/router configuration, perhaps the DSLAM could still supply some form of switching or routing on the user side of the DSLAM. This additional switching or routing need not necessarily be limited to IP packets or ATM cells.

This needs a few words of explanation. Consider a service provider with a frame relay service backbone. Frame relay is a fast packet switching network that uses frame relay switches as network nodes and something called *frame relay access devices* (FRADs) as the user interface to the network. FRADs may be separate devices all on their own, much like a LAN hub, or incorporated into routers as FRAD software modules. Such routers are said to be capable of supporting the "FRAD function" on a given serial port.

Access to a frame relay network is usually over a 64 kbps digital link or even a full 1.5 Mbps T1. When used with frame relay, the link carries frame relay frames (naturally) formatted according to the link access procedure for frame relay (LAP-F) protocol. This link must be provisioned by the local service provider, often at great cost. ADSL can offer a better way to provision frame relay access. Figure 12-4 shows how.

The reason for putting frame relay switching in the DSLAM is both to multiplex and concentrate it for transport over the access network. This access network is typically frame relay, and the service network of interest in this case is a series of frame relay service providers, of course. Note that in this scenario, ADSL can be used to provide frame relay access through the LEC to many other frame relay IXC networks, not just one.

The ATU-R is now either a module in a multiprotocol router implementing a FRAD function or in a separate FRAD device altogether. The whole point of enhancing the DSLAM beyond the simple time division multiplexing of frame relay *permanent virtual circuits* (PVCs) is to make more efficient use of the bandwidth available on the access network. Of course, one characteristic of doing this with ADSL is that the bandwidth is inherently asymmetrical. If symmetrical bandwidth is needed, as would be the case if the SOHO user had a LAN with its own servers on the premises, then the same DSLAM can support users with HDSL or HDSL2, in which case the ATUs become HTU-Cs and HTU-Rs. In fact, almost any DSL can be used, including RADSL, SDSL and IDSL.

Another way that an "expanded role" DSLAM can benefit customers and service providers alike is in the area of a service that is usually called "fractional T1," but that is technically *Nx64* service. With Nx64 service, the link runs at a multiple of 64 kbps (the DS-0 rate of a T1), where in

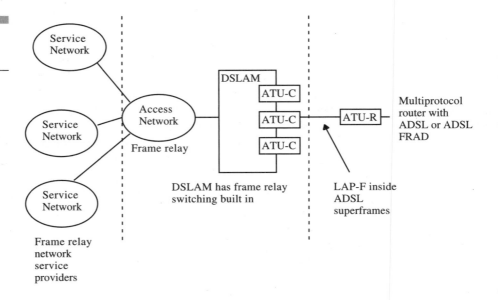

Figure 12-4
ADSL and Frame
Relay

the United States the value can range from N=2 (128 kbps) up to N=24 (which is basically a full DS-1 running at 1.536 Mbps). Fractional T1 services are popular with users, but not so popular with local exchange carriers. Users like having a place to go beyond 64 kbps that is not a full 1.536 Mpbs, but rather can grow there as bandwidth needs and the value of N increase over time.

The reason the LECs are not overly fond of fractional T1 is that there is no real way to provide a piece of a T1 to a single location. Two pairs of wire carries all 24 channels of a full T1. Yet the customer only pays for 384 kbps (N=6), 768 kbps (N=12), or some other piece. This provisioning of 24 channels where only a few are actually used is not very efficient. What the LEC will typically do, therefore, is immediately to take the 64 kbps channels from a number of customers and concentrate them at the wire center in a *digital cross connect* (DCS). The trouble is that this concentration must be done in contiguous channels. That is, the N channels must stay together through the wire center DCS. This is shown in Figure 12-5.

Why should this be a problem? Because the customer knows that fractional T1 enables them to go from 384 kbps to 768 kbps essentially by changing from N=6 to N=12. Suppose Customer #1 in the figure is so successful at whatever the 384 kbps link is used for that the customer wants to upgrade their fractional T1 from N=6 to N=12. The capacity is there, of course, because the link to the customer is still effectively a full T1 anyway. The service provider would like to have the increased revenue, of course. Unfortunately, the only way to go from N=6 to N=12 fractional T1 service is to give Customer #1 contiguous channels 1 through 12 through the DCS. As the figure shows, this is impossible because those channels are currently used by Customer #2. The customer will still get 768 kbps eventually, but only after a great deal of reconfiguration and effort.

Figure 12-5
Concentrating Nx64

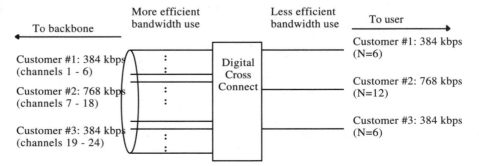

How can xDSL and the DSLAM help? The DSLAM can more intelligently concentrate the Nx64 traffic arriving from the customers and rearrange the channels easily. In this case, HDSL or HDSL2 would be more appropriate because circuits are inherently symmetrical. And now the service can be provisioned on one pair instead of two pairs in a "real" T1. Therefore, provisioning fractional T1 with a DSLAM and xDSL makes a lot of sense.

So the DSLAM can contain a routing function, a frame relay switching function, and even a cross-connect function for delivering fractional T1 services. These functions extend the capabilities of the DLSAM and help to support various modes of ADSL, such as packet mode, but the DSLAM can also be used effectively to deliver ATM services over the ADSL network. As has been mentioned several times, a lot of work done on ADSL and xDSL is aimed at bringing ATM to every possible location.

ATM cells are transported on a variety of physical links, such as T-carrier and SONET. The transport of cells on these diverse physical link types is handled by the ATM *Transmission Convergence* (TC) sublayer of the ATM Physical Media layer. The use of the term "convergence" reflects the fact that there are a variety of the media types must now transport ATM cells inside their respective transmission frame structures instead of their normal circuit-oriented bit patterns.

A potential complication is the presence in ADSL of the "fast" and "interleaved" buffers. It is possible in ADSL to support only one type with ATM, but which one is up to the implementation.

The fast and interleaved buffers have been mentioned before. This is a good place to review their reason for existence and the role they play in ADSL. The interleaved buffer is a way of adding extra protection against transmission errors by using a technique called *interleaving*. This was deemed valuable for video services. The disadvantage is that it adds time delay (latency) to the information stream, as much as 20 milliseconds. Other services, such as voice, cannot tolerate this added delay. However, voice is much less sensitive to transmission errors.

So an ADSL information stream has two possible paths. Data services can be passed through the interleaved path or through the fast (but less error protected) path depending on the type of service. File transfers might need protection, while real-time process control services might need to be fast.

The fast and interleaved buffers in the ADSL modems are associated with each service type. An ADSL system can carry multiple services on one line, and the modem will divert the bits to the appropriate fast of interleaved buffers.

Figure 12-6 shows a possible way of supporting ATM over ADSL with the DSLAM again playing the central role.

In the figure, the ATM transmission convergence is shown for both the ADSL fast and interleaved buffers. Only ATM cells are sent on the ADSL link. No explicit relationship exists between the ATM cells and the ADSL frame structure. The DSLAM performs all ATM layer functions, which includes the key switching aspect of ATM. The presence of the ATM network in the figure is not really an exclusive arrangement. ATM allows for both service and network interworking with other networks types such as IP and frame relay.

On the customer premises, the ATM cells can originate and terminate in an "ATM LAN" running at 25 Mbps, but it will be far more common to have an Ethernet LAN on the premises, in which case the "higher layer functions" in the figure are definitely needed. In the Ethernet case, the premises equipment will need an ATM *segmentation and reassembly* (SAR) sublayer to convert the Ethernet frames to ATM cells and vice versa.

Do-It-Yourself ADSL

This entire chapter has thus far assumed that the DSLAM is located in the service provider's office and that the service provider is a state-certified LEC or CLEC. This section explains that this is not always the case. In fact, the presence of the DLSAM in places other than the LEC local exchange is evidence of a growing "non-telephone company movement" toward ADSL, whether the local service provider embraces ADSL of not.

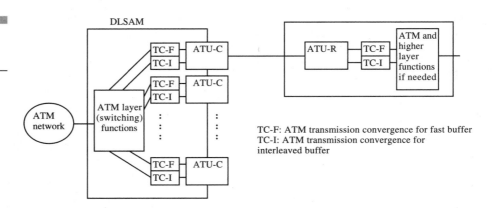

Figure 12-6
Possible ATM over
ADSL Architecture

At first glance, this statement seems astonishing. How can someone obtain ADSL services if their local telephone company does not provide it? The key is understanding that in the newly deregulated local telecommunications marketplace, it is no longer a given that the LEC controls all access to the local loop. Local loop access has officially been "unbundled" to a large degree, and even where it has not, organizations, such as ISPs and groups of users, have been very inventive at gaining access to the local loop in ways the LECs never dreamed of.

After someone gains access to the local loop, all that is technically needed to provide ADSL service is to put an ATU-R at one end and a DSLAM with a group of ATU-Cs at the other. The DSLAM can be a simple time division multiplexer, or add some of the capabilities outlined previously.

In some cases, the LECs are less than pleased at such movements; they are used to having absolute control over their facilities, and who can blame them? The LECs built the physical plant of which the local loops are part at great financial cost and risk. If any other entity is allowed to use the local loop, the LECs want to make sure that they are adequately compensated for it.

In spite of the LEC's concern and sometimes outright opposition, ADSL services are actually appearing without the participation of the LEC. Some organizations, such as ISPs and even groups of ordinary citizens, can bring ADSL—or any xDSL for that matter—to their homes without LEC involvement by using two main methods: this section calls them the "alarm circuit approach" and the "grass roots" approach."

There is little doubt that initial ADSL implementations will emphasize fast Internet access as their premier service. Of course, this type of service will only allow the channels provisioned on the ADSL link to connect two things together: the user's PC and the ISP site's servers/routers. None of the sophisticated ATM switching and broadband services are available in this simple model, but maybe this is all right. Throughout this book, ADSL (and even the other DSLs) have been positioned mainly as a way of getting packet-oriented Internet traffic off the circuit-oriented PSTN. So what if access is limited to the Internet? The whole world is out there on the Internet already.

This scenario does not rule out ADSL use for telecommuters or others. A corporate intranet can just as easily be accessed through the proper ISP as anyway else. The overwhelming concern is security, and the security can be provided in different ways at a variety of steps between user and server. However, this application of ADSL emphasizes the use of ADSL by an ISP for simple Internet access, although it is obviously not limited to this.

Here is how *alarm circuit* ADSL came to exist. An area's ISP wants to offer ADSL services to customers in that area. In addition to the obvious revenue opportunities for the ISP, there are other incentives to do so. Many ISPs may already exist in the area, and the ISP in question might want to distinguish itself from the others. The LEC may be making noises about becoming an ISP themselves (or may already be doing this), and the ISP is seeking to keep ahead of the LEC. And so on.

Of course, providing ADSL service does consist of a little more than having a pool of customers with ATU-Rs and a compatible DSLAM. As has already been discussed, the DSLAM can easily interface with an ISP's internal server/router setup and take users onto the Internet and Web. The ATU-R is customer equipment in the United States and beyond the control of the LEC, so this is not a problem. However, the DLSAM is usually located in the LEC's office space. What about the situation where the DSLAM is owned and operated by the ISP and not the LEC?

In this case, the rules under the Telecommunications Act of 1996 (TA96) take effect. Regardless about how one feels about the overall effectiveness or shortcomings of TA96, a few things in it are fairly explicit. The FCC and state regulators have established rules regarding the *co-location* of other service provider's equipment within a LEC serving office. This applies to all switching locations, but the LEC location is clearly the most important. If an incumbent LEC desires to do certain things under TA96 (such as offer long distance calling), they must offer co-location to competitors. The ISP could, therefore, conceivably pay rent to the LEC and place their DSLAM right inside the switching office, within easy cabling distance of the wire center.

Co-location does have a couple of drawbacks. First, a lot of potential competitors possess a lot of different equipment today. Floor space is at a premium and may not be available. Second, the rent can be very high. Which companies will want to add revenues to their biggest competitor? Third, it is not really a good idea to put such an essential piece of equipment in somebody else's floor space. Access alone could be a problem.

To address these issues, the FCC and many state regulators made co-location easier to do in practice.

It is now allowed for access to an incumbent LEC's equipment when the competitor's equipment is not inside, but is in close proximity to the LEC's local exchange. Usually, this translates to "across the street" but not always or exclusively. The actual cable distance depends on what is actually being done. For some voice services, such as competitive voice mailboxes, a simple channelized voice trunk to the competitor location and back is sufficient. If 600 feet of T3 coaxial cable for a tail-end could not stretch, then fiber optic cable could extend the reach to a mile or so, in some cases. It all depends on the physical limit for the port type in the LEC's equipment being accessed, whether switch, wire center, or something else.

The nice thing about this type of co-location is that it got around the disadvantages of "regular" co-location. Floor space did not matter, rent was just a connection fee (which still might be substantial), and the equipment was now in the competitor's own office space.

With this newer co-location co-location, an ISP with office space close enough to the LEC office obtains a DSLAM from one of a variety of vendors, pays to interconnect with the LEC's wire center or main distribution frame, and distributes compatible ATU-Rs to their customers in the area. The only missing piece is the local loop from LEC to customer—but this is a big piece.

It is now possible to try to buy access to the actual local loop running from LEC to customer. In the normal ADSL configuration, the splitter in the ATU-C and ATU-R keeps the analog voice service going into the voice switch. However, this requires a complex cabling arrangement between ISP and LEC because the DSLAM (and presumably the ATU-C splitters) is now typically across the street from the LEC. This is shown in Figure 12-7.

In the figure, voice and data from a customer splitter and ATU-R enter the wire center. The loop is cross-connected to a cable (perhaps a T1 or T3) leading across the street to the virtually co-located ISP. There, the analog voice is split off at the ATU-C in the DSLAM. The voice is sent back to the LEC via more cables to the PSTN switch ports. The data continues into the ISP network and equipment. The whole system is quite complex and expensive to set up. Also, the cable distance to the ISP counts toward the 12 or 18 kft limit of the xDSL method, and a reasonable wire center signal loss must be factored in, as well. Of course, the cable distance back to the LEC must not exceed the supported distance for the switch port.

Figure 12-7
Co-location of DSLAM
for ADSL Services

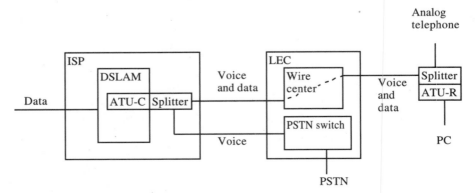

In addition, connector types must match, and so on. All in all, a less than ideal situation.

The whole point of the ISP supporting ADSL, however, is to provide faster Internet access, not voice service, and many people already have second lines used almost exclusively for Internet access, so maybe it is not essential to support the analog voice on the ADSL link at all. In fact, in most cases, all the ISP wants to buy is the actual copper wire and nothing else—not access to the LEC switch, not access to the customer's main telephone, nothing but a copper wire pair from wire center to residence.

Unfortunately, this is sometimes easier said than done. The competitors that are usually allowed to co-locate, virtually or otherwise, typically have to be certified by the state they are located in, which can be a complex and expensive process. Also, the LECs have never totally *unbundled* the local loop from a number of other things that must be bought along with the wire. This is not universally true, but typically is. For instance, the philosophy is that if a competitor wants access to the local loop, the competition is going to offer voice services also, and so the competitor must also buy access to the LEC switch port, and access to directory services, and maybe operator services. And what about reports on usage and billing services? In a totally voice circuit network, all of these things make sense. This is the whole *bundling* approach: sell competitors what they need to compete. When the situation concerns packets and not circuits, however, *unbundling* makes even more sense. Let competitors pick and choose what to buy in almost a totally *a la carte* fashion; this is sometimes done, but rarely. Bundling is more typical, but it is important to note that allowing *any* competitive access to the PSTN is technically known as *unbundling*. The actual issue here is one of degree and extent. Already the topic of court cases, the basic philosophy of TA96 was that if the LEC demonstrates enough competition in local services, then the LEC is permitted to offer long distance services.

So what is an ISP to do if faced with unreasonable (from their perspective) unbundling charges to access the local loop? Why not put in a totally new line and use it exclusively for ISP access? That way, there is no consideration of analog voice support at all, but what about the unbundling? Easy. Just buy a circuit that is tariffed to be the minimal type of service on a pair of wires in the first place. Alarm circuits!

This concept needs some background to explain. Most business burglar alarms are run over simple local loops. No switching is needed, nor dial tone, nor directory assistance, and so forth. All that is needed is a low-quality pair of wires from customer to burglar alarm company. When something trips the alarm at the customer premises, a signal is sent over the wire and triggers something at the alarm company's premises, such as a siren and a flashing red light on a console. The alarm service may be a private security firm or the local police. No matter, the concept is the same.

Alarm circuits can be very low-quality. No voice bandwidth is needed, although it is usually there. The alarm circuit can be a point-to-point leased line, although some have dial tone on them. In any case, the point is that the alarm circuit is functional, inexpensive (about $25 per month), and usually carries none of the "extra" unbundling features the ISP could do without.

Several ISPs have tried this approach already. Order a bunch of alarm circuits for their customers, have them terminated at the ISP site, give ATU-Rs to each user, and away they go. However, the LECs have recently begun to fight back over this use of what are usually called *local-area data service* (LADS) lines. Other names include *dry pairs*, *burglar alarm wire*, *modified metallic circuit with no ring down generators*, and *voice grade 36 circuits*.

The ISPs argue the restricted local loop access and unwanted unbundling features make this method the only efficient way to get ADSL services to their customers. The LEC argue that this use of the lines circumvents regulations and floods lines intended for low bit rates with high speed data. Some LECs have even stopped selling LADS lines to ISPs.

Both sides are playing tough. The ISPs threaten legal action over "anticompetitive" practices when LECs claim there are no more wire pairs available to a given ISP location. Such a claim on the part of a LEC is a little strange when dial-up service is replaced with ADSL, but the recently terminated dial-up service pairs seem to mysteriously disappear. The LECs are correct in saying that they do not have tariff approval to offer these lines for ADSL, and it is also true that running ADSL signals in cable bundles can disrupt other services in the same bundle. This usually happens not when the ADSL signals in the downstream direction are running from the same origin to the same destination, but just the opposite. In other words, if ADSL "downstream" from the PSTN switch is opposed to an ADSL "downstream" from the ISP is the same bundle, the crosstalk and interference might knock out both lines.

This should not happen often. Few complaints have been made about the use of ADSL over LADS lines and other burglar alarm-type services. But the LECs are both concerned and nervous. Some of the LECs have even been accused of placing or keeping loading coils on these circuits to prevent their use for ADSL. Whether LEC opposition to ADSL on alarm circuits is out of concern for the network or concern for potential competition to the LEC's own ADSL plans, there is another way that ADSL services can be provided without LEC involvement.

The biggest problem with the alarm circuit approach is that it does not scale well. That is, it is impractical and uneconomic for widespread services. The ISP is still paying retail and not wholesale costs for the lines, for instance. This will therefore remain only a niche ADSL market.

The "grass roots" approach is really just another variation on the alarm circuit theme. In this scenario, local users themselves get together to purchase access to their local loops. The same unbundling issues apply, but with significant differences. Some state regulators have been more sympathetic to users buying their own loops than LEC competitors buying access to the loops. The LECs themselves seem less concerned about citizens buying the loops than competitive LECs or ISPs. And even in cases where certification is required, the process for such "community groups" in often quick and inexpensive.

Some states have set initial costs for direct user access to the local loop very low, even lower than the alarm circuit price of about $25 per month. The LECs are claiming that such low rates are below the costs of providing the loop, but there is little hard data to go on, and so the same battle lines exist, but often on a much smaller and less intense scale.

In the *grass roots* approach, users serviced by the same LEC switch organize and approach the LEC about buying access to their local loop—or even buying them outright. They can then rent office space according to the virtual co-location rules, site their DSLAM as before, and tie the DSLAM into a recognized ISP. Alternatively, then can become their own ISP themselves. In this case, the voice trunks back to the PSTN switch might be absolutely necessary if continued voice support is needed. Perhaps in the near future the Internet will supply all such needs on its own.

In any case, the grass roots approach appeals to the pockets of highly technical users around the country. Most already have their own LANs at home. They do not hesitate to use new technology to help their situation. Usually, they want faster access not only to the Internet and Web, but to their own corporate Intranet when telecommuting. In fact, these groups do not even need a router in many cases. A simple ADSL bridge between home LAN and company LAN is all they want or need.

DSLAMS and SONET Rings

Most service providers interested in xDSL services are local telephone companies or LECs. After all, the LECs have control of the analog local loops that ADSL enhances for high-speed digital access.

Many service providers also have another distinctive technology available to them, the *Synchronous Optical Network* (SONET) ring, which transfers standard transmission frames on fiber optic cable rings. Outside of the United States, these are more properly called *Synchronous Digital Hierarchy* (SDH) rings. Whenever a service provider mentions the "backbone" links in their network today, this usually refers to the SONET/SDH rings. The rings help to prevent service outages because if a signal traveling one way around the ring is interrupted by a cable break or failure, which is the cause of 75 percent of service outages in the 1990s, the signals re-route themselves the other way around the ring. This rerouting occurs in about 60 milliseconds, which is faster than the blink of a eye (about 100 milliseconds). SONET/SDH rings also offer enormous fiber bandwidths for aggregate traffic between switching offices, well into the multiple gigabit-per-second ranges (a Gbps equals 1,000 Mbps). Perhaps there is a way to interface ADSL—or any xDSL technology—with SONET/SDH rings, and so maximize the benefits of both. In fact, such a method is shown is Figure 12-8.

The central feature in the figure is the SONET ring itself, which may span many miles and link switching offices to special *remote digital terminal* (RDT, or just RT) devices as part of a *digital loop carrier* (DLC) architecture, serving what are called *carrier serving areas* (CSA). CSAs are typically employed in office parks and housing developments to minimize the need to run several point-to-point wires over several miles to a serving office. Of course, the whole CSA concept and terminology is also employed in all local loop situations today, whether an RDT is present or not. The ring provides protection switching, as well. The RDT may be a small cabinet of modest size and cost.

Figure 12-8
xDSL and SONET
Rings

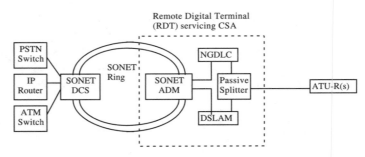

ADM: Add-drop Multiplexer

CSA: Carrier Serving Area

DCS: Digital Cross-connect System

NGDLC: Next Generation Digital Loop Carrier

Inside the RDT, a SONET *add/drop multiplexer* (ADM) interfaces with the ring. Although the figure shows only one such device, there would in reality be many RDTs on a ring. Note the location of the DSLAM in this architecture: inside the RDT. What is it doing there, of all places?

This usage of the DSLAM is related to the use of a *Remote Access Multiplexer* (RAM) outlined in the general DSLAM architecture discussed at the beginning of this chapter. To understand the relationship between a pure RAM and distributed DSLAM (which is what is shown in the figure), consider that in most DSLAM representations, the ADSL services are basically "home-run" arrangements. That is, the loops connect directly from the home to the local exchange. In a DLC architecture, however, which is by some accounts now nearly 30 percent of the total installed access lines in the United States, the end of the ADSL service is not in the local exchange office, but in the DLC RDT itself. A RAM can multiplex various loops to a DSLAM. A distributed RAM in this architecture actually *is* the DSLAM, as well. Usually, a RAM handles up to 100 lines or so. A distributed DSLAM can do whatever a local exchange-based DSLAM can handle, maybe up to 1,000 lines or so.

As shown in the figure, a passive splitter is located at the end of the local loop. Now, the analog voice still enters the "NGDLC" on the SONET rings, but the ADSL digital information now passes through the distributed DSLAM onto the SONET ring on its own. The SONET ring contains enough bandwidth to accommodate both numerous analog voice channels as DS-0s running at 64 kbps (the digitization is a function of the NGDLC) and just as many ADSL channels running at higher and asymmetrical speeds. No matter, the flexible configuration capabilities of SONET rings allows for this.

Inside the RDT, along with the familiar DSL access multiplexer, is a device called a *Next Generation Digital Loop Carrier* (NGDLC). Simply put, a NGDLC is a DLC optimized for use on a SONET ring. A DLC is frequently used today to support many analog local loops in a CSA, but is not appropriate for ADSL in many cases. The NGDLC will have more flexible and remote configuration capabilities, and have the traffic carrying capacity to tolerate the presence of ADSL on the loops. Most important of all, the NGDLC will have an interface designed for SONET ring connectivity. A passive splitter combines analog and digital signals to and from the ATU-Rs. All analog voice is digitized by the NGDLC for transport on the SONET ring.

In the switching office, a SONET *digital cross-connect system* (DCS) splits off the digitized voice traffic from the ADSL traffic. Voice still goes into the PSTN switch, and other traffic still finds its way into an IP router or ATM switch. In fact, the switches and routers could be located almost anywhere on the SONET ring where a DCS is employed, even in a competitor's service location.

This use of ADSL and SONET rings gives maximum service outage protection and carries enough traffic that thousands of ADSL downstream traffic flows could be supported at once. However, it may be a while before such architectures become common.

Issues still to be resolved are the following: Is the RDT fed by a SONET ring now, or is it just a point-to-point link? Is there space in the remote cabinet or area to hold the remote DSLAM? Will enough customers want the ADSL (or other DSL) service to justify the remote DSLAM expense?

A related issue is the relationship between the remote DSLAM or RAM and the DSLAM in the local exchange. A pure RAM, which just aggregates xDSL traffic, might still need a DSLAM back at the local exchange. A true remote DSLAM, which is still sometimes called a RAM, might be able to dispense with this central DSLAM altogether, especially if the remote DSLAM has some switching or routing capabilities (or both) built in or added on.

A remote DSLAM can be made small enough today to fit into almost any DLC site. There are DLCs in small, pale green cabinets or huts (something called a *controlled environment vault* (CEV)) in urban areas or even in the basement of an office building. A key issue here is the availability of power, battery backup, heat dissipation, and other mundane operational factors. Most DLC sites have power and battery backup already. Space in a RDT cabinet is another concern, since they were never designed or intended to house extra equipment like ADSL.

The penetration of DLC and CSA arrangements has increased over the past 15 years or so. A lot of this occurred when the CSA rules were introduced in 1983. By limiting new loops to 12 kft with 22 or 24 gauge wire, and 9 kft with 26 gauge wire, and forbidding loading coils, the CSA rules virtually guaranteed the need for DLCs for new shopping malls, condominium and town house units, and office parks. Under CSA rules, bridged taps were allowed, but were limited to 2,500 feet total, with no single bridged tap exceeding 2,000 feet. The older rules continued to allow loading coils, of course, but with restricted distances that mandated new local exchanges if DLCs were not used with CSAs.

The CSA system has become quite popular around the world, and as DLC systems increase and SONET rings multiple, the remote DSLAM and NGDLC arrangement will become even more popular.

13

DSL Migration Scenarios

As has been pointed out numerous times in this book, DSL comes in many flavors. The emphasis for the last few chapters has been on ADSL, but that does not mean that ADSL is the end of user and service provider migration strategies. All of the various forms, such as HDSL (High-bit-rate), SDSL (Symmetric), MDSL (Moderate Speed), and IDSL (ISDN), could play a role in the total xDSL picture. However, it is undeniably true that ADSL (Asymmetric) and VDSL (Very High-speed) seem to occupy most of the interest of service providers and technical writers. This is no accident. Most equipment vendors and service providers have focused on ADSL and VDSL for two reasons: First, ADSL appears to offer the broadest base of near-term services (fast Internet access, for example) to the widest possible range of local loop conditions. Second, VDSL appears to offer the bandwidth needed for almost any broadband service. If this is the case, then not only is ADSL the most promising member of the xDSL family, but VDSL might well be the future of ADSL.

In fact, there are a number of possible migration paths that a customer might be offered and a service provider might propose to furnish ADSL and VDSL. The ADSL Forum has spent a lot of time and effort on this topic and mention should be made of their work. This chapter relies heavily on documentation from the ADSL Forum. Along these lines, several DSL migration paths have been outlined, as listed below.

- **From Analog Modems to ADSL:** Customers using analog modems migrate to simple ADSL services, such as faster Internet access.

- **From DLC to ADSL:** Customers currently serviced by a *digital loop carrier* (DLC) feeder (not necessarily fiber-based) migrate to simple ADSL services such as faster Internet access.

- **From ISDN to ADSL:** Customers currently using ISDN services for Internet access migrate to simple ADSL services, such as faster Internet access.

- **From ADSL to Next Generation DLC (NGDLC):** Customers currently using ADSL, in some form, migrate to next generation DLC systems that can provide a wider range of services in a more cost-effective manner.

- **From ADSL to VDSL:** Customers currently using ADSL (or NGDSL-based ADSL) migrate to VDSL, which offers the highest speeds and widest range of broadband services.

Many more migration paths than those listed here exist, but these are the ones most likely to be offered to the vast majority of customers and are probably the most important. Each migration scenario has its own set of issues, benefits, and hazards for customers and service providers alike.

DSL Migration Architecture

A number of possible migration paths have been outlined to furnish ADSL and eventually VDSL. Not all of them are discussed in this chapter, but the main and most important ones are all treated in detail. These DSL migration scenarios presented above are shown in Figure 13-1 as a number of different paths through the total DSL migration architecture. Many service providers and equipment vendors envision migration from an analog plain old telephone service world to ADSL, and then eventually on to a VDSL network.

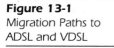

Figure 13-1
Migration Paths to
ADSL and VDSL

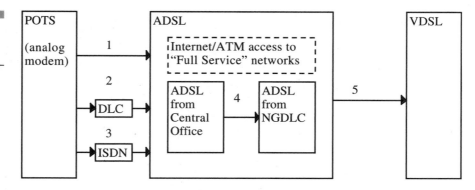

xDSL migration scenarios can include technologies such as HDSL, HDSL2, and others; however, these are the xDSL technologies and migration strategies that the equipment vendors and service providers have emphasized most often to supply ADSL and VDSL services. The xDSL migration scenarios shown in the figure do not mention HDSL2 or any other xDSL technology. However, almost any of the three paths shown in the figure could include other xDSL family members. For example, the ISDN path could potentially include IDSL, and digital loop carriers could employ HDSL or HDSL2. The key idea is that the migration will probably move from an analog POTS world to ADSL services, and then ultimately on to a full VDSL network.

More information on the five paths indicated in the figure are listed below.

1. **From Analog Modems to ADSL:** The nice thing about customers using analog modems is that they are fairly familiar with the technology. Much of the ADSL marketing strategy emphasizes ADSL devices as just a new type of modem. Because most early ADSL services are expected to be nothing than faster Internet access, there is a natural attraction and real incentive to perform this migration.

2. **From DLC to ADSL:** ADSL makes maximum use of twisted-pair copper wire, but not everyone is at the end of a single pair of copper wires. It has been mentioned several times that more customers are serviced by a DLC feeder (not necessarily fiber-based), especially when following the guidelines from the CSA concept. These users must be able to migrate to simple ADSL services such as faster Internet access as well.

3. **From ISDN to ADSL:** It came as a shock to some telephone companies that customers wanted ISDN not to make telephone calls or to access telephone company switch-based ISDN services, such as they were. Even when ISDN switches were used for simple point-to-point links at 64 or 128 kbps for Internet access, the link still looks like ISDN more than anything else. So customers currently using ISDN services for no more than faster Internet access need a migration path to simple ADSL services that can provide faster Internet access with much less stress and strain on the voice network.

4. **From ADSL to Next Generation DLC (NGDLC):** After customers in sufficient numbers get on ADSL, and the whole CSA concept becomes even more prevalent than it currently is, it might be a good idea to migrate to a more distributed architecture. Now ADSL could be provided on SONET rings, DSLAMs could be distributed, as well, and the whole process of furnishing ADSL services would be more efficient. In this scenario, ADSL is a prerequisite, but customers currently using ADSL in some form who are migrated to next generation DLC systems can be provided with a wider range of services.

5. **From ADSL to VDSL:** ADSL is quite fast, no doubt about it. But ADSL will still struggle to supply the bandwidths needed to support true broadband applications such as high-definition television and the like. Customers currently using ADSL, either on home-run local loops or as NGDLC-based ADSL, will be migrated to VDSL at some point in the future. VDSL, of course, offers the highest speeds to support the widest range of broadband services.

Note that some of the paths involve conscious decisions on the part of customers. For instance, customers must choose when and if the transition from analog modem to ADSL should be made, but other migration paths are totally independent of and transparent to the user. Any decision on the part of the service provider to furnish ADSL services on DLC systems, for example, is not only invisible to the customer, but totally out of the user's hands. Always remember that other possible migration paths exist, but are not shown in the figure.

This last point is especially true within the ADSL network itself. Note that within the ADSL network itself, the migration from simple, LE-based ADSL to providing ADSL services from NGDLC systems is essentially transparent to the users and requires no changes, or even awareness, on the customer's part.

The migration architecture also points out that the intent of ADSL is to provide either Internet (that is, Transport Control Protocol/Internet Protocol, or TCP/IP) or ATM network access to a *full*-service network that provides customers with all the services they need. Whether this network is the Internet with more broadband capability or an ATM network with realistic broadband services is not essential to ADSL itself. ADSL will perform in either situation.

Migration Urgencies and Priorities

More is needed than just an outline of possible migration paths to furnish ADSL and VDSL. All of the migration paths, even those not considered in this survey, can be characterized in terms of a *starting network* and a *target network* scenario. For instance, the first migration path listed, from analog modem to ADSL, could have "analog modem Internet access" as the starting network and "ADSL for Internet access" as the target network.

The five major migration paths considered in this chapter are shown in Table 13-1. Most of the information presented here is based on ADSL Forum documentation. Keep in mind that the starting and target networks shown in the table are not all that exist, but these seem to be the ones on which most service providers and equipment vendors are currently concentrating. Each migration scenario is assigned an *urgency* and *priority* based on service provider and equipment vendor assessments of just how crucial each one is for the success of ADSL. The other migration scenarios are mainly concerned with variations of these major themes, and none of those omitted can be considered to have immediate urgency and high priorities.

Table 13-1

ADSL migration paths

Starting Network	Target Network	Urgency	Priority
Analog modem Internet access	ADSL for Internet access	Immediate	High
Analog Modem on DCL	ADSL delivered service(s)	Immediate	Medium
ISDN for Internet access	ADSL for Internet access	Immediate	High
ADSL from Local exchange	ADSL from NGDLC	Future	High
ADSL delivered service(s)	VDSL delivered service(s)	Future	High

Each migration scenario has been assigned an urgency and priority for standardization and implementation. These are detailed as follows, using the starting and target terminology instead of numbered transition paths:

- **From analog modem Internet access to ADSL for Internet access:** The urgency is immediate, the priority is high. The pressure placed on the PSTN switches and trunks from packets-on-circuits Internet access make this a critical migration scenario.

- **From analog modem on digital loop carrier (DLC) to ADSL delivered service(s):** The urgency is immediate, the priority is medium. The special challenges of DLC systems (e.g., limited bandwidth, remote configuration concerns, and so forth) for ADSL make this an important item; the priority is only medium, though, given the current distribution of DLC systems. This priority may increase in the future.

- **From ISDN for Internet access to ADSL for Internet access:** The urgency is immediate, the priority is high. The concerns are basically the same as those for analog modem Internet access.

- **From ADSL in the local exchange (LE) to ADSL in the Next Generation DLC equipment:** The urgency is future, the priority is high. Next Generation DLC systems will not be constrained by bandwidth and may be integrated with SONET rings and even ATM switches. This will further decrease traffic aggregation woes at the LE. However, NGDLC systems are not in the immediate future of ADSL.

- **From ADSL delivered service(s) to VDSL delivered services:** The urgency is future, the priority is high. Most see VDSL as a natural evolutionary path for ADSL, but probably only until NGDLC fiber-based systems become common.

Each of these starting and target network migration scenarios has its own set of pros and cons for both the customers and service providers. These are considered in the sections that follow.

From Analog Modem to ADSL

The most common migration scenario for customers and service providers venturing into the wonderful world of ADSL is most likely to be from a starting network scenario of limited Internet access speeds with 56k or slower analog modems, although 56k is already said sometimes to be "half way to ADSL," due to the presence of digital downstream signals and the fact that 56k modems are still 33.6 kbps upstream. The target network would be to higher Internet access speeds with ADSL. Of course, the limited analog modem speeds offer a ready market for ADSL services in the first place, which should explain the anticipated popularity of this migration scenario.

The whole point would be to speed up current long waits for Web graphics and files. It is not unusual today to wait 30 seconds or longer for a full Web page of writhing graphics. File downloads at 33.6 kbps can take about six minutes per megabyte on a good day and time, and twice as long during congested intervals. The shorter wait is the greatest benefit to ADSL Internet users.

The greatest benefit to the service provider is that it moves long holding time Web sessions off of the voice network, and this is true even if the service provider is not a telephone company. After all, even an ISP has to use the PSTN to reach their customers at the end of the local loop in the first place, and so some ISPs have proved to be avid early adapters of ADSL. Of course, some ISPs have accused the LECs of blocking their attempts to deploy ADSL for the ISP's customers because the LECs plan to offer their own ADSL services at some point. However, many of the delays with early ADSL deployments can be traced to the simple growing pains of any new technology, and not necessarily the sinister actions of incumbent service providers.

If the service provider happens to be a telephone company, an added benefit of this ADSL migration scenario is that it does not require expensive ISDN upgrades to PSTN switches, which can cost about $500,000 dollars, not to mention the need to coordinate and manage the upgrade process.

This migration plan also gives the telcos an answer to the possibility that competing cable modem services from the cable TV companies could become popular. This is not the place to rehash cable modems, but it should be pointed out that much ADSL activity seems to be a direct response to early cable modem trials and developments. Ironically, many telephone companies considered buying and even did buy cable TV companies to deliver higher bandwidth services than the copper loop could deliver, but many telephone companies that have experimented with becoming cable TV companies, and even offering voice services on these cable networks, seem to be abandoning their early efforts. It seems that customers like watching TV on cable, but talking and accessing the Internet through a separate telephone network.

Finally, all of this makes even better sense were more telephone companies to become ISPs, as seems to be the case in the current deregulated environment in the United States.

From DLC to ADSL

Some 15 percent of all local loops are currently serviced with some kind of pairgain system, which multiplex many telephone customers onto a single carrier system (thus, gaining pairs that can be used for other purposes). The vast majority of these pairgain systems are *digital loop carrier* (DLC) systems, as were detailed in a previous chapter. DLCs pose problems for even simple ADSL deployment scenarios.

Most troublesome is the fact that current DLCs will not pass 1.1 MHz ADSL bandwidths, so the service provider end of the ADSL loop is not at the local exchange, but at the DLC remote digital terminal because the most common DLC systems, such as SLC-96 (*Subscriber Loop Carrier* carrying 96 analog signals on a group of four T1s), simply digitize the analog voice in the 4 kHz baseband range and filter out the other frequencies. The voice passband is digitized at 64 kbps, essentially limiting any ADSL deployed at the end of a DLC system to this speed downstream (and upstream as well, but the downstream speed in ADSL is much more critical, of course).

However, in such DLC-to-ADSL migration scenarios, it might be possible to derive a 64 kbps channel from the total ADSL bandwidth, which means that the analog voice passband is digitized at the DLC and can still enter the DLC equipment as before. In this case, the ADSL signals above and beyond the analog voice channel must find their way to the local exchange some other way, perhaps on new wire pairs installed expressly for this purpose. This would preserve the DLC line for voice service, but calls into question the entire ADSL strategy of employing existing copper pairs.

Alternatively, it might be possible to shift non-ADSL customers onto the DLC. In other words, when a customer serviced on a DLC system wants to use ADSL services, a nearby customer employing a simple analog loop can be shifted onto the DLC, freeing up this loop for easy ADSL deployment. Naturally, the feasibility of this would depend of the DLC/straight loop service mix, resource commitment on the part of the service provider, and several other factors.

Another possibility is to provide another local loop pair for ADSL alone and forget about analog voice support altogether. This should not be a serious liability because most people would be using ADSL for Internet access and not to make telephone calls, just as they use ISDN today.

Of course, the most likely way will be to eventually deploy a DSLAM at the RT and backhaul the traffic on SONET/SDH fiber optic cables.

From ISDN to ADSL

The limitations of analog modems when used almost exclusively for Internet access have been apparent to users for years. The situation was even worse several years ago when the fastest analog modems in common use were only 14.4 kbps. As recently as 1992, a Hayes external 9600 bps modem cost $799. Throughout the 1990s, therefore, users who have been especially sensitive to analog modem limitations for Internet access (Internet site designers and other networking personnel working from home) have employed ISDN lines where available in an attempt to speed up their Internet access.

ADSL provides much higher speeds than the 128 kbps (achieved by bonding the two 64 kbps bearer channels) that can be used over the residential and small business 2B+D *Basic Rate Interface* (BRI) of ISDN. Also, using ISDN almost exclusively for Internet access unnecessarily ties up switches and trunks for long periods of time when the line is inactive. Also, because ISDN is a switched service, a call setup delay occurs when accessing an ISP with ISDN. Newer versions of ISDN, such as "always on" ISDN, seek to get around this problem, more or less successfully.

The real problem with ISDN and ADSL is that ISDN is a Digital Subscriber Line (DSL) in its own right. ISDN lines might already support analog telephones at the premises through ISDN *terminal adapters* (TA). In addition, some telephones could be compliant ISDN devices, and more exotic ISDN terminals might be present for services such as videoconferencing. Any migration from ISDN to ADSL must address all of these possibilities. One possible "merging" of ISDN and ADSL is shown in Figure 13-2.

As shown in the figure, it might be possible to carry ISDN BRI service and analog POTS under ADSL. However, this seems to be asking a lot from supposedly inexpensive ADSL equipment. In this way, both POTS

Figure 13-2
ISDN and ADSL

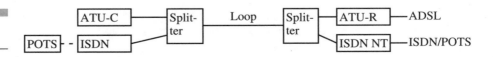

and ISDN bandwidths are located somewhere under the starting ADSL frequency of 25 kHz. Furthermore, both POTS and 2B1Q are essentially baseband signals (analog baseband for POTS and digital baseband for 2B1Q, but baseband nonetheless).

Perhaps it could be possible to embed ISDN inside some range of the total 1.1 MHz ADSL bandwidth. That way, analog phones would still be handled by normal splitter arrangements, while the 2B1Q signals traveled elsewhere. In this arrangement, the option is to embed the 2B1Q somewhere "inside" the 25 kHZ to 1.1 MHz range and modulate the 2B1Q exactly as if it were just another digital input. Neither option is ideal, but some consideration must be given to ISDN users in an ADSL scenario.

Regardless of the approach considered, all ISDN-to-ADSL network scenarios must address ISDN standard issues. ISDN has been an international standard for years. Many ISDN documents specify that some interfaces operate, be provisioned, and be administered in a certain standard fashion. Therefore, violations to the ISDN standards in the name of ADSL should be approached with caution.

There is an even more important issue with regard to a migration from ISDN to ADSL. ISDN is a standard that defines more than DSL access to an ISDN switch. ISDN defines services and applications that can be accessed over the ISDN DSL. ISDN video servers and videoconferencing services are in place and in use already. Now replace an ISDN BRI line with ADSL. How do ISDN applications make the leap to ADSL?

A few variations on the whole ISDN and ADSL theme should be mentioned. These issues come up because both ISDN and ADSL are forms of DSLs, as was mentioned previously. One of the variations of the xDSL family mentioned earlier in this book was something called IDSL. This form of "ISDN over ADSL" is not really the same as the migration from previously installed ISDN to ADSL considered here. In the case of IDSL, there is no ISDN switch to contend with. The ISDN link runs at 128 kbps (usually), and the D-channel is essentially ignored (again, usually). IDSL allows the continued use of this speed and structure through the DSLAM by using the proper termination units at each end.

In many countries, especially outside of the United States, it is also common to deliver two analog telephone lines over one copper pair. This is done by digitizing the voice signals and transporting each individual conversation by means of a 64 kbps digital channel.

This concept is known as *two channel pairgain* or digitally added main line (DAML). Manufacturers of this equipment were quick to notice that existing ISDN DSL chips offered an economic means of carrying the two 64 kbps channels over the copper pair. A DAML system will therefore typcially utilize ISDN transmission chips, and so the signals on the wire appear identical to an ISDN signal. However, the service is quite different from ISDN.

One ADSL equipment vendor already supports ADSL with DAML "underneath" the ADSL instead of analog voice. It is, therefore, possible to deliver 2B1Q (two binary, one quaternary) coded ISDN-like service "under" ADSL.

Inside ADSL: Using IP for "Full Services"

ADSL is more than just a new way to digitize the analog local loop. As should be readily apparent, ADSL is part of a mostly complete network architecture. ADSL may be used and is indeed intended to be used to access a wide array of broadband, multimedia, and graphical services. Several of these services, which mostly remain potential services for ADSL, have been relatively routine for many ISDN users for years. However, if ADSL is to be a "better ISDN than ISDN," then what happens to the services?

It is undoubtedly true that most users will employ ADSL initially for fast Internet access, and probably be happy with that. Sooner or later, though, users will want to do more, especially if those users have previously employed, or been exposed to, a complete range of ISDN services. Perhaps there will soon be a need to provide a full service network with ADSL, not just faster Internet access.

However, it is also true that TCP/IP-based servers are the most common in the world. Many of these already serve up video and other broadband services (although in a limited capacity, to be sure). Nevertheless, most ADSL users will be running TCP/IP as client for their initial Internet access service.

Perhaps someday soon all broadband services delivered over ADSL networks should also be accessible through TCP/IP. These services would ultimately include video-on-demand, broadcast TV services, and so on; they would be served up from TCP/IP-based servers, maybe still on the Internet, but not exclusively. For reliability and logistical reasons, the servers could just as well be in the service provider's local exchange location.

Of course, this move from Internet access to broadband services would require massive changes to existing TCP/IP protocols. The fact is, and nobody denies it, the current TCP/IP Internet protocol stack is woefully inadequate for delivering more demanding voice and video bit streams.

Some changes are being made, but others will take years to complete. It is one thing to say that TCP/IP should have quality-of-service parameters, have low and stable delays, and be as robust as it is now, but it is quite another thing to figure out how to do it.

The move to a TCP/IP and Internet-based full-service broadband network would require massive changes to existing Internet structure, as well. The Internet backbone, as it is currently constructed, leaves a lot to be desired for handling massive flows of delay-sensitive and error-sensitive information. Detailing all the limitations cannot be done here, but a good example is the common "site not reachable" message that result from congestion rather than a server being down. This obviously will not do for 911 voice services or even for routine television watching. Some changes are being made in this area, but much work remains. Furthermore, it is not yet clear who gets to build, pay for, and run this Internet of the future.

Using ATM for "Full Service" ADSL

Because ADSL will most likely be used for Internet access initially, it might be beneficial to try to deliver broadband services exclusively with TCP/IP. This approach, which was strongly suggested previously, would require massive changes to the existing TCP/IP protocols and underlying Internet infrastructure. In fairness, many of these changes are underway, but they will take years to complete. Early versions of "reservation" protocols on the Internet have left a lot to be desired.

An alternative to IP-based ADSL "full-service" networks is *asynchronous transfer mode* technology. ATM seems to be a better fit for "full-service" deployments, especially because ATM is part of the full *Broadband ISDN* (B-ISDN) protocol stack, which outlines the form that most of these broadband services should take. So ATM is specifically designed for broadband services, not just a data delivery protocol with some voice and video features added on.

Internet proponents have started saying things along the lines of "with a lot of hard work and effort, TCP/IP could be as good as ATM is." This statement is startling for a number of reasons, but the aspect that startles when it comes to ATM and ADSL is the implication in the statement that ATM *today* is as good as TCP/IP *might be* tomorrow—but only after a lot of hard work and effort expended on TCP/IP. Why bother? Why not just use ATM instead of TCP/IP in ADSL (or any xDSL) and be done with it?

Things are never that simple. The fact is that there is no ATM hardware or software on the vast majority of customer premises. The network protocol on the premises is most commonly TCP/IP from the Internet protocol suite, now bundled inside the majority of new PCs. The common argument is that the protocol of choice for ADSL must be TCP/IP, and maybe this will be true forever.

The situation, however, is not totally bleak for ATM on xDSL supporters. IP packets can be carried and switched inside a stream of ATM cells. In fact, this is one of the allowed ADSL transport modes of operation, as shown in an earlier chapter. If there is more than one TCP/IP device or PC on the premises, then the ATM Forum's *LAN emulation* (LANE) might be needed to shuttle the IP packets from user client to service provider server effectively. The ATM Forum also has a specification known as *multiprotocol over ATM* (MPOA), which essentially makes a multiprotocol router out of an ATM switch. Such a switch can route not just IP, but other protocols.

If IP is the only protocol of concern, then a group of protocols known collectively as *IP switching* can be used very effectively. Although implementations of IP switching have been proprietary to date, at least six major router and ATM switch vendors have released their own versions of IP switching software, which have generated a great deal of interest.

Be that as it may, the use of ATM as the basis for "full-service" network deployment has been seen as the only viable, long-term broadband solution. This is due mostly to ATM's roots in the B-ISDN service arena.

Moreover, using ATM for this purpose requires only relatively minor changes to B-ISDN/ATM standards. Furthermore, many of these would merely involve redefining the speeds and media at which B-ISDN services and ATM networks are allowed to function.

It is worth noting that some of the architectures for implementing such full-service networks start with IP routers and others start with ATM switches. Consider the ATM switch first. ATM cells flow in and out on all ports. Now add LANE and MPOA (LANE is a kind of prerequisite for MPOA). The result is a device that sends and receives ATM cells and routes IP and other protocols. If only IP matters, add only IP switching.

Now start over again with an IP router. Add ATM adapters for input and output ports, mainly for speed and delay reasons. Now add LANE and MPOA, as before. The addition can still be made because both LANE and MPOA are simply standard documents that can be implemented in almost any programming language on almost any hardware platform. The result is a device that sends and receives ATM cells and routes IP and other protocols (maybe), but this is almost exactly what the ATM switch became! So which is it, an IP router with ATM switching, or an ATM switch with IP routing? It does not matter because the result is the same.

Perhaps the apparent collision course between IP and ATM, Internet and telephone company, and router vendor and switch vendor is not an impending disaster for either. Perhaps they are *both* needed and wanted in all types of networks.

From ADSL to NGDLC

Whether the PSTN is used with analog modems or ISDN, the transport of packets on circuits makes congestion a problem. The problem actually runs deeper than this: At the root of the entire switch and trunk congestion issue is the very *local exchange* (LE) concept upon which the PSTN is founded. Although the official term is "local exchange," which is what ISDN and B-ISDN documentation calls them, these switching offices are almost always called *central offices* (COs) in the United States. Local exchanges or central offices, by definition, aggregate traffic for switching and trunking purposes. Even if the local exchange is optimized for packets, if there are enough packets, congestion will result. After all, the Internet is optimized for packets, and it gets congested all the time, as telephone companies with lots of circuits are fond of pointing out.

Local exchanges (or central offices, the point is the same) exist mainly for historical reasons. Aggregating traffic makes sense when equipment is fragile, expensive, or both. Local exchange switches could not simply be plunked down anywhere and left alone. Even today, when PSTN switches are more robust than ever, they remain a huge expense to the service provider. These large, central switches do, however, provide the focal point for access to services.

Today, PSTN switches closely resemble large computers. The advances in modern computing power and methods make distribution, rather than centralization, attractive. Robust modern switches can be made smaller, consume less power, and require less local maintenance than ever before. A local exchange switch used to be one huge device handling 10,000 local loops and hundreds of trunks. Today, the local exchange switch might be a grouping of same units handling 1,000 loops each, linked together by a 100 Mbps Ethernet LAN or fiber ring. Trunks may all emanate from a specialized trunk unit. In addition, the services no longer need to be located right next to the switch if the trunking network is reliable and fast enough.

Obviously, a local exchange switch on a LAN has already taken a big step from centralized to distributed. Why not distribute the services and their servers around the LAN, also? Why stop there? And the possible move from central to distributed access to services works hand-in-hand with the move to more widespread DLC systems. Bellcore estimates 45 percent of all loops will be on DLC systems by the year 2000, due mainly to changing customer living patterns (they live further and further away from obvious local exchange sites) and the falling cost of fiber-based DLC equipment. As was pointed out earlier, most of the newer DLCs will access some form of SONET ring through a small SONET *add / drop multiplexer* (ADM) to support this robust *Next Generation DLC* (NGDLC) equipment.

The next step, to be taken sometime after the year 2000, seems clear. Why not move SONET ADM, NGDLC, and DSLAM technologies to a remote NGDLC site? Some switching functions could be included to avoid aggregating traffic into a funnel, which would lead to a congested local exchange or single-point-of-failure site. After all, how do you call to report that your "full-service" network is now acting empty?

Perhaps even ATM switching and some services will follow. After all, why bring the cells all the way back to the central location just to switch them? Maybe there should be not only *community* Web site servers in the local exchange, but *neighborhood* Web site servers for a few hundred homes, tops. Whether this is feasible and whether this distributed approach occurs quickly or slowly all depends on how quickly local exchanges can be converted to participate in this new, distributed architecture.

From ADSL to VDSL

In the next 10 or 20 years, it may be common for the public network to include fiber feeder systems, as a matter of course, to offer widely used ATM/B-ISDN services and to consist of many SOHO users. Such a network will be common in the near future as end device computing power continues to grow. Naturally, applications will demand more and more bandwidth, as they always have.

The point is that internal network computers and applications must keep up with the demand placed on the network by "external" user computers and applications—this is the reason for VDSL.

Given these simple network growth factors, VDSL seems as rational as the progression of Ethernet from 10 Mbps to 100 Mbps to 1 Gbps. However, where the migration from 10 Mbps to 100 Mbps Ethernet took some ten years and is still not a given, migration from ADSL to VDSL could take place more rapidly.

Furthermore, VDSL speeds align very nicely with ATM/B-ISDN speeds. That is, the top VDSL speed of 51.84 Mbps (although VDSL can run marginally faster) is exactly the same as the OC-1 SONET interface that can be used for ATM networks. However, please note that 51.84 Mbps technically is not a speed supported by B-ISDN itself at present, and so perhaps VDSL could someday be an allowed option for SONET transport in a B-ISDN full-service network scenario.

By now it should be apparent that the relationship between ADSL and VDSL is not quite the same as the relationship between ADSL and other xDSL technologies because of all the alphabet soup of DSLs, only VDSL is one of the ADSL target networks. This only makes sense. It would not be of any benefit to anyone to take existing ADSL links and turn them into HDSL, SDSL, or anything else. It does, however, make sense to convert existing ADSL links and turn them into VDSL. Maybe the time has come to take an even closer look at VDSL itself.

14

Very High-speed Digital Subscriber Line (VDSL)

Very High-speed Digital Subscriber Line offers the highest upstream and downstream rates of any DSL designed to date. The downstream speeds range from about 13 Mbps to 55 Mbps, depending on distance. The upstream speeds start at 1.5 Mbps (where ADSL essentially tops out for longer distances) and end at about 26 Mbps. It should be noted that in some forms, VDSL is symmetrical rather than asymmetrical. Current VDSL schemes use *frequency division multiplexing* (FDM) for line code signals, but echo cancellation may be needed ultimately for symmetrical or very high bit rates.

Figure 14-1 repeats the basics of the VDSL architecture. With one obvious exception, the figure could easily be detailing ADSL instead, but the presence of the *optical network unit* (ONU) shows that this is not the ADSL world at all. Note that if the ONU were a *digital loop carrier* (DLC), plain and simple, then the figure would simply be a variation on the ADSL theme, one that was mentioned previously in the ADSL migration scenarios. However, the ONU is not just a DLC. It is a special DLC, one that must employ fiber-optic cable to the local exchange. And this use of fiber takes the fiber closer to the home than other architectures, an important point. Of course, if the fiber were in a ring configuration, the ONU could also potentially become a NGDLC, but for the purposes of this discussion, the ONU is similar to but not quite a NGDLC. The continued presence of the splitter is to support previously existing analog voice in the newly installed ONU/DLC world. Certainly nothing in VDSL mandates the presence of the analog voice service if digitization has already occurred.

Two issues about VDSL should be addressed: First, how can VDSL offer such high speeds when other DSLs seem to struggle to achieve 1.5 Mbps and (with Asymmetric DSL (ADSL)) a mere fraction of that speed upstream? Second, is all of this blazing speed and its associated expense necessary for faster Internet access?

With regard to the first issue, VDSL is intended for the widespread fiber feeders or DLC systems of the near future. In other words, VDSL starts out by assuming that there will be fewer and fewer straight runs of twisted-pair copper from local exchange to customer premises, and that fiber-based DLC systems will become more prevalent—these are not bad assumptions. Some 15 percent of all local loops are currently serviced by DLC arrangements, as has already been mentioned several times, and a sharp increase is expected due to the combined effects of decreasing fiber costs and increasing condominium, apartment complex, and housing development construction in many parts of the country. With the current

Figure 14-1
The VDSL
Architecture

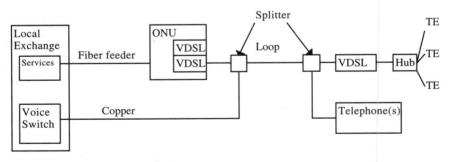

Downstream speeds:	12.96 - 13.8 Mbps	4.5 Kft
	25.92 - 27.6 Mbps	3 Kft
	51.84 - 55.2 Mbps	1 Kft
Upstream speeds:	From 1.5 Mbps to equal to the downstream in some cases.	

CSA rules limiting local loop runs and gauge changes, the only alternative to a world with many DLCs is a world with many small local exchanges every few miles or so. And so VDSL anticipates no more than about 4,500 feet of copper on a loop, with the rest of the loop being fiber.

All of the upstream and downstream VDSL speeds are shown in the figure. Downstream speeds range from 12.96 through 13.8 Mbps at 4,500 feet, to 25.92 through 27.6 Mbps at 3,000 feet, to 51.84 through 55.2 Mbps at 1,000 feet. In the upstream direction, speeds range all the way from a modest 1.5 Mbps to 26 Mbps, and in some cases could be equal to the downstream speed. Keep in mind that this is just a design parameter. Many technical obstacles must be handled before we can expect inexpensive customer premises equipment to function effectively in the 10 Mbps ranges, even at these restricted distances.

Regarding the second issue—the purpose and use of these appreciable VDSL speeds—the intent of VDSL is not for mere Internet access. It is intended for widespread *asynchronous transfer mode* (ATM)/*Broadband ISDN* (B-ISDN) service deployment. Whereas ADSL allows ATM to be used to provide services, for example, VDSL more or less expects ATM to be used for this purpose. These broadband services would include a wide range of multimedia and video services, including broadcast TV services. The key is that when used in conjunction with current compression technologies, which will play an important role in VDSL specifications, VDSL might remove loop bandwidth restrictions once and for all.

Also, VDSL is intended for widespread small office/home office (SOHO) premises networks. As housing spreads into the "exurbs" and formerly rural areas, the commuting pressure puts added stress on the highway infrastructure and adds to concerns about air pollution, energy consumption, and so on. Perhaps the best way to deal with this issue is to tell everyone to stay home, but still go to work.

All in all, VDSL is designed and intended for the network of the next 10-20 years in many countries around the world. Oddly, countries with less developed copper infrastructures than in the United States may turn out to enjoy an edge when it comes to VDSL.

VDSL and ATM

VDSL is intended to take advantage of the presence of fiber in modern pairgain systems, the need for more and more bandwidth for "converged" residential services, and the promise of ATM cell transport as a unifying transmission method for mixed-media broadband services. The whole package is neatly tied together. What keeps it from falling apart? Each component is a vital part of the whole. If one is weak, it weakens the whole basis of VDSL.

On first glance, the "weak link" in this plan may seem to be asynchronous transfer mode. ATM has struggled to gain acceptance in all but a few well-defined markets, and depending on ATM to popularize VDSL might seem like a mistake. This is not all that VDSL offers, however.

VDSL documentation defines five main transport modes across the basic copper/fiber VDSL architecture. Most of these closely mirror their ADSL cousins. These VDSL transport modes are shown in Figure 14-2.

In its simplest form, VDSL offers *synchronous transfer mode* (STM), or what most people would just call *time division multiplexing* (TDM). This is the same type of "bit pipe" that bit sync mode in ADSL supports. Here, bit streams to and from different services and devices are assigned fixed bandwidth channels for the duration of the interaction.

VDSL also supports packet mode. In this transport mode, all bit steams to and from different services and devices are organized into individually addressed packets of varying lengths. All packets are sent on the same "channel" of maximum bandwidth. Naturally, the packets would be expected to be IP packets, but IP packets are not tied to VDSL in any way.

Additionally, VDSL supports ATM mode. In fact, ATM can be used in three distinct ways in a VDSL network. ATM mode is similar to packet mode in the sense that each bit unit is individually addressed and sent on the unchannelized line, but in this case the packets form small, fixed-length units called *cells* rather than variable length packets. ATM can be

Figure 14-2
VDSL Transport
Modes

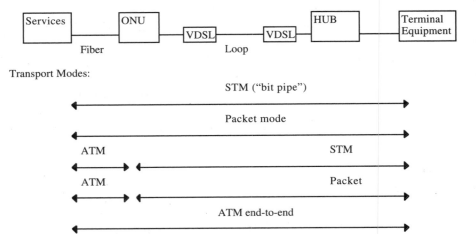

used to access service network in combination with STM and packet mode VDSL. At its logical limit, VDSL forms an end-to-end transport between ATM devices.

So VDSL is also allowed to function with ATM at the service side to the ONU and STM (that is, TDM) on the loop, which makes sense because service providers are more likely to employ ATM switches and servers than users are. In fact, ATM is an ideal solution here, where mixing voice, video, and data services is common; and the ONU can take care of any conversion chores. A combination of ATM services to the ONU and packets on the loop is also supported; this combination takes advantage of the widespread use of IP packets on most types of distribution networks.

Please note that most implementers of early VDSL pilot systems intend to focus on the ATM end-to-end transport mode, which seems rather astonishing at first. However, the more mixing of services and types of traffic there is, the more ATM makes sense, both technically and financially, and VDSL is certainly intended for this mixture of traffic types.

VDSL/ADSL Downstream Speeds and Distances

ADSL is designed for a world where most analog local loops consist of simple "home-run" pairs of copper wires extending from local exchange to customer premises. Fifteen percent (and up to 45 percent in the near future) of this local loop plant, however, is already on some form of feeder system, such as a DLC. More often, the DLC pairgain portion of the loop is carried on fiber-optic cable. This is most likely SONET, but not necessarily in a ring configuration. Figure 14-3 shows the speeds and distances for typical ADSL and VDSL loop lengths.

Sometimes there is debate about carrying the fiber all the way to the home (called *fiber to the home*, or FTTH). It is a little known fact that this is supposed to be called *fiber to the basement* (FTTB) when a small office building is involved. Anyway, this premises deployment of fiber to all homes and businesses is probably too expensive and not absolutely necessary, and so there are variations for carrying the *fiber to the curb* (FTTC) and *fiber to the neighborhood* (FTTN, about 10 to 100 homes). The whole point is that VDSL is designed for FTTN, but nothing rules out the use of VDSL for other forms of fiber feeder networks.

In fact, it could be said that VDSL takes over where ADSL tops out. The figure compares the VDSL speeds and distances with those of ADSL.

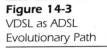

Figure 14-3
VDSL as ADSL
Evolutionary Path

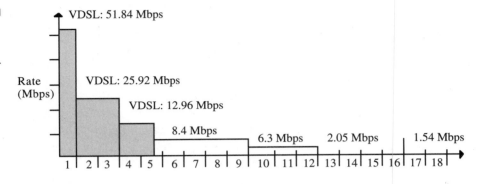

Distance in thousands of feet

Note that VDSL may be used in conjunction with ADSL, depending on twisted-pair distance. Various standards groups have suggested that a combination of VDSL and ADSL offers the best opportunity to provide access to almost any PC server-based application, with interactive TV services on the same system.

As mentioned previously, VDSL can be employed in three general downstream ranges, as shown in the Figure 14-2:

- 12.96-13.8 Mbps to 4,500 feet (1,500 meters)
- 25.92-27.6 Mbps to 3,000 feet (1,000 meters)
- 51.84-55.2 Mbps to 1,000 feet (300 meters)

VDSL upstream rates also fall into three categories but are not restricted by distance. These are 1.6 to 2.3 Mbps, 19.2 Mbps, and upstream speed equal to that of downstream. The boundaries between ADSL speeds and distance in the figure are not always fixed and absolute, but they are certainly representative.

VDSL in Detail

This last section takes a look at VDSL in more technical detail. It presents VDSL in terms of projected capabilities, underlying technology, and even a few of the outstanding issues. The section includes a survey of current VDSL standards activity and ends with the suggestion that ADSL and VDSL together can give network providers a very good combination of technologies for evolving a full-service network while offering access to most PC applications and interactive TV applications at the same time and while the total network develops naturally.

Just as in ADSL, both data channels in VDSL will be separated in frequency from bands used for analog voice and ISDN in order to allow service providers to overlay VDSL on top of existing voice or ISDN services, which is always a good idea. For now, the two high-speed channels will also be separated in frequency. If the need for greater bandwidth in the upstream direction arises for higher-speed upstream channels or symmetric rates, then VDSL systems will presumably need to use echo cancellation.

Also as in ADSL, VDSL must transmit compressed video downstream to users. Compressed video is a real time signal that requires low and stable delays and is not well-suited for the usual retransmission methods of error control used in traditional data communications networks. In order to give compressed video the error rates needed, VDSL will also have to incorporate *forward error correction* (FEC) with interleaving that would be sufficient to correct all of the errors created by impulse noise of some specified duration on the line. Of course, this interleaving introduces delay, and there is a definite relationship between the delay and correctable impulse noise interval that the line can recover from. In turns out that the delay is on the order of some 40 times the maximum length of the correctable impulse, a considerable interval.

Signal distribution in VDSL can be accomplished in many ways. The simplest method is to broadcast the information in the downstream direction to every device on a customer premises. Alternatively, information can be transmitted to a separate hub device that distributes information to addressed devices based on cell or straight time division multiplexing within the information stream itself.

In the upstream direction, multiplexing will be more difficult in VDSL. Systems that use a passive *network interface device* (NID) must insert information onto a shared medium, namely the local loop itself. This can be done either with a form of time division multiplexing known as *time division multiple access* (TDMA), often used in wireless telephone systems, or some form of FDM. The TDMA used in VDSL may use a type of token passing called *cell grants* passed in the downstream direction from the VDSL modem (VTU) in the ONU modem.

However, other methods of upstream multiplexing, such as straight contention or even tokens and contention, can be used. For instance, contention could be used for unrecognized (previously inactive or unconfigured) devices and cell grants for recognized (previously active or configured) devices. On the other hand, an FDM scheme would give each customer premises device its own separate channel. This does away with the need for a shared channel upstream, but also would either limit the information rate available to any one customer device or require dynamic allocation of the bandwidth and even inverse multiplexing at each piece of customer premises equipment. For VDSL systems using active NIDs, the upstream information aggregation problem is most easily solved by using a logically separate hub that would use Ethernet (most typical) or ATM (less common) protocols for the upstream multiplexing.

The twin considerations of migration scenarios and inventory considerations might mean that VDSL units should be capable of operating at various speeds, perhaps at *all* of the speeds, with automatic recognition of a newly connected device to a VDSL line or a change in operating speed. Passive network interface devices need to have a capability known as *hot insertion*, where a new VDSL premises unit can be placed on the line without interfering with the operation of other modems.

VDSL Technology

It is obvious that VDSL technology resembles ADSL technology to a great extent. This is not necessarily a bad thing and may actually be good. What is not so obvious is that ADSL systems must deal with much larger dynamic ranges in operation. As a result, ADSL is considerably more complex than it could be under more controlled conditions. VDSL must be lower in cost and lower in power than ADSL, and premises VDSL units might have to implement some form of physical layer *media access control* (MAC) for multiplexing upstream data, as outlined above.

A key aspect of any xDSL technology, from HDSL2 to ADSL, is the line code employed. VDSL is no exception. In fact, as many as four line codes have been proposed for VDSL:

1. **Carrierless Amplitude/Phase modulation (CAP).** This is the same carrier suppressed QAM employed in ADSL, but translated to VDSL. With passive NIDs, CAP would use quadrature phase shift keying (QPSK) in the upstream direction and a form of TDMA for the multiplexing. Actually, CAP does not rule out an FDM approach to the upstream multiplexing; it simply makes more sense to use TDMA.

2. **Discrete Multitone (DMT).** This is also the same technique as in ADSL systems. DMT is a multiple carrier system that uses *dis-*

crete Fourier transforms (no more will be forthcoming here on DCTs) in order to create and demodulate the individual carriers. For passive NID configurations, DMT would use FDM for upstream multiplexing, although DMT, in the reverse of CAP, does not rule out a TDMA approach to upstream multiplexing.

3. **Discrete Wavelet Multitone (DWMT).** This is another form of multicarrier system. DWMT uses *wavelet transforms* to create and demodulate the individual carriers. Like DMT, DWMT also would use FDM for upstream multiplexing, but of course it also allows for TDMA.

4. **Simple Line Code (SLC).** This is a type of four-level baseband signaling. SLC filters the baseband signal and restores it back to the original at the receiver. For passive NID configurations, the SLC would probably use TDMA for upstream multiplexing, like CAP, although FDM is possible.

This list is presented mainly to show the research efforts into VDSL. Today, DWMT and SLC have been ruled out as viable VDSL line codes. Only DMT and CAP/QAM remain in the debate.

It is anticipated that early implementations of VDSL will use FDM to separate downstream from upstream channels, and both of them, in turn, from the analog voice and ISDN, if present. Echo cancellation will almost certainly be required for later systems implemented with symmetric information rates. In order to produce very simple and cost-effective analog voice splitters, a fairly large distance in terms of frequency will need to be maintained between the lowest information channel of the VDSL services and the analog voice. Now, normal practice (as in ADSL) would dictate that the downstream channel be located above the upstream channel. However, the digital video specification *Digital Audio-Visual Council* (DAVIC) reverses the usual order of the upstream and downstream channels in order to allow for the distribution of VDSL signals over a coaxial cable TV system.

The FEC used in VDSL will use the same type of optional interleaving to correct bursts of errors caused by impulse noise as in ADSL. Although the structure will be very similar to ADSL as defined in T1.413, an outstanding issue to be resolved is whether the FEC overhead (in the range of 8 percent) will be taken from the information payload capacity or added as a totally out-of-band signal. The first approach reduces the total information payload capacity but keeps the standard reach of the signal. The out-of-band approach keeps the normal payload capacity but requires a small reduction in the signal's reach. As has been mentioned, ADSL puts the FEC overhead out of band in a special part of the ADSL superframe.

In the discussion of VDSL line codes, mention was made of upstream multiplexing in a passive NID arrangement, as is common in the United States. Note that if the premise's VDSL unit employs an active NID, then the multiplexing of upstream cells or information channels from more than one customer premises device into a single upstream flow becomes the responsibility of the premise's network because an active NID is part of the service provider network. In this case, the premise's VDSL unit simply transfers raw bit streams in both directions. With an active NID, one type of premise's network to accomplish this involves the use of a star configuration to connect each premise device to a multiplexer or switching hub. The hub could, of course, be integrated into the premises VDSL unit.

In a passive NID configuration, which is more common in the United States, each piece of a customer's premise equipment is associated with a VDSL unit. please consider that a passive NID does not necessarily rule out the use of multiple customer devices with each VDSL unit. However, in this case the issue of active or passive NID becomes a matter of ownership, not a matter of the wiring topology and multiplexing method.

There are two major strategies for handling upstream contention due to the fact that the upstream channels in VDSL for each customer premises equipment device must share a common wire. Although a LAN-like collision detection system could be used, the need for guaranteed bandwidth in VDSL would be better served by one of two solutions. One involves a form of cell-grant protocol in which downstream frames generated at the ONU or even further up in the VDSL network contain a few bits that grant access to a specific piece of customer equipment for a specified time interval that begins after the frame is received . A "granted" piece of customer equipment in allowed to send one upstream cell during this time interval. To do this, the transmitter at the premises must turn itself on, send a preamble signal to condition the ONU receiver, send the cell, and then turn itself off. One early version of this protocol uses 77 octet intervals to transmit a single 53 octet cell.

A second way to accomplish this divides the upstream channel into frequency bands and assigns one of the frequency bands to each piece of customer premises equipment. This method has the benefit of doing away with any media access control and its associated overhead, although a multiplexer has to be built into the ONU, of course. The drawback is that this method either restricts the bit rate available to any one piece of equipment (due to the banding) or requires the use of a type of dynamic inverse multiplexing method that allows one piece of customer equipment to send more than its share of bits for a given time interval. This inverse multiplexing approach would look very much like a media access control protocol, only without the bandwidth penalty from the operation of the contention scheme.

VDSL Issues

VDSL has not achieved anywhere near the maturity of even ADSL, which is arguably immature itself. In fact, VDSL is still only in the definition stage. Few preliminary VDSL products even exist. However, a lot of work needs to be done on local loop, noise effects, upstream multiplexing techniques, and other aspects of VDSL operation to establish a set of firm standard properties. The biggest and most important VDSL question mark involves the maximum distance that VDSL can operate reliably for a given bit rate. This may sound odd for a new technology such as VDSL, but it really reflects the unknowns of real local loops at VDSL frequencies. Furthermore, short bridged taps or unterminated extensions in homes that have no effect at all on analog voice, ISDN, or even ADSL, may have very disruptive effects on VDSL.

VDSL also is susceptible to radio interference, just like ADSL. A long local loop acts as both an antenna and a transmitter at these frequencies. If VDSL signal levels are low, the antenna effects inject amateur radio noise into the line. If the signal levels are increased, then the transmitter effects begin to dominate and could interfere with the radio signals.

Another issue involves the services that VDSL is designed to deliver. VDSL more or less needs ATM cell streams at some as-yet unspecified bit rates to carry information. What about non-ATM information and packet formats? VDSL, like anything else at this low level, cannot be completely independent of upper layer protocols. This is particularly true in the upstream direction where multiplexing information from potentially many pieces of customer equipment may require VDSL to be aware of a variety of protocols, ATM or not.

A third VDSL issue also is a concern in ADSL—the area of the premises distribution network and the interface between the telephony portion of the network and customer premises equipment itself. There are many considerations, including cost, that favor a passive network interface device with VDSL on the premises installed in a customer premises equipment package. In this form, the upstream multiplexing is handled very much like multiple access in LANs. However, other factors exist, including system management, reliability, regulatory constraints, and migration considerations that would favor an active VDSL network interface. The active VDSL interface could act just like a LAN hub and use either point-to-point or shared media distribution to multiple premises equipment devices over premises wiring. This wiring would be physically isolated and totally independent from the network wiring.

However, as with most things, cost may be the overwhelming concern. The fact is that there are only a few customers serviced by small ONU arrangements, at least compared to *hybrid fiber-coax* (HFC) cable TV systems. The common equipment costs in the ONU, such as fiber links, interfaces, and equipment cabinets, must therefore be spread over a relatively small number of customers. VDSL has a much lower target cost than ADSL because ADSL can be supplied directly from a wire center in a local exchange. The bottom line seems to be that passive NIDs for VDSL could be more expensive than passive ones, but even this is not a given. In spite of the obvious benefits of the active NID arrangement for VDSL, eliminating responsibilities for other premises network electronics make the passive NID for VDSL very attractive from a cost point of view.

VDSL and Standards

There are five different standards organizations and forums working on the VDSL standard. Into the first category go ANSI T1E1.4, the European Telecommunications Standards Institute (ETSI), and Digital Audio-Visual Council (DAVIC). In the second category are the ATM Forum and the ADSL Forum. Each is discussed here:

1. **T1E1.4.** In the United States, the ANSI T1E1.4 standards group has started a project for VDSL. This is looking first at overall VDSL system requirements and should evolve into a stable system and protocol definition.

2. **ETSI.** In Europe, an ETSI has a VDSL standards project, as well, originally under the overall title *High Speed* (metallic) *Access Systems* (HSAS), but now just as VDSL. They have compiled a list of objectives, problems, and requirements for VDSL systems. ETSI has already called for active NIDs and compatibility with SONET in its international standard form of *Synchronous Digital Hierarchy* (SDH), which is hardly surprising. Note that ETSI works very closely with ANSI T1E1.4 and the ADSL Forum.

3. **Digital Audio-Visual Council (DAVIC).** DAVIC was quite interested in VDSL due to its video delivery capabilities, as might be

expected. They have called for media access control for upstream multiplexing based on TDMA over shared wiring. DAVIC had initially specified VDSL for only a single downstream rate of 51.84 Mbps and a single upstream rate of 1.6 Mbps over 300 meters or less of copper. DAVIC has favored use of a passive NID and assumes that premises distribution from the NID will be over new coaxial cable or new copper wiring.

4. **ATM Forum.** Although unrelated to VDSL, the ATM Forum has defined a 51.84 Mbps interface for private ATM network UNIs. They have also considered the issue of premises distribution and delivery of ATM cells all the way to premises over VDSL.

5. **ADSL Forum.** The ADSL Forum also has started to work on VDSL. The ADSL Forum is initially looking at the major network, protocol, and architectural aspects of VDSL. The emphasis will be on the applications, and the Forum is leaving the line code and other low level issues to T1E1.4 and ETSI. Other higher-layer protocols will be left up to the ATM Forum and DAVIC.

As a last word on VDSL, consider the relationship between ADSL and VDSL once more. VDSL has some real technical similarities to ADSL, but VDSL achieves bit rates some ten times greater than ADSL. However, ADSL is the more complex transmission technology, mostly because ADSL must deal with much greater variations in local loop performance characteristics than VDSL. But the two technologies, at heart, are essentially twins with regard to the whole xDSL family. For instance, ADSL uses advanced transmission techniques such as forward error correction to achieve bit rates from 1.5 to almost 8 Mbps over twisted pair ranging to 18,000 feet. VDSL uses basically the same advanced transmission techniques and forward error correction to achieve bit rates from 13 to 55 Mbps over twisted pair ranging to 4,500 feet.

ADSL and VDSL can be considered a matched pair forming a set of transmission techniques capable of transporting about as much information as theoretically possible over varying distances using analog local loop wiring. VDSL is very well-suited for a full-service network with up to two channels of HDTV at the highest VDSL bit rates.

Of course, service providers cannot deploy ONUs overnight, and all of the technology needed is not yet available in products. At least ADSL, in spite of its limitations, offers services over local loops that exist and are used today. And many of the new services that are being talked about today can now be delivered at speeds at or below T1/E1 rates, such as decent video conferencing, Internet access, video on demand, remote LAN access, and others. For these services, ADSL and VDSL provide an ideal mixture for network migration. On very long loops, ADSL only delivers a single channel, but as the loop length shrinks, either due to natural closeness of sites to the local exchange or due to new deployment of fiber-based access nodes, both ADSL and VDSL offer more channels and higher capacity for services, such as HDTV, that require bit rates above T1/E1.

So initial broadband service offerings will be based on ADSL in most cases. However, as services, networks, and the physical infrastructure evolves, the future may indeed belong to VDSL. Whatever the future of ADSL, it seems most likely as time goes on that it will include at least portions of VDSL.

Outstanding DSL Issues

So far, all of the major components of the xDSL network family in general and ADSL in particular have been explored in detail, but the approach taken has been mainly descriptive, without mentioning many of the numerous controversies and issues that divide xDSL proponents. A notable exception was the ADSL line coding controversy, where CAP and DMT solutions were covered in some detail. However, the same type of debate goes on over many other aspects of DSL technologies. The time has come to consider some of these issues in more detail. The answers to these questions are crucial in determining just what DSL technologies will look like, how they will perform, and how much they will cost.

This chapter considers the current issues and controversies surrounding xDSL technology in four main areas. First explored are overall DSL network issues. Issues in this category include items that affect the entire operation and functioning of an xDSL network. Some equipment issues are addressed next. Issues in this category include items that affect the various product packages in which DSL equipment will be available. Then some service issues are investigated. Issues in this category include items that affect the various service packages that DSL systems will deliver. Finally, the chapter looks at the issue of active and passive network interface device or network termination arrangements.

DSL Network Issues

Any Digital Subscriber Line technology, from HDSL to VDSL, is a relatively new technology. As with any new technology, from ATM to Gigabit Ethernet, a good number of network issues should be resolved before anyone grabs a couple of boxes and begins plugging them together. These issues address overall networking concerns. The following lists some of the more critical network issues regarding DSL. Each issue is then explored in more detail.

1. How should loading coils and *Digital Loop Carrier* (DLC) local loops be addressed?
2. What will be the network of choice on the service side of the *DSL access multiplexer* (DSLAM)?
3. How are DSL links to be tested, repaired, and managed?
4. What should be done with non-DSL telco solutions?
5. How should premises issues about splitters and wiring be addressed?
6. How should varying DSL bit rates be billed over time and among customers?

How Should Loading Coils and Digital Loop Carrier (DLC) Local Loops be Addressed?

Obviously, because some 15 to 20 percent of local loops have loading coils, and 15 percent are on DLC systems (this could rise to 45 percent by the year 2000), something must be done about these loops before any DSL can be deployed, but what? In order to use any DSL on local loops with loading coils, the loading coils must be removed. This was a big stumbling block to ISDN deployment. The process of removing the loading coils is labor intensive and time consuming. It will be no easier with xDSL. The trouble with DLCs is that they only digitize and pass along the analog voice passband of 300 to 3.3 KHz, so any xDSL cannot take advantage of the 1.1 MHz usually used on the xDSL twisted pair without doing something about the DLC system. The simplest answer would be to run something else to these loading coil and DLC customers, but this only shifts the issue. Perhaps these users' best bet would be satellite services or cable modems, although this sounds strange to telephone company service providers.

Actually, in spite of the issues addressed here, there are clear answers to the loading coil and DLC issues today. First, loading coils will be dealt with as they were in ISDN: they will be removed. This is not an ADSL show-stopper. It's just a pain. Second, with regard to DLC, the ADSL equipment will be installed at the remote terminal (RT) and separate trunks will handle the extra loads. Another unavoidable pain in the neck.

What Will Be the Network of Choice on the Service Side of the DSL Access Multiplexer (DSLAM)?

Basically, what this boils down to is the choice of using either the *Internet Protocol* (IP) or ATM. Both positions have merit and neither can be totally ruled out, even on grounds of economics. ATM may be initially more expensive, but perhaps the performance aspects of ATM networks and services would justify this approach. On the other hand, it is a TCP/IP world, especially where the Internet and Web are concerned. TCP/IP is evolving to a more adequate transport for voice and video, but at least part of this is due to the increased use of ATM in the Internet itself. This will be a very interesting battleground over the next few years, to say the least.

How are DSL Links to be Tested, Repaired, and Managed?

It took a long time for outside plant personnel to be trained and comfortable with ISDN DSL links. Personnel took a while to become familiar with *Basic Rate Interface - Terminal Equipment* (BRI-TE) cards and so forth. It was quickly apparent that this was no longer an analog world and that radically different testing, repair, and management techniques were needed. By the same token, try naming a test box specifically designed to test ADSL or other xDSL links. They exist, but are neither common nor well-known. Other questions along these lines exist. What is the standard procedure for splicing a break in an ADSL link? Where are the uniform network management packages for multivendor DSL environments? Any DSL might prove to be too immature for the constant day-to-day pounding that working networks demand.

It is easy to take a pessimistic attitude regard this issue, but it need not be. HDSL links are being readily maintained today. And adding xDSL electronics at the local loop might actually help with management, due to the increased intelligence and capabilities of the devices.

What Should be Done with Non-DSL Telephone Company Solutions?

DSL is not the only game in town, which is painfully obvious at many trade shows and conferences. In fact, xDSL solutions are not always the solution of choice within the telephone companies in spite of xDSL's promise to maximize use of existing twisted-pair copper wire. One reason might be that the telephone companies fought for years for the right to build and operate cable TV networks. At least one large telephone company has decided practically to abandon their twisted-pair infrastructure in favor of building the whole thing over again as a full-service *hybrid fiber-coax* (HFC) network that will be almost indistinguishable from any other cable TV network. The *multichannel multipoint distribution system* (MMDS) is a telephone company initiative. What is the future of these technologies in what might yet turn out to be a DSL world?

How Should Premises Issues about Splitters and Wiring be Addressed?

Exactly where should the splitter be located and how should it be packaged? When packaged with the ATU-R, as the vast majority are, the concern is maximized that ATU-R failures will affect analog telephone service, although there are ways to address this possibility. Unfortunately, these vary from vendor to vendor. If the splitter is packaged elsewhere, there may be long parallel runs of wire that cross-couple noise from the analog voice, which, of course, is always a major concern. How should those entrusted to run the premises wiring be monitored for standard compliance? Inspections of wiring are rare outside of new commercial construction, and even here the inspection is often limited to a visual inspection of splices, connectors, and other non-electrical characteristics of the wiring system. Although these physical features affect electrical performance of the wiring, more precise electrical testing should be conducted.

Of course, CDSL makes a lot of these issues easier to address. Nothing could be simpler than using splitterless ADSL, and in fact this will be very desirable from the customer point of view in the majority of cases.

How Should Varying DSL Bit Rates be Billed Over Time and Among Customers?

Money is always a crucial issue both to customers and service providers. Because DSL line rates vary, what are customers really paying for? Obviously, it is not pure bandwidth. Also, packet-based services are bursty, so a customer might not want to pay for pure connection time either (or else customers might want their old circuit back!). Just how should DSL be billed? After all, if a modem connects at 14.4 kbps on a noisy local loop, a customer pays the same as a neighbor that consistently connects at 33.6 kbps. What if neighbors with ADSL or RADSL have service that varies even more drastically? What if a customer can get only 640 kbps while a neighbor consistently gets 1.5 Mbps, about three times more bandwidth for the same price? Service providers can try to justify equal charges, but it is debatable whether the customers will stand for it.

DSL Equipment Issues

The technology exists today to combine a server, LAN hub, and router all in the same package. Many reasons to do so exist because most of these devices are used together in most cases anyway. The result would be about the size of a PC anyway, but at a potentially much lower cost. Why aren't these devices all over the place? There are many reasons, among which are that people like having separate devices for all these functions, different vendors have different strengths, and so on. There are similar issues regarding DSL equipment, as shown in the following list. Each issue is explored in more detail below.

1. How should different DSL coding methods and technologies relate to each other?
2. How should premises equipment be packaged to ensure customer acceptance?
3. What types of customer equipment interfaces should DSL devices support?
4. What is the best ATM or TCP/IP interface for DSL access multiplexers (DSLAM)?
5. Where are the end-to-end, turnkey DSL systems?
6. How much will all this equipment cost?

How Should Different DSL Coding Methods and Technologies Relate to Each Other?

If both carrier amplitude/phase modulation and discrete multitone technology become common xDSL line coding methods, what would or should happen if an ADSL Terminal Unit-Central Office is DMT-based and the ADSL Terminal Unit-Remote (ATU-R, which is, after all, customer premises equipment) is CAP-based? Is it logical to expect that xDSL equipment will eventually support both? Some vendors, faced with trying to distinguish their products in an increasingly standards-based, commodity market, have announced duel CAP/DMT support in an effort both to separate themselves from the crowd and to allow the maximum use of their products in all situations. Should this duel CAP/DMT support be part of all ADSL equipment? What will this do to costs in the crucial deployment phases?

How Should Premises Equipment be Packaged to Ensure Customer Acceptance?

The plain old telephone service splitter can be located in one of several places to support the analog voice in ADSL. Which splitter form and position will users prefer? Which will they tolerate? Which will they reject? The ATU-R itself can be a PC card, a set-top box, a stand-alone device, or something else entirely. How does a vendor or service provider decide which packaging the customer will buy and use? This issue involves considerations of economics, marketing, psychology, and just plain consumer intangibles. The art of packaging is hard enough to do just right for packaging food, let alone technology!

CDSL is betting that customers will embrace splitterless arrangements. So far, this seems more or less guaranteed for Internet access, and perhaps even other xDSL services.

What Types of Customer Equipment Interfaces should DSL Devices Support?

Customers do not really care about the technical aspects of their equipment; they just want to plug things in and get working. But what kind of plug is being plugged in? The more types and choices on the customer premises, the higher the potential cost of the device. Just give the users an Ethernet port? In what form? A 10Base-T RJ-45 only, as many ATU-R vendors do now? Or something else, as well, such as what is known as the DIX Ethernet connector? And what about the *Consumer Electronics Bus* (CEBus), which sends signals over the building electrical wiring? Newer PCs now support the *Universal Serial Bus* (USB), a sort of "serial SCSI" interface with multiple fast serial devices hooked up to a PC. Most vendors just put in the 10Base-T RJ-45 and that is that, but this means users have to buy an Ethernet card just to be able to plug in to the xDSL modem.

What is the Best ATM or TCP/IP Interface for DSL Access Multiplexers (DSLAM)?

This is not a question of ATM or TCP/IP, which was already addressed. It is rather a question of what kind of ATM or TCP/IP interface or device should be employed. Look at the IP choices first. Should the IP router be accessed at 10 Mbps? 100 Mbps? Gigabit Ethernet? Is even more speed needed? Should the service provider consider something more exotic such as Fibre Channel, which offers speeds into the multiple gigabit ranges? Should the DSLAM support the *Small Computer Systems Interface* (SCSI)? Something else here as well? What about the *High Performance Parallel Interface* (HPPI), which runs close to 6 Gbps. ATM interfaces are pretty much in the same boat, but choices could be more restricted due to ATM Forum and B-ISDN standards.

Where Are the End-to-end, Turnkey DSL Systems?

Customers may like simply to plug it in and run it—whatever the "it" may be—but so do service providers, believe it or not. Why buy equipment from four different vendors and spend months performing quality assurance and interoperability testing? Just buy everything from the ATU-R to the DSLAM to the backend distribution network to the ATM server (or IP server) from one vendor or *value-added reseller* (VAR) and go and sign up users. These are called *turnkey* solutions. Common enough in Web site, LAN, and even some ATM applications, common turnkey solutions seem to be totally lacking in xDSL. There seems to be a precedent in MMDS systems. Telephone companies do not have the in-house expertise to engineer and build MMDS networks so they rely on outside contractors to turnkey the whole thing for them. Why limit this to MMDS?

How Much Will All This Equipment Cost?

This is the stickiest issue. Perhaps subsidies will be needed to get DSL off the ground, a process that worked quite well with cellular phones, at least in the early stages of deployment. Early cellular phones that cost about $700 were commonly sold for $350 of less. How could this be? Simple. The cell phone had to be used with one service provider or another. Whoever signed up the user would rebate a portion of the subscriber's monthly bill to the manufacturer. When the difference in cost and consumer price was even, it was all the same in the long run. Could ATU-Rs be marketed the same way? Perhaps, but this is pure speculation.

DSL Service Issues

People make money from technology in two ways. The first is to package and sell the technology itself, such as with automobiles or VCRs. The second way is to package and sell the service that the technology represents, such as leasing a limousine for the evening or renting videos. Obviously, the two approaches can be combined to some degree (such as buying a modem, the technology itself, or accessing the online service, which the modem technology represents). When it comes to DSL technologies and computer/network technologies in general, people are not so much impressed by the technology (which remains a semi-mystical, black box device to users and even to many in the industry), but are more often impressed by the service ("I need that and I want that now!"). Many of the service issues surrounding DSL concern pricing and packaging, but not exclusively. The issues are listed here, and more detail is provided below.

1. How should DSL services be priced to maximize use and minimize payback periods?

2. What marketing mechanisms exist to identify DSL target customers?

3. Will DSL services simply shift bottlenecks elsewhere in the network?

4. What impact will DSL services have on other provider revenue streams?

5. Will slow deployments mimic the glacial ISDN deployment pace?

6. Can I change my voice provider and keep my DSL service provider?

How should DSL Services be Priced to Maximize Use and Minimize Payback Periods?

A *payback period* is the length of time that it takes a service provider to recover the initial costs necessary to start offering a new service from scratch. The new service may involve totally new technology, but not always. The initial AT&T coast-to-coast, direct-dialed calls (DDD, or *direct distance dialing*) cost $12 per minute (in 1960 dollars). This amount was too expensive for most people to afford, but based on the cost of the equipment, prorated over a certain payback period, AT&T figured that this was a fair price to charge. No one used it initially. AT&T dropped the price to $6 per minute and made money hand over fist. Long distance is now deregulated, but the LECs that are considering DSL services are still closely watched by the state regulators. The point is that the same type of cost, revenue, and payback period curves for DSL have yet to be widely publicized, and there is debate over the ones initially released.

What Marketing Mechanisms Exist to Identify DSL Target Customers?

Marketing is the art of creating a perceived need for a product. The key term here is *perceived*. How many people have found themselves saying, "I really shouldn't have, but it was such a bargain?" To be most effective, marketing efforts should focus on those who are most likely to recognize the product's benefits. Why waste time marketing to users who have no interest in the service or could not use it even if they wanted to, such as people without home PCs? The benefits could be real or imagined, but good marketing makes them known in either case. However, outside of obtaining a list of current Internet users (and where exactly would that be?), how are DSL service providers to decide whether one neighborhood has more potential buyers than another?

Will DSL Services Simply Shift Bottlenecks Elsewhere in the Network?

DSL will alleviate the bottlenecks caused by restricted local loop bandwidths, true enough. What about other potential bottlenecks in the network? Will the DSLAMs now be flooded with traffic, or the IP routers or ATM switches? Will the backbone network now grind to a halt, which has been predicted for the Internet for some time now? Will the servers be able to keep pace with the traffic demand? It is one thing for a heavily visited Web site to have to stream out information over a 100Base-T Ethernet LAN to a router feeding 64 kbps downstream to 10 users (640 kbps aggregate), but will the same server be able to handle the load over the same LAN to the same router when users now get 1.5 Mbps downstream (15 Mbps aggregate)?

What Impact Will DSL Services Have on Other Provider Revenue Streams?

Service providers, especially telephone companies, do not often put all of their eggs in one basket. A variety of voice services are available to fit almost any user need. Services include call forwarding, call waiting, caller ID, and so on, all of which complement each other nicely. Users do not need a wireline phone or a cellular phone—they need both! But DSL is different. If DSL becomes common for SOHO usage, what will that do to

T-carrier revenues? What about users of frame relay, *Switched Multi-megabit Data Service* (SMDS), or even ATM services? The success of DSL could cause a drastic drop in other revenue streams.

Will Slow Deployments Mimic the Glacial ISDN Deployment Pace?

The ISDN has been around for 20 years, and yet, as recently as October 1996, there was no ISDN available in New Mexico, Alaska, Washington, D.C., nor in several other major portions of the United States. The kiss of death for almost any technology has frequently been the familiar "we will deploy services based on customer demand" refrain from the service providers. Customers do not often demand services, any more than people demand bus stops. The bus does not just cruise the boulevards and look for knots of people waiting on corners ("better put a bus stop here?"). However, if the bus stops at a convenient location, and there is a bus stop sign there, people will climb aboard. Today, of course, there are even competitive buses to hop onto.

Can I Change My Voice Provider and Keep My DSL Service Provider?

Deregulation in many countries enables customers to shift service providers almost effortlessly. However, if DSL is delivered to a user with a POTS splitter, how can a customer change DSL providers and keep their analog voice service provider? Not many telephone service providers with DSL plans have even considered this possibility. If the market for DSL users heats up the way that the market for Internet providers has heated up, however, not to mention the market for long-distance services, such "slamming" tactics for DSL customers might become more common.

The LECs, naturally, see the Internet customer as a way to help hold and protect their voice customers.

Active and Passive NIDs

The *customer premises equipment* (CPE) remote unit in many DSL arrangements (especially ADSL) supports both newer services and analog voice services. The CPE in ADSL consists of an ATU-R and POTS split-ter, sometimes in the same device, but not always. The boundary between the CPE and the network is determined by a *network interface device* (NID). Sometimes the term *network interface unit* (NIU) or even *Network Termination* (NT) is used in the same context. It all depends on who or what is talking about this network access point. In any case, there must be some form of NID involved in the DSL link, which is the term used in this chapter.

As it turns out, there are two major types of NID: active and passive. The differences are mainly due to differing regulatory environments and have little to do with straightforward technological issues. These are sometimes referred to as the "wet wire" NID and "dry wire" NID, respec-tively. The terms *wet* and *dry* refer to the ability and willingness of the service provider to deliver electricity to the DSL device over the copper loop itself. In a wet arrangement, electricity is part of the service pack-age, but in a dry arrangement, electricity must be provided by the cus-tomer separately.

In the United States, where deregulation is considered a worthwhile goal and widely pursued, the NID is almost always a passive, "wires-only" interface, which means that the NID delivers no power to the premises and works by just plugging things in to the proper jacks on the NID. There are exceptions, but not using a passive NID for a service in the United States often involves the permission and cooperation of the regulating au-thority, typically the individual state in DSL arrangements.

It is more common in a deregulated environment, particularly in the United States, for the ATU-R to be the CPE and contain all of the active DSL electronics. That is, the customer can purchase the ATU-R from any approved vendor, install it themselves, and the service provider has no control over the situation at all.

In many other countries, however, where deregulation is not as preva-lent, both the NID and ATU-R are part of the network. This is an active NID because the electronics can be split between the NID and ATU-R. In these countries, the ATU-R can be bundled with the NID device. The cus-tomer must purchase or lease the ATU-R from the service provider. The service provider arranges for installation, and so the service provider ex-ercises a fair amount of control over the situation.

In many other countries deregulation is more of a future goal and less actively pursued. In such countries, both the NID and ATU-R are consid-ered to be part of the network. In these countries, the ATU-R might be bundled with the NID device. That is, the customer must purchase or lease the ATU-R from the service provider, the service provider arranges for installation, and the service provider exercises a fair amount of control over the situation.

The question is, how should DSL vendors package their products for these environments? Should a vendor manufacture passive NIDs and active CPEs for sale in the United States and other deregulated countries? Should the vendor then repackage the product as an active NID/ATU-R combination for sale elsewhere? This is not an ideal situation, but there seems to be no way around it.

Figure 15-1 shows how both the passive and active NID arrangements work. In the passive NID scenario, as will be typical in the United States, the ATU-R is packaged separately from the NID. There is a "wires-only" interface for DSL signals to and from the ATU-R unit, which has its own separate alternating current (AC) power supply (or shares power with another device, such as a PC or set-top box). The NID sits at the end of a dryî (no electrical power) DSL link.

In the active NID scenario, also shown in the figure, the ATU-R can be bundled within the NID itself. There is no need for a wired DSL interface anywhere else. After all, the user interfaces on the ATU-R are not governed or determined by DSL standards. These non-DSL interfaces might be 10BASE-T, CEBus, or something entirely different. This arrangement is quite attractive for that very reason. The service provider is not responsible for the non-DSL interfaces. The active NID/ATU-R unit sits at the end of a possibly wet (powered) DSL link, although power distribu-

Figure 15-1
Passive and active
NID arrangements

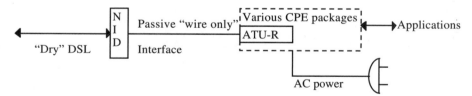

Passive NID/Active ATU-R Arrangement (United States)

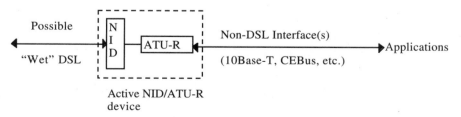

Active NID/Active ATU-R Arrangement (many countries)

tion methods are subject to more than just active/passive considerations (for example, battery power is an option).

Passive NID/Active CPE Issues

Although mandated by deregulation initiatives, the passive NID is simple and minimizes service provider involvement, which can be seen as a direct benefit to both the customer and service provider. The passive NID fits in very well with the traditional "modem" purchase and installation philosophy. That is, a customer can buy whatever modem they like; if it complies with certain common standards, the customer can use it with confidence, without concern for service provider support and compatibility.

The passive NID delivers no power to the CPE. However, if packaged as a PC board, the DSL CPE needs no additional power supply. In addition, even if packaged as an external "modem" with separate power supply needs, the DSL CPE can be deployed as the customer pleases, without regard for PC location.

One distinct benefit to vendors exisits in the passive NID environment: There is ample opportunity for vendor distinction based on feature set. In other words, because customers have a wide range of choices for DSL CPE equipment, competition might be fierce, and prices will drop rapidly at the same time the features become richer.

Active NID/Passive CPE Issues

Regulated countries can use a bundled, active NID and ADSL Terminal Unit-Remote. When the active NID/ATU-R issues are considered, this arrangement guarantees maximum service provider control over the entire network loop. The service provider might have an easier time configuring, troubleshooting, and maintaining the loop, although this is not necessarily a given.

This arrangement means that the service provider has no premises wiring concerns at all when it comes to DSL. The interfaces on the ATU-R are governed by standards other than DSL and are beyond the scope of DSL.

Also, there is no ATU-C-to-ATU-R (or any DSL devices) interoperability concern at all in this configuration. Interoperability can be a problem in deregulated environments, where there is no guarantee that a customer's CPE will function properly with the service provider's equipment if they come from different vendors. Of course, if they both comply with all relevant standards, they are supposed to work together flawlessly. However, this is not a perfect world.

The active NID presents a good opportunity for the service provider to perform effective remote management and configuration. Knowing exactly what the device is at the premises will help this situation. Moreover, the active NID/ATU-R presents a simple and familiar interface(s) to users (such as 10BASE-T). There is no need to worry about whether premises wiring will carry DSL signals because there will be none within the premises itself.

Naturally, future deregulation trends in many countries could alter the applicability of active NID solutions. Many nations see deregulation as an ultimate goal, although projected timelines vary greatly.

Please note that in some circumstance, even in the United States, an active NID configuration arrangement is still possible. It all depends on the willingness of the relevant regulating bodies to go along with service providers' plans and tariff agreements. In fact, some initial ADSL offerings will be based on this active NID/passive CPE model.

16

International Issues and xDSL

ADSL and all xDSL technologies in general are not intended exclusively for use within the United States, of course. The goal of maximizing use of the existing copper local loop plant is a worthy one, and not only a concern of service providers in the United States. All developed countries, telecommunications-wise, have a relatively extensive infrastructure of copper local loops. Not all are as old as the loop infrastructure in the United States, but there are still challenges in terms of usage and increasing network access speeds on these loops. Many of the concerns about local loops are truly global in nature.

This is not to say that there are not significant differences between the copper local loop infrastructure within the United States and throughout the world. Some appear to be simply cosmetic, such as the difference between measuring distances in kilometers instead of miles and kilofeet, but others are more profound and should be addressed. This chapter considers the issues surrounding the international use of xDSL, including not only the obvious differences in wire gauges and distances, but also considering the degree of pre-existing wiring infrastructure.

For the more extensive the copper local loop infrastructure is in a given country, the more interest there will be in xDSL products, equipment, and services. These differences are certainly worth pointing out in a chapter of their own.

Before looking at local loops in particular, a few general comments about telephone networks around the world may be in order. First and foremost, tariffs for telephony services vary widely around the world. Many countries do not have monthly flat rate services, but they may have flat rate service on a per-call basis, which is at least similar in concept. Some countries have a monthly maximum on measured service, which amounts to a kind of price cap. The point here with regard to DSL is that the switch clogging effects of Internet access due to flat rate services might not be as severe around the globe as they are in the United States. It is worth noting, however, that the traffic engineering practices (such as the use of Erlang B tables) used to design telephone networks is universal.

Also, 911 is not a universal emergency number. In business environments, activation digits for special services or outside lines may be different. For instance, an outside line in many countries is obtained by dialing a 0 and not a 9 as is common in the United States.

Cable TV, which is almost as universally available in the United States as telephone service, is not found all over the world. Much of the world's television is delivered by way of government-controlled networks. Most countries only have three or four television stations anyway, making 60-channel cable TV networks somewhat useless. However, satellite systems are becoming popular in many countries with developed telecommunications infrastructures, which is an important point because cable modems are frequently mentioned as an alternative to DSL technologies, but when the marketing considerations broaden to include non-U.S. markets, the applicability of cable modems as a broadband access solution falls off dramatically.

Oddly, ISDN varies widely around the world. The United States follows something known as *National ISDN* (NISDN) whereas many other countries have followed a more international flavor of ISDN. As a result, much of the ISDN deployed in the United States is incompatible with other countries' forms of ISDN. Even when it comes to connectors, the world is not totally on the standard modular connectors known as RJ-11 or RJ-45 as the United States is. There may be an RJ-11 in addition to the "national connector" on the telephone handset, and so even the simple act of plugging something in can be a challenge.

There are other differences, of course, but the differences of most concern here involve the local loop itself.

Local Loops and the ROW

Naturally, there are many types of local loops outside of the United States that do not necessarily correspond to the common local loop architectures developed and deployed in the United States. All other countries must adapt local loop technology for their local environments, generally categorized as the *Rest of the World* (ROW) to easily distinguish them from the United States. Some of the changes are more or less straightforward, such as the changes from feet to kilometers or from American wire gauges (AWG) to international wire gauges, measured in millimeters, but other differences are more profound. Because all countries are interested in higher-speed access lines and xDSL in general, this topic is an important one.

It is always good to keep in mind when dealing with international situations that the telecommunications infrastructure in the United States is somewhat unique. After all, the telephone was invented in the United States, and many other countries are not as developed as the United States given this built-in head start.

One of the most significant differences in local loop infrastructures is the fact that not all countries make extraordinary efforts to reach distant customers. Although it is true in many countries—especially developed telecommunications countries like the United States—that governments and telephone companies have a firm commitment to the concept of universal service, where everyone who can pay for a telephone can get one, this is not always the case.

If a potential subscriber cannot generate enough revenue to cover the cost of extending the local loop to the premises, so much the worse for the customer. Many other countries' local exchanges therefore serve more compact areas than in the United States. This makes the use of loading coils almost unheard of except on long trunk circuits between local exchanges. Oddly enough, trunk loading coils are never used in the United States today, due mainly to the nearly universal use of digital trunks in the United States.

Of course, in the ROW, mixed gauges and bridged taps still sometimes exist. In fact, mixed gauges are even more common in many countries than in the United States because the ROW compensates for signal loss with creative wire gauge changes rather than loading coils in most cases. But most European countries do not have bridged taps at all.

Some countries, especially developed telecommunications countries, have a high percentage of digital access lines. Some of this is due to ISDN deployment, but not always. Germany has about 10 percent of its access lines digitized already, and 50% of the world's ISDN lines, because backward compatibility is not so much an issue as new deployment after reunification. South America has many digital access lines, but few ISDN switches. And so on.

Given the enormous variability of local conditions for telecommunications infrastructure in the ROW, it is hard to make valid generalizations. One of the main things to look for is to divide the countries into emerging, developing, and developed telecommunications infrastructures, as does the United Nations and ITU. This grading is based mainly on the infrastructure maturity of the country in question. The internationally accepted infrastructure maturity indicator has been *telephone penetration*: the ratio of the number of telephones to the population of a given country.

International Wire Gauges

In the United States, the copper wire used in the local loop is manufactured according to the AWG sizes. Naturally, the rest of the world feels no urge to conform with this specification, and indeed other countries do not. It is universal outside the United States to measure wire gauges in simple terms of millimeters (one thousandth of a meter, which is 3.28 feet).

Table 16-1 shows the differences between wire gauges used in the United States and in the ROW. The AWG is shown in the left column. The center column translates the AWG to an individual wire's diameter in millimeters. The rightmost column shows the closest international gauge to the AWG in question.

Table 16-1

International Wire Gauges

American Wire Gauge (AWG)	Diameter (mm)	Closest International Gauge (mm)
19	0.912	0.9
22	0.643	0.63
24	0.511	0.5
26	0.404	0.4
28	0.320	0.32

The most common gauges in use internationally are 0.4 mm and 0.5 mm, which corresponds to 26 and 24 gauge wire in the United States. However, it is common is many local loop configurations to see 0.62 mm (22 gauge) or 0.9mm (19 gauge) wire in the ROW, as well. Even 0.32 mm (28 gauge) wire is used on some local loops. Keep in mind that the ROW can be much more creative than the United States when it comes to mixing wire gauges and constructing bridged taps.

Kilofeet to Kilometers

The United States is one of the few countries left in the world that insists on using the traditional English system of measurement (inches/feet/miles). The rest of the world uses the more rational metric system, based as it is a tens and hundreds instead of twelves and other multiples and fractions of this basic factor. There is nothing critical about technology when seeing DSLs in terms of feet instead of meters. It is just quite awkward to try to convert units mentally.

Table 16-2 shows the most common local loop lengths used in loading coil situations in the United States. The rightmost column translates these figures to kilometers. Keep in mind that The United States and one small African nation are the only countries in the world that are not on the metric system.

For local loop purposes, the table shows local loops of 12,000 feet (3.66 km), the point at which creative steps are taken to extend its reach. The table also lists 18,000 (5.49 km), long considered the ISDN limit. Even 30,000 feet (9.15 km) is listed to show the extent of some local loops in some parts of the United States (legends of 20-mile (105,600 feet) local loops running on barbed wire fences are common in the Far West).

For all other conversion purposes, keep in mind that 1 km equals 3.28 kft or 3,280 feet. For going the other way, 1,000 feet equals 0.305 km.

Table 16-2

Kilofeet and Kilometers

Feet	Kilometers
12,000	3.66
14,000	4.27
16,000	4.88
18,000	5.49
30,000	9.15

Some Sample ROW Local Loops

Those concerned with ROW local loops are often faced with a bewildering array of local practices and variations in local loop construction. Fortunately, standards organizations, long concerned with testing methods for compliance among equipment vendors, have established a set of standard international local loops in common use around the world. For instance, a set of eight test loops is included even in the ANSI T1.413-1995 specification for ADSL. The four samples shown in Figure 16-1 are based on this document.

In its simplest form, a ROW local loop might consist of 2 to 4 km of either 0.4 or 0.5 mm twisted-pair copper wire. This is the first configuration shown in the figure.

The second configuration shows a mixed gauge situation where up to about 2.5 km of 0.4 mm wire is followed by 1.5 km of 0.5 mm wire. This is quite common. The 0.4 mm section can vary from 1.5 to 4.55 km in length under most conditions, extending the reach to about 6 km (18,000 feet), which is about the ISDN limit.

The third configuration shows a more complex mixture of gauges. It is common to take 0.32 mm (28 AWG) wire about 0.2 km (only some 656 feet)

Figure 16-1
ROW Local Loops

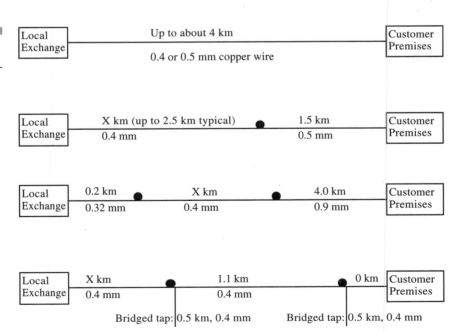

from the local exchange and then travel from 1 to 5 kms on 0.4 mm wire. At that point, 0.9 mm (19 AWG) wire travels the remaining distance of up to 4.0 km to the customer premises. Note that this mix extends the reach of the loop without the need for loading coils, at the cost of reduced voice quality.

Finally, a bridged tap configuration is shown. In this case, all of the wire is 0.4 mm. However, the bridged taps can extend up to 0.5 km (about 1,600 feet) from the loop itself. Note also that it is common to have one of the bridged taps right at the customer premises.

The other four configurations not shown are basically variations of gauge and distances on the third loop type in the figure. Those in need of more detail are referred to the ANSI document itself.

ROW Local Loops in General

It is hard to summarize local loop architectures outside of the United States in the so-called ROW. Considerable variation exists due to local conditions, both regulatory and with regard to existing infrastructure. Nevertheless, it is possible at least to attempt to make some valid statements about ROW local loops.

First, local loops in excess of 5 km are relatively rare, with 2 to 4 km as the typical local loop length. This was one reason why the maximum access line length for ISDN was established at 5.49 km (18,000 feet). In most parts of the world, this is more than adequate.

In spite of the high degree of modest length local loops, signal quality on local loops in many places around the world is quite poor. One reason is that loading coils are rare, and efforts are made to extend loops with creative gauge mixing, often with poor results. Oddly, loading coils are often used on analog trunks in the ROW, due to their longer distances between local exchanges. This practice is no longer done in United States.

Also, mixed gauges are even more common in the ROW than in the United States, which is at least partly because of the effort to extend loop distance without the need for loading coils.

In the ROW, there are usually fewer switches and access lines per unit area. For example, the United States has 22,000 switches of all types (some 13,000 of them local exchanges) to serve about 266.5 million people. This works out to about 12,000 people per switch. In contrast, Australia, considered to be as developed a country as the United States with regard to telecommunications, has some 5,000 switches to serve 18.2

million people. This works out to only 3,600 people per switch. Part of the difference is because of regulation, but another part is due to the smaller reach of the serving area.

As a final note, some countries are not seeking to preserve their copper local loop infrastructure, but to more or less abandon it. These countries are moving right into wireless local loop or fiber-based loop services.

Infrastructures Around the World

xDSL technologies are of interest to service providers not only in the United States, but around the world, and yet each individual country has its own distinct approach to handling their telecommunications infrastructure. These differences reflect local conditions that vary from country to country and are to be expected. For example, a country without a concept of eminent domain to pre-empt private rights-of-way for public use might favor undersea cables for coastal trunks, whereas a country comprised of islands might favor as much wireless links as possible.

Whatever the circumstances, it turns out that there are three recognized categories of nations with regard to telecommunications: emerging, developing, and developed. This structure is established by the United Nations and ITU.

- *Emerging* nations have only rudimentary telecommunications infrastructures and services are typically offered only in major metropolitan areas.
- *Developing* countries have more mature telecommunications infrastructures based on advanced technology, and services are typically offered beyond the major metropolitan areas.
- *Developed* countries have fairly state-of-the-art telecommunications infrastructures, and services are typically offered almost everywhere.

The point is not to be critical of various countries, but rather that countries have much more in common within each of these three categories than they have with other countries in the same geographical area that are not in their category. And one of the most commonly used indicators of telecommunications category is a measure known as service penetration, which is the ratio of a service device to a number of people. The faults of this type of measure have been frequently pointed out. It is true that raw math cannot take into account many of the widespread variations in coverage and local conditions. However, this has not stopped this form of statistic from being used almost universally, especially by the countries that it concerns the most.

Table 16-3 shows the penetration rate of basic telephony service for a variety of countries around the world. Note that telephony penetration closely follows other technologies such as television, but there are exceptions (Germany, for example). In fact, countries such as Germany point out the hazards of using pure numbers like this to establish categories along these lines. Of course, this stops no one from using statistics of this nature and such categorization is done all the time.

Table 16-3

Penetration Rate of Telephones and Television

Country	Population (M)	TVs (M)	Penetration Rate	Telephones (M)	Penetration Rate
United States	266.5	215	80.7%	182.5	68.5%
Canada	28.8	11.5	39.9%	15.3	53.1%
Germany	83.0	85.0	100%	44.0	53.0%
Australia	18.2	9.2	50.5%	8.7	47.8%
Malaysia	20.0	2.0	10%	2.5	12.5%

xDSL and International Conditions

Due to the large degree of variation among countries when it comes to the telecommunications environment, sweeping generalizations are difficult to make. Nevertheless, some statements still apply to countries that share a category of telecommunications infrastructure maturity.

The United Nations considers a nation developed if the penetration rate for telephony is 45 percent or more. The nation is developing if the penetration rate is between about 20 to 30 percent, and emerging if the penetration is less than 20 percent or so. Many nations fall into this emerging category. Others are considered to be developing. Few have telecommunications infrastructures mature enough to be considered developed.

What has this to do with xDSL? It is mostly developing and developed telecommunications nations that are most interested in xDSL due to the presence of a mature infrastructure. Other nations have more to explore from a telecommunications perspective than the need to preserve existing copper-wire local loops.

Also, wireless services are popular in some environments, especially in countries consisting of islands or those with densely forested interiors or uplands. For these countries, xDSL becomes less of an issue except in major metropolitan areas.

Regardless of category or geographical circumstance, all countries are interested in deregulation and broadband access services. Both are desirable from an xDSL perspective. Deregulation spurs competition and service deployments, whereas broadband access is always a given for things such as video and Internet access.

Using HDSL/HDSL2 Around the World

There is lot of interest in using HDSL and HDSL2 in many countries around the world. Without HDSL, providing high-speed digital links at 1.5 Mbps or 2.0 Mbps can be an expensive proposition.

A potential drawback for HDSL until now was the fact that 2.0 Mbps service required the provision of two pairs of wires (and even three pairs before 1995). Of course, the 2.0 Mbps E1 service is widely used around the world except in the United States, Japan, and the Philippines. The full HDSL2 specification for 2.0 Mbps service on one pair of wires is greatly awaited for this reason.

Regardless of the number of pairs, HDSL on existing wire pairs offers numerous advantages over installing new infrastructure. A study reported in *Data Communications* magazine (October 1997, p. 75) shows why.

In order to provision E1 2.0 Mbps service on new wires, streets must be dug up. It can take two technicians up to 30 days to provision the service. The cost to the service provider is about $4,000 US for a 5 km (16,000 feet) link.

Now consider the same E1 2.0 Mbps service using HDSL/HDSL2. Only six person-hours are needed, which could be one person for six hours or two people for three hours. The cost to the service provider is now only $1,800 US for the same 5 km link. This lower provisioning cost can also keep down the price of the service to the customer.

Using ADSL Around the World

Although ADSL might appeal to almost every country, most of the interest in ADSL services to date has come from compact, developing/developed countries. This is not surprising because many emerging countries do not have the infrastructure in place to be overly concerned about preserving existing local loops, although there are exceptions.

This section briefly mentions three examples of ADSL initiatives taken by countries outside of North America. In Singapore, Singapore Telecom Limited has ADSL equipment in 27 local exchanges. More equipment is to be added, and, in 1998, 100 percent of the population of Singapore will have ADSL service available to them.

In Finland, the Helsinki Telephone Company has a pilot ADSL service trial in progress. Although limited in number, the participants will have Internet access through ADSL available to them in 1998.

And just to show it is not all an ADSL world, Hong Kong Telecom Limited is seriously looking at running ATM cells on VDSL links. The bandwidth and services offered will be unparalleled but limited to access lines of about 1 km or so. Luckily, the one kilometer limit is no problem at all in Hong Kong.

NID Issues Revisited

Although most of the issues regarding network interface devices have been addressed more fully previously, this chapter might be a good place to review them in summary.

Countries that have a regulated telecommunications environment can use a bundled, active NID and ATU-R. When the active NID/ATU-R issues are considered, this arrangement guarantees maximum service provider control over the entire network loop. The service provider might have an easier time configuring, troubleshooting, and maintaining the loop, although this is not necessarily a given, as has been pointed out previously.

This arrangement means that the service provider has no premises wiring concerns at all when it comes to DSL. The interfaces on the ATU-R are governed by standards other than DSL and are beyond the scope of DSL.

Of course, there is no ADSL Terminal Unit-Central Office (ATU-C) to ATU-R (or any DSL devices) interoperability concern at all in this environment. Interoperability can be a problem in deregulated environments, where there is no guarantee that a customer's CPE will function properly with the service provider's equipment if they come from different vendors. Of course, if they both comply with all relevant standards, they are *supposed* to work together flawlessly.

The active NID presents a good opportunity for the service provider to perform effective remote management and configuration. Knowing exactly what the device is at the premises will help this situation.

Also, the active NID/ATU-R presents a simple and familiar interface(s) to users (such as 10BASE-T). There is no need to worry about whether premises wiring will carry DSL signals because there will be none within the premises itself.

Future deregulation trends in many countries could alter the applicability of active NID solutions. Many nations see deregulation as an ultimate goal, but the projected timelines to deregulation vary greatly from country to country.

Appendix A

xDSL Services

It is somewhat risky to give a listing of available xDSL services in a book because hardly a week goes by without some new announcement of xDSL service trials or availability. Even if an announcement is made, not all of the details about pricing and the like are always available. Nevertheless, the exercise should be attempted.

Most of the information here comes from the October 27, 1997 issue of *Network World* magazine. The tables in question appear on pages 60 and 62 in a slightly fuller version than presented here. Those lucky enough to live in one of the areas listed are urged to contact the service providers directly. It should be noted that information such as this quickly becomes dated, and the tables make no real attempt to distinguish between ambitious, well-funded roll-outs and tiny trials of perhaps a few dozen lines. However, there is still value even with these limitations. The contact information, also taken from the article tables, is shown in Table A-1.

Some words of explanation are in order before presenting the service tables themselves. First, the abbreviation *TBA* means "to be announced" and indicates that the service provider has no yet made firm decisions about their exact plans in one area or another. Next, a *blank* indicates that no information about a particular area is available or that the item is not applicable. Finally, keep in mind that there may be variations that are not documented.

Even in this brief form, some general conclusions about xDSL services can be drawn. Note that there are five xDSL service areas listed: ADSL, HDSL, IDSL, RADSL, and SDSL. This is impressive given the newness of these technologies. Only VDSL has yet to make an appearance on more than an experimental basis. Nine vendors are already present on the lists: Alcatel, Amati, Ascend, Copper Mountain, Netspeed, Paradyne, PairGain, Tut Systems, Westell. This will only increase with time.

Note that many of the service providers include the remote termination unit with the service package, more than half in the case of ADSL alone and almost all of the other xDSL types. This would seem to contradict United States deregulation rules but is to be expected in the early days of xDSL. ADSL equipment is expected some day to become as interchangeable as modems are now. In the meanwhile, users want the service to work correctly.

Table A-1

xDSL service
providers

Company	Telephone Number	Web Site
Ameritech	(800) 832-6328	www.ameritech.com
Aspen Internet Exchange	(970) 927-0336	www.aspn.net
Dakota Services Limited	(414) 221-9212	www.dslnet.com
GTE Corporation	(800) 483-4926	www.gte.com
HarvardNet	(800) 772-6771	www.harvard.net
Intelecom Data Systems	(888) 885-4437	www.ids.com
InterAccess	(312) 496-4400	www.interaccess.com
Pacific Bell	(888) 884-2375	www.pacbell.com
Signet Partners	(512) 306-0700	www.signet.com
Southwestern Bell Telephone	(888) 792-3751	www.swbell.com
Transport Logic	(503) 243-1940	www.transport.com
US WEST !nterprise	(888) 634-2879	www.uswest.com/interprise
UUNET Technologies	(800) 488-6384	www.uu.net
WebWave Resources	(714) 497-8611	www.webwave.net

A few more comments about the entries. When the service provider has chosen a marketing name for the service, it is included in parentheses after the company name. Where dates are given, they apply to the trial areas and then the service deployment plans. Where there is a variety of speeds listed, they generally apply to different loop lengths and geographical areas. The service providers should be consulted directly for details.

The applications listed belong to one of three categories. First, the service is intended for fast Internet access. Second, the service can be used for LAN interconnections. In this case, there will be no Internet component of the service. Finally, the service can be used for remote access. Usually, this would mean that a remote PC or small cluster of PCs links to a corporate LAN site, but not always.

At least four more service providers are intending to offer xDSL services. The tables do not include them, but they are noted here for the sake of completeness. These are Network Access Solutions of Herndon, VA; Vitts of Manchester, NH; North Point Communications of San Francisco, CA; and Covad Communications of Santa Clara, CA (who have just announced firm plans). Most of these companies have achieved CLEC status in a number of states and plan to resell local loops with xDSL as an added value to their services.

Finally, keep in mind that this list is concerns only the United States. Bell Canada's agreement with Alcatel for ADSL equipment, for example, is not included. This exclusion is not to slight international efforts, but to reflect the emphasis of the original article. International trials based on information from the ADSL Forum is given in tables following the first five tables.

The ADSL Forum lists several companies not included in the article cited here, but they are mentioned here for the sake of completeness. Except where noted, these are all ADSL services. Advanced Corporate Solutions is involved with Transport Logic throughout the Pacific Northwest. ALLTEL, a CLEC and ISP, offers service to Dalton, Georgia and Hudson, Ohio. AUSNet Services is an ISP in Portland, Oregon, as is Branch Internet Services in Ann Arbor, Michigan. Western Network Management is in western Colorado and eastern Utah, and ioNET is an ISP in Oklahoma City and Tulsa, Oklahoma.

Some larger and smaller telephone companies are experimenting with ADSL services, as well. Bell Atlantic has some services in northern Virginia, Pittsburgh, and Boston. BellSouth has service in Atlanta, Georgia. Cincinnati Bell offers services in Cincinnati. The LEACO Rural Telephone Cooperative offers service in Lovington, New Mexico, and Northwest Iowa Telephone (with Northwest Iowa Power) offers service in partnership with MCI in Detroit, Iowa, and New York.

ADSL Services

Company	Trial Area	Service Area	Speed Upstream	Downstream	CPE Vendor	Included?	User Cost	Application	Cost Installation	Monthly
Advanced Corporate Solutions	Pacific NW 4/97	Pacific NW	up to 1 Mbps	up to 2.5 Mbps				Internet, LAN, other		
ALLTEL	Dalton, GA Hudson, OH Harrison, AR	TBA	64 kbps	1.5 Mbps				Internet, LAN		
Ameritech	Wheaton IL 10/96	TBA	64 kbps	1.3 Mbps	Alcatel service, Westell trial	TBA	TBA	Internet, LAN, remote access	TBA	TBA
Bay Junction	San Jose, CA 12/97	?	1.5 Mbps					Internet, LAN		
Bell Atlantic	Northern VA Fairfaxe, VA Boston, MA 1998	64 kbps	64 kbps 1.5 Mbps 64 kbps	1.5 Mbps 1.5 Mbps				Internet Video on demand Internet, LAN		
Bell Atlantic & CMU	Pittsburg, PA 2/98	64 kbps	1.5 Mbps					Various		
BellSouth	Atlanta, GA 1998 Birmingham, AL	640 kbps	64 kbps ?	6 Mbps			?	Internet, LAN		
Concentric	Northern CA 10/97		384 kbps	up to 1.5 Mbps				Internet, LAN		
Covad	SF Area, CA 12/97		?	144 kbps – 1.5 Mbps				Internet, LAN		

Company	Trial Area	Service Area	Speed Upstream	Downstream	CPE Vendor	Included?	User Cost	Application	Cost Installation	Monthly
DNAI	SF Area, CA 12/97		384 kbps	up to 1.5 Mbps				Internet, LAN		
GTE	Irving, Plano, TX 10/96	TBA	64 kbps 160 kbps	1.5 Mbps	Westell Amati	Yes		Internet, LAN, remote access	TBA	TBA
	Redmond, WA 8/96	TBA	64 kbps 160 kbps 384 kbps 640 kbps	1.5 Mbps 6.0 Mbps	Westell Amati	Yes		Internet, LAN, remote access|	TBA	TBA
	Durham, NC; West Lafayette, IN	TBA 10/97	64 kbps 160 kbps 384 kbps	1.5 Mbps 4.0 Mbps	Westell Amati	Yes		Internet, LAN, remote access	TBA	TBA
LEACO	SE New Mexico 7/97							Internet		
MCI	Iowa 8/97	1998	64–640 kbps	1.5–6 Mbps				Internet, LAN		
Northland Comm.	NY 1997		64 kbps	1.5 Mbps				Internet, LAN		
SBC	Houston, TX		640 kbps	6 Mbps				Internet		
South-western Bell (FasTrak)	Austin, TX Q4 97	TX	384 kbps	384 kbps 1.5 Mbps	Alcatel	No	$435-$660	Internet, remote access	$125	$150–250
Pacific Bell (FasTrak)	LA, SF, San Jose, CA Q4 97	CA	384 Kbps 1.5 Mbps	384 kbps	Alcatel	No	$435-$660	Internet, remote access	$125	$150–250

continues

ADSL Services (Concluded)

Company	Trial Area	Service Area	Speed Upstream	Downstream	CPE Vendor	Included?	User Cost	Application	Cost Installation	Monthly
Signet Partners	Basic	Austin, TX 1/97	620 kbps	247 kbps	Netspeed	No	$1295	Internet, LAN, remote access	$1250	$535
	Managed	Austin, TX 1/97	620 kbps 960 kbps 1.6 Mbps 2.24 Mbps	247 kbps 408 kbps 680 kbps 942 kbps	Netspeed	Yes	$1295	Internet, LAN, remote access	$1325	$695-5695
	Alternative-Full	Austin, TX 1/97	2.24 Mbps	942 kbps	Netspeed	No	$1295	Internet, LAN, remote access	$1250	$535+
Slip.net	SF Area, CA	1998	384 kbps	up to 1.5 Mbps				Internet LAN		
Vitts	NH	?		up to 6 Mbps				Various		
WebWave Resources	Laguna Beach, CA 6/97	TBA	64 kbps	1.6 Mbps	PairGain	Yes		Internet access	$1600	$200-800

HDSL Services

Company	Trial Area	Service Area	Speed Upstream	Speed Downstream	CPE Vendor	CPE Included?	User Cost	Application	Cost Installation	Cost Monthly
Aspen Internet Exchange	None	Roaring Fork Valley, CO 6/96	768 kbps 1.1 Mbps	768 kbps 1.1 Mbps	PairGain	Yes		Internet, LAN	$1250–1500	$100–500
Dakota Services Limited	TBA 11/97	Milwaukee, WI TBA	1.5 Mbps	1.5 Mbps	Paradyne	No	Varies	Internet, LAN, remote access	Varies	Varies
US WEST Enterprise (MegaHome, MegaOffice)	Boulder, CO, Arizona 5/96 & 6/97	Utah 10/97	192 kbps 320 kbps 704 kbps	192 kbps 320 kbps 704 kbps	PairGain	Yes		Internet, LAN, remote access	$215	$40, $65, $125
WebWave Resources (MegaService)	Laguna Beach, CA 5/97	Laguna Beach, CA 6/97	768 kbps	768 kbps	PairGain	Yes		Internet, LAN	$1600	$200–800

IDSL Services

Company	Trial Area	Service Area	Speed Upstream	Speed Downstream	CPE Vendor	CPE Included?	User Cost	Application	Cost Installation	Cost Monthly
UUNET Technologies (Preferred Access)	San Jose CA 6/97	LA, San Diego, Silicon Valley, NY, 117 cities 1998	128 kbps	128 kbps	Ascend	No	$745	Internet, LAN, remote access	$3500	$150-200 plus 650–750 Internet usage charge

RADSL Services

Company	Trial Area	Service Area	Speed Upstream	Downstream	CPE Vendor	Included?	User Cost	Application	Cost Installation	Monthly
AUSNet	Portland, OR 2/97	92–972 kbps	1.5 Mbps					Internet, LAN		
Branch Internet	Ann Arbor, MI	1 Mbps	up to 2 Mbps					Variety		
Cincinnati Bell	Cincinnati, OH 1/98	150–400 kbps	1.5–6 Mbps					Internet, LAN		
Dakota Services Limited (DSLNet Express)	Milwaukee 5/97	Wisconsin 7/97	1.0 Mbps	1.5 Mbps	Paradyne	No	$1595	Internet, LAN	Varies	$650
easy.net	Denver, CO 8/97	272 kbps -	640 kbps - 1 Mbps	2.5 Mbps				Internet, LAN		
GTE Comm.	South CA 11/97		256-384 kbps	680 kbps-1.5 Mbps				Internet, LAN		
HarvardNet	Boston, MA, Portland, ME 9–10/97	Boston, Cambridge, Waltham, MA, 11 NE cities 12/97	768 kbps	1.0 Mbps	Paradyne	Yes		Internet, LAN, remote access	$1600	$299–699
Intelecom Data Systems (IDS Fastlane)	Rhode Island 4/97–Q198	Rhode Island 4/97–Q1 98	1.06 Mbps	2.56 Mbps	Paradyne	Yes		Internet, LAN, remote access	$2000	$600–1500

Company	Trial Area	Service Area	Speed		CPE Vendor	Included?	User Cost	Application	Cost	
			Upstream	Downstream					Installation	Monthly
InterAccess Company (DSL Internet)	Chicago 7/96	Chicago and suburbs 9/96 & Q1 98	1.08 Mbps	2.0 Mbps+	Paradyne	No	$650 900	Internet access	$200–	$150–1200
Interastate	Westpoint, GA	1 Mbps	7 Mbps					Internet		
ioNET	OK	OK & TX	1 Mbps	7 Mbps				Internet		
Transport Logic (DSL Access)	Portland, OR 3/97 & 10/97	Portland, OR 3/97 & TBA	1.08 Mbps 1.0 Mbps	2.5 Mbps 2.0 Mbps	Paradyne Tut Systems	No	Varies	Internet, LAN, remote access	$300	$675+
WebWave Resources (Mega-Service)	Laguna Beach, CA 12/97	Laguna Beach TBA	1.0 Mbps	3.0 Mbps	PairGain	Yes		Internet access	$1600	$200–1600

SDSL Services

Company	Trial Area	Service Area	Speed Upstream	Downstream	CPE Vendor	Included?	User Cost	Application	Cost Installation	Monthly
HarvardNet (SDSL Service)	Boston 11/97	13 cities in MA, ME, NH, and RI, 2/97 & 2/98	128 kbps 384 kbps 768 kbps	128 kbps 384 kbps 768 kbps	Paradyne, PairGain, Copper Mountain	Yes		Internet, LAN, remote access	$1600	$299–699
UUNET Technologies (Preferred Access)	TBA	LA, San Diego, Silicon Valley, & NY 117 cities 1998	768 kbps	768 kbps	Ascend		TBA	Internet, LAN, remote access	TBA	TBA

xDSL Services in Canada (ADSL Forum)

Company	Trial Area	Service Area	Speed Upstream	Downstream	Application
British Columbia Telephone (RADSL)	Greater Vancouver and Victoria, BC, 11/96	Greater Vancouver and Victoria, BC, 9/97	64-640 kbps	1.5 to 8 Mbps	Internet access, LAN
Bell Canada	Kanata, Ontario and St. Bruno, Quebec, 9/96	Ottawa, Hull, and Quebec city areas 10/97	1.0 Mbps	2.2 Mbps	Internet access, LAN
CADVision (RADSL)	Calgary, Alberta	Calgary, Alberta 11/96	1.0 Mbps	2.5 Mbps	Internet access
SaskTel	Regina and Saskatoon	Regina and Saskatoon 11/96	64 kbps	1.5 Mbps	Internet access
New Brunswick Telephone	Moncton, New Brunswick 12/96	Moncton, New Brunswick Q2 97	64 kbps	1.5 Mbps	Internet access
Manitoba Telephone System	Winnipeg, Manitoba 11/96	Winnipeg, Manitoba Q3 97	64 kbps	1.5 Mbps	Internet access
Maritime Telephone and Telegraph	Novia Scotia 4/97	Novia Scotia Q3 97			
Telus Comm.	Edmonton and Calgary, Alberta 3/96	Edmonton and Calgary, Alberta 10/97	64 kbps	1.5 Mbps	Internet access, LAN
UUNET Canada	Toronto, Ontario 6-12/96	Toronto, Ontario, Q1 97	64 kbps	1.5 Mbps	Internet access

xDSL Services in Europe (ADSL Forum)

Company	Trial Area	Service Area	Speed Upstream	Downstream	Application
AMUSE (European Commission program)	Milan, Italy (Telecom Italia) early 1997		640 kbps	8.2 Mbps	Video on demand, Internet access
Belgacom (Belgium)	9/95				Video on demand
British Telecom (UK)	Colchester and Ipswich 6/95	TBA 1998	384 kbps	2.0 Mbps	Video on demand, Internet access
Clinet OY (Finland)	Finland 1/97	Finland 9/97	750 kbps	2-6 Mbps	Internet, Intranet, and Extranet access
Deutsche Telecom AG (Germany)	Nuremburg, North Rhine-Westphalia 97	TBA			VOD, Home shopping, Internet access
France Telecom	Brittany 11/96		640 kbps	8 Mbps	Video on demand
Helsinki Telephone Co. (Finland)	Helsinki 8/95	Helsinki 2/97	9.6 kbps	2 Mbps	VOD, remote work, Internet access
Kingston Comm.-Hull (England)	Hull, Fall 97	TBA			Video on demand
Swiss Telecom PTT (Switz.)	Grenchen 9/95		9.6 kbps	2 Mbps	VOD, Internet access
Telecom Eirann (Ireland)	TBA		2 Mbps	2 Mbps	Internet access, LAN
HDSL Telecom Finland	TBA				Internet access, LAN
Telecom Italia (Italy)	Turin, early 97	Plans 1.5 million users by year 2000	272 kbps - 1 Mbps	640 kbps - 2.24 Mbps	Internet access, video conferencing
Telefonica Espana (Spain)	TBA 1997				
Telenor (Norway)	1/96				Video on demand
Tellia AB (Sweden)	Stockholm 9/95	Stockholm Summer 97			Internet access

xDSL Services in Middle East/Africa (ADSL Forum)

Company	Trial Area	Service Area	Speed Upstream	Downstream	Application
Bezeq (Israel Telecom)	Tel Aviv, Jerusalem, 4/96		9.6 kbps	2 Mbps	Video on demand

xDSL Services in Asia/Pacific (ADSL Forum)

Company	Trial Area	Service Area	Speed Upstream	Downstream	Application
Agricultural Assoc. of Ina City on Broadcasting and Telephone Services	Ina City, Nagano Prefecture, Japan 9/97	TBA	1 Mbps	8 Mbps	Internet access, LAN
Chunghwa Telecom (Taiwan)	Central Taipei 12/96		9.6 kbps	1.5 Mbps	Near VOD, remote access
Hongkong Telecom (VDSL)	Hongkong, 12/96	250,00 users by year 2000	1.5 Mbps	51 Mbps	Video on demand
Japanese UNIX Business Assoc. (UBA) (RADSL)	Ina City, Nagano prefecture, Japan 9/97	TBA	Up to 1 Mbps	Up to 2.2 Mbps	Internet access, LAN
Korea Telecom	Pusan and 5 cities, 8/96	3.5 million users by year 2000	128 kbps	4 Mbps	Internet access, VOD, education, shopping, and so on
Nippon Telegraph & Telephone (NTT)		Spring 1998?			
Singapore Telecom	5,000 homes 2/96-6/97	80,000 users by end 1998	168 kbps	5.5 Mbps	Internet access, VOD
Telstra (Australia)	Melbourne 4/10/96	Melbourne 2nd half 97	—	2 Mbps 8 Mbps	Broadcast, Video on demand

Appendix B

The ADSL Forum

The ADSL Forum is the vendor consortium concerned with the advancement and completion of ADSL standards. The Forum and their week have been mentioned several times in this book, and suffice it to say that this book could not have been written without the ADSL Forum and its members.

Rather than trying to reproduce information about the ADSL Forum, it might be better to let the organization speak for itself. The rest of this appendix is from the official ADSL Forum Web site at **www.adsl.com**.

Welcome to The ADSL Forum

Copper telephone lines are everywhere. There are over 700 million now, and the next century will begin with a number closer to one billion. Today, they connect telephones, fax machines, and computers at very slow speeds—28.8 kbps for modems and 128 kpbs for ISDN. Tomorrow, with ADSL, they will connect computers and televisions at speeds up to 9 megabits per second, which is 300 times faster than modems and 70 times faster than ISDN. ADSL will remove the last bottleneck to high-speed access for the Internet, corporate LANs, video on demand, video education, and myriad applications we can only imagine today.

The ADSL Forum was formed in late 1994 to help telephone companies and their suppliers realize the enormous market potential of ADSL. Our assistance comes in two forms—technical and marketing. The Forum's marketing programs attempt to uncomplicate ADSL's inherent technical complexity and spread the news. The Forum's public output, therefore, mixes the tutorial with the promotional. Our web site, for example, offers short explanations of all DSL technologies, the system environment for ADSL, and the basic features of the ADSL market. It also follows market dynamics—trials, vendors, tariffs when they happen, and opinion essays designed to give texture to the market's general frame. Forum members speak at numerous conferences and trade shows, and the Forum now has an active public relations program. Next year the Forum will be presenting one-day conferences before each of its four meetings to offer— both to the public and its members—up-to-date information on ADSL, megabit applications, and next-generation access networks. At present, the Forum's formal marketing work falls into four major projects:

1. Web Site
2. Conferences
3. Public Relations
4. Competitive Analysis (of, for example, cable modems)

ADSL is, in essence, a pair of modems on either end of a twisted-pair copper wire. To deliver megabit services, telephone companies must deploy new access systems that embody ADSL. These systems require protocols and connections for home networks and terminals, access networks that concentrate traffic and route signals to appropriate destinations, and network management to install, configure, maintain and migrate ADSL systems. The Forum's technical projects focus on these system features, without which ADSL could not perform its transmission triumphs.

At present, the Forum's formal technical work divides into seven major projects:

1. Interfaces
2. Access Networks
3. Packet Mode Protocols
4. ATM Mode Protocols
5. Migration
6. Network Management
7. Testing and Spectral Management

Each project develops technical reports through a working technique called *Working Texts*, which are web site documents that capture and organize work in progress. As work completes, the Working Text becomes a *Technical Report*, subject to membership approval, which is then made public and distributed by the Forum to interested parties. In many cases the Forum identifies a particular requirement for ADSL and then advises another standard body, such as T1E1, of this requirement, in the hopes that it will take suitable action. Indeed, The ADSL Forum finds itself working ever more closely with adjacent standards groups—T1E1.4, ETSI, DAVIC, the ATM Forum, and TR41.

The ADSL Forum has begun to realize that its role as a "forum"— a place to meet and discuss—has as much value as its technical and promotional work. We gather four times a year, twice in the U.S. and twice in Europe. With more than 200 members now (April 1997) the Forum includes most significant members of the telecommunications community. Our gatherings, therefore, create opportunities for informal exchanges afforded by spontaneous conversation and meetings around the coffee service.

It must be stated, in this context, that The ADSL Forum is an association of competing companies and must therefore abide by various anti-trust rules. These rules bar the Forum from engaging in market projections or discussions of issues such as costs, market shares, telephone company RFQs, or any other area that could incubate possible charges of price fixing. Although many such discussions are not in violation of anti-trust, separating those that are from those that do not often requires investigation into the source of information and its intent. The Forum therefore takes a so-called "green-line" approach of avoiding the issues altogether (as do most other similar associations). Doing so does not affect our basic work

How To Join

The ADSL Forum is a Corporation operated for the benefit of its members. Recognizing that its members compete, the Forum strictly observes applicable anti-trust laws. Within these laws The ADSL Forum intends to promote the market for ADSL and facilitate development of interoperable ADSL-based network components.

MEMBERSHIP GUIDELINES
PRINCIPAL MEMBERSHIP—$5,000 USD

Principal Membership in the ADSL Forum entitles the member company to attend all annual, general, and special meetings of the Forum, as well as all committee meetings of the Forum. Principal Members have one vote each on all Forum issues. Principal Members have access to all working documents, contributions, technical reports, and meeting minutes. Principal Members are eligible to run for Board of Directors and Committee officer positions. Principal Members are eligible to submit Technical Committee and Marketing Committee contributions. Principal Members may subscribe to all ADSL Forum email mailing lists.

NOTE: *The Bylaws of the ADSL Forum bars two divisions of the same company or any affiliates of a Principal Member from also being Principal Members. Other divisions and affiliates may be Auditing Members.*

TO BECOME A PRINCIPAL MEMBER:

— Complete and submit a membership application

— Sign and return the last page of the Bylaws (contact the ADSL Forum office for a copy of the Bylaws)

SMALL COMPANY PRINCIPAL MEMBERSHIP—$2,500 USD

Small Company Principal Members are entitled to the same privileges as Principal Members. Small Company Principal Members must have had annual revenues under $10 million USD in the previous year to qualify for this membership class.

TO BECOME A SMALL COMPANY PRINCIPAL MEMBER:

— Complete and submit a membership application

— Sign and return the last page of the Bylaws (contact the ADSL Forum office for a copy of the Bylaws)

— Provide sufficient evidence that your company meets the qualification to become a Small Company Principal Member

AUDITING MEMBERSHIP—$1,500 USD

Auditing Membership in the ADSL Forum entitles the member company to attend all annual, general, and special meetings of the Forum. Auditing Members have access to all working documents, contributions, technical reports, and meeting minutes. Auditing Members may subscribe to all ADSL Forum e-mail mailing lists. Auditing Members may NOT attend committee meetings of the Forum. Auditing Members may NOT vote on any Forum issues. Auditing Members are NOT eligible to submit Technical Committee or Marketing Committee contributions.

TO BECOME AN AUDITING MEMBER:

— Complete and submit a membership application

— Sign and return the last page of the Bylaws (contact the ADSL Forum office for a copy of the Bylaws)

You may also contact the Forum office for a Membership Application to be mailed to you.

The ADSL Forum
39355 California Street
Suite 307
Fremont, CA 94538
Phone: +1.510.608.5905
Fax: +1.510.608.5917
adslforum@adsl.com

ADSL Forum Members

December 17, 1997

The links provided by this homepage to other homepages are being provided as a service by The ADSL Forum. The ADSL Forum, however, neither has nor assumes any responsibility or liability, either expressed or implied, for any such links or the accuracy of any information or service provided by any homepage or company to which this homepage is linked. The use of the links provided by this homepage is on an "as is" basis and at the users sole risk. The links provided by this homepage are not intended to be and are not an endorsement of any service, product, or company.

PRINCIPAL CLASS
75 Members

3Com

ADC Telecommunications

AG Communication Systems

ATM Ltd.

AT&T Laboratories

Advanced Fibre Communication

Alcatel Telecom

Amati

Ameritech

Analog Devices

Applied Innovation

Ariel

Ascend Communications

Atlantech Technologies

Bay Networks

Bell Atlantic

BellSouth Telecommunications

British Telecom

Cabletron Systems

Cayman Systems

IBM TJ Watson Research

Interphase

Italtel

Lucent Technologies

Metalink

Microcom

Microsoft

Mitsubishi Electric Information
Technology America

Motorola Semiconductor

NEC

NetSpeed

Newbridge Networks

Nokia Telecommunications

Nortel

Orckit Communications

P.T.T. Telecom B.V.

PairGain Technologies

Paradyne

Performance Telecom

Cisco Systems

Communications Technology

Copper Development Association

Copper Mountain Networks

DSC Communications

ECI Telecom

ELSA GmbH

EPL Ltd

Efficient Networks

Ericsson Austria AG

France Telecom

Fujitsu Network Communications

GTE Corporation

Global Village Communication

GlobeSpan Technologies

Gorham & Partners

Hayes Microcomputer Products

Xyplex Networks

Promatory Communications

Pulsecom

Rad Data Communications

RedBack Networks

SGS-Thomson Microelectronics

Samsung A.I.T.

Siemens Stromberg-Carlson

Sourcecom

Southwestern Bell Technology Resources

Standard Microsystems Corporation (SMC)

Sun Microsystems

Telecom Italia

Telesis Technologies Laboratory

Telstra

Teradyne

Texas Instruments

Westell, Inc.

AUDITING CLASS
218 Members

3M

AKM

ATI Technologies

ATRI

AT&T Wireless Services

Accelerated Networks

AccessLan Communications

Addtron Technology

Advanced Micro Devices

Aptis Communications

LG Information & Communications

Level One Communications

Livingston Enterprises

MMC Networks

Madge Networks

Matsushita Electric Industrial

MediaLight

Mitel Semiconductor

Mitsubishi Electric Corporation

Motorola Information Systems Group

Aspen Internet Systems

AVIDIA Systems Inc.

Aware

Belgacom

Bell Canada

Bellcore

Bermuda Telephone Co.

Bosch Telecom

Broadband Technologies

Broadcom

Burr-Brown Corporation

C-DOT

CS Telecom

Cable & Wireless

Cadence Design Systems

Case Technology

Cellware GmbH

Centillium

CETECOM

Charles Industries

Cheng-Hwa Communications

Chrontel

Cincinnati Bell Telephone

Cirrus Logic

Coastcom

Comatlas

CommHome Systems Corporation

Compaq Computer

Conin Co., Ltd.

Conklin Instrument

CopperCom

MuLogic

NEC Australia

National Semiconductor

Natural MicroSystems

NeoNET LLC

NETMANSYS

Network Communications

Next Level Communications

Novell

OKI America

Objective Systems Integrators

PC-Tel

PMC-Sierra

Panasonic Singapore
Laboratories

Personal Systems Group,
Fujitsu Ltd.

Philips GmbH

Pivotal Networking

Predictive Systems

Premisys Communications

Pulse

Q&I Networks

Racal Research

Raychem

RELTEC

Robertson, Stephens & Co.

Rockwell Semiconductor Systems

Ryan-Hankin-Kent

Rythms NetConnections

SESSI

SPEA Software GmbH,
Diamond Multimedia

Creatix

CrossComm Corporation

Crystal Semiconductor

Cypress Semiconductor

DNA Enterprises

DTI

DAGAZ Technologies

Dallas Semiconductor

Davicom Semiconductor

DeLoitte & Touche Consulting Group

DSET Corporation

Design of Systems on Silicon (DS2)

Diamond Lane Communications

Digi International

Digital Link

Digital Technology MA

ETRI

EXFO Electro-Optical

Eicon Technology

Elantec

Electronic Techniques (Anglia) Ltd.

Epigram

Equator Technologies, Inc.

Ericsson Components AB

Ericsson Telecom

Fantastic Corporation

FORE Systems

Filtran

Flowpoint

Fluke

Fujitsu Microelectronics

Sage Instruments

Satronix

Schmid Telecommunication

Schott Corporation

Sharp Laboratories of America

Shenzhen Huawei Tech. Co.

Shiva

Siecor

Silicon Automation Systems India

Silicon Spice

Singapore Telecommunications

Sony Semiconductor Europe

Southern Methodist University

Specialix International Ltd.

SpellCaster Telecommunications

Sprint

Stellar One

Stentor Resource Centre

Sumitomo Electric U.S.A.

Superior Telecommunications

TTC

TYAN Computer

Tainet Communication Systems

Team Accountants

Tektronix

Telamon Corporation

Telco Systems

Tele Danmark

TeleChoice

Telecom Finland

Telecom New Zealand Ltd.

General DataComm

General Signal Networks

GenRad

George Washington University

GlobaLoop

Hanchang Group

Harris & Jeffries

Harris Corporation

Helsinki Telephone Company

Hewlett-Packard Cerjac

Hewlett-Packard Company

Hi/fn

Hitachi

Hitachi Cable America

IIS - Software, Inc.

INET

IPM Datacom

ISR Global Telecom

ITD/CIA

ITK Telecommunications

ITRI

IX Micro

Integral Access

Integrated Device Technology

Integrated Network Corporation

Integrated Technology Express

Intel

International Copper Association

InterSpeed

Intracom S.A.

JNA Telecommunications

Japan Aviation Electronics Ind.

Teledata

Telefication

Telefonica de Espana S.A.

Telenetworks

Telenor

TeleSoft International

Telia AB

Telindus

Tellabs OY

Telmax Communications

Telrad

Tennoz Initiative

Toko America, Inc.

Tollgrade Communications

Toshiba

Toucan Systems

Transcend Access Systems

TranSwitch

Tut Systems

University of New Hampshire

VLSI Technology CPG

VLSI Technology EPD

VSIS

VTT Electronics

Vacuumschmelze GmbH

Valor Electronics

Verilink

Viacom

Wandel & Goltermann A.T.E. Systems

Web Silicon

Westell International

Whitehead YCL USA

Jersey Telecoms

Katron Technologies

Kingston Communications
(Hull) PLC

Korea Telecom

LEP (Laboratoire d'Electronique
Philips)

Wilcom

Winbond Electronics

XAVI Technologies Corporation

XZ Group

Xecom

BOARD OF DIRECTORS

Chairman and President
Hans-Erhard Reiter
Ericsson Austria AG
Pottendorferstrasse 25-27
Vienna A-1121 Austria
Phone: +43.1.81100.5421
Fax: +43.1.81100.5381
100106.35@compuserve.com

Secretary
David Greggains
Gorham & Partners Ltd
61 Coleherne Court
Old Brompton Road
London, England SW5 0EF
Phone: + 44.171.370.1263
Fax: + 44.171.370.5375
davidgreg@cix.compulink.co.uk

Dawn Diflumeri-Kelly
GlobeSpan Technologies
8545 126th Avenue North
Largo, FL 33773
Phone: +1.813.530.8141
Fax: +1.813.530.8038
dkelly@globespan.net

Vice President and Treasurer
William Rodey
Westell Inc.
75 Executive Drive
Aurora, IL 60504
Phone: +1.630.898.2800
Fax: +1.630.898.4859
brode@westell.com

Martin Jackson
ATM Ltd.
Mount Pleasant House
2 Mount Pleasant
Huntingdon Road
Cambridge CB3 0BL
Phone: +44 1223 577241
Fax: +44 1223 566915
mjackson@atml.co.uk

Gavin Young
British Telecom
Martlesham Heath, B67/107
Ipswich, Suffolk
England IP5 7RE
Phone: +44.147.364.5963
Fax: +44.147.361.4580
youngg3@boat.bt.com

Debbie Sallee
Motorola
7755 South Research Drive
Suite 110
Tempe, AZ 85284
Phone: +1.602.641.0492
Fax: +1.602.641.8105
rzep90@email.sps.mot.com

Dan Arazi
Orckit Communications
38 Nahalat Yitzhak
Tel Aviv
ISRAEL
Phone: +972.3.6962121
Fax: +972.3.696.5678
danny@orckit.com

Kamran Sistanizadeh
Bell Atlantic
1310 North Court House Road
3rd Floor
Arlington, VA 22201
Phone: +1.703.974.3269
Fax: +1.703.974.0613
kamran.sistanizadeh@bell-atl.com

Tom Starr
Ameritech
2000 West Ameritech Center Dr.
Room 3C52
Hoffman Estates, IL 60196
Phone: +1.847.248.5467
Fax: +1.847.248.3775
tom.starr@x400gw.ameritech.com

John Shelnutt
Alcatel Telecom
1967 Lakeside Parkway
Tucker, GA 30084
Phone: +1.770.270.8328
Fax: +1.770.493.4856
Email: **john_m_shelnutt@aud.alcatel.com**

Executive Director
Karen Moreland
Association Management
Solutions
39355 California Street
Suite 307
Fremont, CA 94538
Phone: +1.510.608.5902
Fax: +1.510.608.5917
kmoreland@adsl.com

Association Coordinator
Cindy Carrillo
Association Management
Solutions
39355 California Street
Suite 307
Fremont, CA 94538
Phone: +1.510.608.5909
Fax: +1.510.608.5917
ccarrillo@adsl.com

Public Relations Representative
Ann Jansen
Jansen Communications

1475 SE Brookwood Avenue
Hillsboro, OR 97123
Phone: +1.503.648.3545
Fax: +1.503.640.1456
Email: **annjansen@compuserve.com**
or **jansen@aol.com**

European Public Relations
Representative
Carol Friend
Pielle Consulting
Museum House
25 Museum Street
London, ENGLAND WC1A 1PL
Phone: +44.171.322.1587
Fax: +44.171.631.0029
Email: **info@pielle.co.uk**

Association Coordinator
Karen Campo
Association Management
Solutions
39355 California Street
Suite 307
Fremont, CA 94538
Phone: +1.510.608.5921
Fax: +1.510.608.5917
kcampo@adsl.com

COMMITTEE CHAIR LISTING
MARKETING COMMITTEE

Chair
Marc Zionts
Westell
750 N. Commons Drive
Aurora, IL 60504
Phone: +1.630.898.2800
Fax: +1.630.898.4859
mzion@westell.com

Public Relations Chair
Dawn Diflumeri-Kelly
GlobeSpan Technologies
8545 126th Avenue North
Largo, FL 33773
Phone: +1.813.530.8141
Fax: +1.813.530.8038
dkelly@globespan.net

Vice Chair
Franz Starnberger
Ericsson Austria AG
Pottendorferstrasse 25-27
Vienna A-1121 Austria
Phone: +43.1.81100.6652
Fax: +43.1.81100.5381
franz.starnberger@sea.
ericsson.se

TECHNICAL COMMITTEE

Chair	Vice Chair
Nigel Cole	Frank Van Der Putten
Orckit Communications	Alcatel
110 Blue Ravine Road, Suite 160	Francis Wellesplein, 1
Folsom, CA 95630	B-2018 Antwerp, BELGIUM
Phone: +1.916.351.5600	Phone: +32.3.240.7081
Fax: +1.916.351.5700	Fax: +32.3.240.9920
nigel@orckit.com	**puttenf@sebb.bel.alcatel.be**

TECHNICAL COMMITTEE WORKING GROUPS

ATM End-to-End

Chair	Vice Chair
Jean-Francois Van Kerckhove	Bernard Dugerdil
DSC Communications	Motorola
Phone: +1.707.792.5722	Phone:+41.22.799.1385
jf_kervan@optilink.dsccc.com	**tte004@email.sps.mot.com**

NOTE: *From time to time during the two-day meeting period of this very large working group, the group will divide into two break-out groups as follows:*

- *"Network Architecture Group"—Chair: David Allen, Nortel*
- *"Access Transport Group"—Chair: Jean-Francois Van Kerckhove, DSC*

CPE & CO PHYSICAL CONFIGURATIONS & INTERFACES

Chair
Ken Martinez
Paradyne
Phone: +1.813.530.2306
martinez@eng.paradyne.com

NETWORK MANAGEMENT

Chair
Chi-lin Tom
Advanced Fibre Communications
Phone: +1.707.792.6244
chi-lin.tom@fibre.com

PACKET END-TO-END

Chair
Dave Krinsky
Aware
Phone: +1.617.276.4000
dkrinsk@aware.com

Vice Chair
Shuang Deng
Aptis Communications
Phone: +1.617.229.6863
shuang@aptis.com

OPERATIONS & TESTING

Chair
Fadi Daou
GenRad
Phone: +1.508.287.7296
fhd@genrad.com

TECHNICAL COMMITTEE FORMAL LIAISON OFFICERS

IETF
David Allan
Nortel
Phone: +1.613.763.6362
dallan@nortel.ca

ANSI T1E1.4
Frank Van Der Putten
Alcatel
Phone: +32.3.240.7081
puttenf@sebb.bel.alcatel.be

ATM Forum
Subra Ambati
AG Communicaitons
Phone: +1.602.581.4421
ambatis@agcs.com

ETSI TM6
Bernard Dugerdil
Motorola
Phone: +41.22.799.1385
tte004@email.sps.mot.com

Appendix C

ADSL Vendors

ADSL SEMICONDUCTOR COMPANIES

ATM Limited

Advanced Micro Devices

Alcatel Telecom *

Analog Devices

Broadcom Corporation *

Burr-Brown Ltd.

Elantec

GlobeSpan Technologies

Harris Semiconductor

Hi/fn

Integrated Device Technology

Integrated Technology Express

Intel

Metalink

Motorola Semiconductor

NEC Australia

NEC Electronics

Orckit Communications

PairGain Technologies

Rockwell Semiconductor Systems

SGS-Thomson Microelectronics

Web Silicon

NOTE: *Company has announced ADSL chip products.*

ADSL MODEM SUPPLIERS

3Com

ADC Telecommunications

AG Communication Systems

ATM Limited

Addtron Technology Corporation

Alcatel Telecom

Amati

Ariel

Aware

Cabletron Systems

Cayman Systems

Compaq Computer

DSL Networks

ECI Telecom

EPL Ltd.

Efficient Networks

Ericsson

Global Village Communication

GlobaLoop Ltd.

Intel

Interphase

interspeed

Orckit Communications

NEC Australia

PairGain Technologies

Paradyne

Performance Telecom

Pulsecom

RNS

SpellCaster Telecommunications

Tainet Communication System

Telrad Telecommunications

Teltrend

Westell Inc.

Westell INternational

@ux:XAVI Technologies Corporation

ADSL ACCESS NETWORK SUPPLIERS

3Com

3M

AG Communication Systems

ATM Limited

Addtron Technology Corporation

Alcatel Telecom

Amati

Ameritech

Aptis Communications

Ariel

BOSCH Telecom GmbH

BellSouth Telecommunications

BellSouth Telecommunications

BellSouth Telecommunications

Cayman Systems

Cellware GmbH

Compaq Computer

DSC Communications

DAGAZ Technologies

Diamond Lane Communications

ECI Telecom

EPL Ltd.

Efficient Networks

Ericsson

GlobaLoop Ltd.

Integrated Network Systems

Interphase

interspeed

Microcom

NEC Australia

Orckit Communications

PairGain Technologies

Performance Telecom

Promatory Communications

Promptus Communications

Pulsecom

Tainet Communication System

Telrad Communications

Teltrend

Westell Inc.

Westell INternational

XAVI Technologies Corporation

NETWORK/SERVICE PROVIDERS

Ameritech

Bell Atlantic

BellSouth Telecommunications

British Telecom

Cincinnati Bell

Kingston Communications

Korea Telecom

Sprint

TERACOM Svensk Rundradio AB

Telecom Finland

US West

TESTING/MANAGEMENT SUPPLIERS

3Com

3M

Atlantech Technologies

Digital Technology MA

EPL Ltd.

Fluke Corporation

Kingston Communications

TTC

University of New Hampshire

CONSULTANTS & OTHERS

3M—Consultant

Cadence Design Systems—Consultant & ADSL Design Services

Copper Development Association—Trade Association—Copper Industry

DIGI International—Remote Access Supplier

Digital Technology

General DataComm—DSL, HDSL, SDSL

Filtran Ltd.—Manufacturer of "POTS Splitters" and related magnetic
   components, transformers, inductors, and so forth

Global Village Communication—Consultant

GlobaLoop Ltd.

Gorham & Partners—Consultant

Harris & Jeffries—Frame Relay & ATM Software Source

Intel—PC Industry Player

Kingston Communications—Consultant

Predictive Systems—Consultant

Ryan-Hankin-Kent

SESSI—Consultant & Software Systems Engineering

Schott Corporation—ADSL Transformer Supplier. Interfaces with semiconductor chipsets.

Sharp Laboratories of America

Siecor—Copper Connections & Protection Supplier

Tainet Communication System

Toucan Systems—Consultant

University of New Hampshire—ADSL Consortium

VLSI Technology—Semicustom & Standard Semiconductor
Manufacturer

Web Silicon—ADSL ASIC Core Supplier

GLOSSARY

ADSL is a technology that firmly has one foot in the world of telephony and the other in the world of the Internet. Unfortunately, many potential readers of this book may be quite familiar with the concepts of the public-switched telephone system but at a loss when it comes to the World Wide Web. Others may have personal Web pages but have no idea what goes on inside the telephone network when an Internet service provider is contacted. Still others may find both worlds as mysterious as the inside of their television set—they just know that they turn it on and it works.

All of the terms and concepts presented here are detailed in much greater depth in the book itself. However, some of these terms and concepts may be so unfamiliar—and so numerous—that readers just cannot keep track of them all, especially through the early chapters. So this glossary represents a set of basic essentials for gaining a better understanding of what ADSL is all about. Telephony and Internet terms and concepts are mixed in a purely alphabetic sequence. The days are gone when these ideas could safely exist separate from one another. Terms and concepts cross-referenced in this glossary are presented in *italic* type.

Access lines.   Another term usually used for *local loop*. These are links used to allow users to access the *central office* (or *local exchange*) switch and have been optimized for voice use. Access lines have restricted bandwidth, known as the passband, that make it difficult to run at *broadband* speeds.

Analog.   Continuously varying over time. Opposite of *digital*, which has only discrete values. Voice is the prime example of analog information and a text file on a computer is a good example of digital information. However, either can be sent over a telecommunications link with an analog or digital signal.

Asynchronous.   One of the most overworked terms in telecommunications. In the context of this book, asynchronous can either stand for the usual operation of a PC *modem* or the method used in a *multiplexer* in a *cell* switching technology known as *Asynchronous Transfer Mode* (ATM). The meaning must usually be gathered from the context.

Bandwidth.   A good measure of the carrying capacity of a link, but not the only measure. Bandwidth just indicates how many bits per second a link can carry, but says nothing about the *delay* through the network. This is important for *delay bound* applications such as voice which need low and stable network delays to function properly.

Bandwidth bound.   An application which will not necessarily benefit from lower *delay* in a network, but can only run properly with a minimum amount of *bandwidth* at their disposal. A bulk file transfer is a good example of a bandwidth bound application.

Bell system.   Before 1984, the local telephone companies that belonged to AT&T were commonly grouped together as the Bell system. All others were independents. After 1984, it became common to speak of the entire telephone network as the *public switched telephone network* (PSTN).

Blocking.   Whenever bits cannot make their way from an input port to an output port in a *network node*, they are considered to be blocked. In the voice network, the call will not go through. In a data network, the bits may be stored in a *buffer* or discarded, depending on the situation.

Bridged tap.   An extension to a *local loop* (or *access line*) generally used to attach a remote user to a *central office* (or *local exchange*) switch without having to run a new pair of wires all the way back. Bridged taps branch off from a main line. The main line is not cut to establish the bridge. Bridged taps are fine for voice, but severely limits the speed of digital information flow on the link.

Broadband.   Strictly speaking, a telecommunications link that runs at more than 1.5 Mbps in the United States and more than 2 Mbps almost everywhere else. This is the primary rate of *ISDN*. Today, most people consider broadband speeds to be much higher, perhaps as high as 5 or 10 Mbps. Includes elements of both *bandwidth* and *delay*, neither of which can be ignored.

Brownout.   In the context of this book, a situation which occurs when a *central office* (or *local exchange*) cannot handle all of the calls attempted and even disrupts calls in progress. Also called a "brown down."

Browser.   A universal *client* for accessing information on the *Web* portion of the *Internet*. The development of the browser directly led to the explosion of interest in the Web and indirectly to the current crisis in *local access* speeds that ADSL addresses.

Buffer.   A storage area in a computer or other processor's memory dedicated for telecommunications purposes. The whole art of network design is a balancing of the need to buffer bits in order to store and process them and the need for adequate *bandwidth* and *delay* to actually send the bits somewhere. Bigger buffers can compensate for slower links and *network nodes* that experience *blocking*, but at the risk of offending waiting users.

Central office.   The term most often used in the United States to describe the basic public switched telephone network's *network node*. Used to attach *local loops* (or *access lines*) to users. See also *local exchange*.

Cell.   A fixed length unit of information. Most other data units can vary in length, but a cell is fixed in size. This helps to cut down of network delays and most importantly variations in the *delay* through a network. ATM is the international standard way of building networks that employ cells. Contrast with *packet* and *frame*.

Channel.   A portion of a total communication link's capacity. The channel may be established by an *analog* or *digital* technique using a *multiplexer*. The channel can also be physical or virtual. If a channel represents the entire carrying capacity of a physical link, it is said to be *unchannelized*, or (more properly) *non-channelized*.

Circuit.   A path through a network from source to destination (and usually back). In a circuit-switched network, this path uses a fixed route and fixed amount of bandwidth for the duration of the connection between end points. Circuits are efficient for voice use. Contrast with *packet* and *cell*.

Client/server.   A model for computer interactions. One half of the typical interaction between user and information content on the *Internet* and *Web* is called the client. The other half is the server. Usually, a user runs a client process to obtain and process information present on a remote server. That is, the client "talks" and the server "listens" and responds to client requests. The interaction between client and server is usually asymmetrical, which means that more bits flow from server to client than from client to server. ADSL reflects this difference, assuming most servers are *not* located in homes and small offices. This may not prove to be a valid assumption.

Codec.   The interface device required to carry *analog* information such as voice across a link using *digital* signaling. Contrast with *modem* and *DSU/CSU*.

Convergence.   The concept that at some unspecified time in the future, all information will be *digital*, all networks will one (no more separate TV, voice, or data networks), all user devices will just be various forms of computers, and there will be peace on earth.

Crosstalk. The interference caused by signals on adjacent *circuits* in a network. Annoying enough on the *analog* voice network, crosstalk is a hazard that limits distance and speed on *digital* networks.

CSU. See *DSU/CSU.*

Cyberspace. Meaningless term to describe the past, present, and most of all the future of the *Internet* and *Web.* Extreme Internet and Web zealots often claim that cyberspace forms a different level of existence or reality ("virtual reality") than that which the computers and networks themselves occupy, a somewhat odd thought.

Delay. A contributing measure of the carrying capacity of a link, but not the only measure. Delay just indicates how long it takes bits to find their way through a network, but says nothing about the *bandwidth* through the network. Delay can be zero, but the network useless if it only delivers one bit per hour. This is important for *bandwidth bound* applications such as bulk data transfers which need adequate bandwidth to function properly.

Delay bound. An application which will not necessarily benefit from more *bandwidth* in a network, but can only run properly with a minimum and stable *delay* at their disposal. A voice telephone call is a good example of a delay bound application. Adding more bandwidth beyond what it needs will not make the voice call any better.

Digital. Having only discrete values, such as 0 or 1. Opposite of *analog*, which is continuously varying over time. A text file on a computer is a good example of digital information and voice is the prime example of analog information. However, either can be sent over a telecommunications link with an analog or digital signal.

DSU/CSU. Data Service Unit/Channel Service Unit. The interface device required to change one form of *digital* signal to another. Many of the devices used in xDSL technologies are basically advanced forms of DSU/CSU, such as the HTU (HDSL Termination Unit). Contrast with *modem* and *DSU/CSU.*

Echo. The reflecting of a signal back to its source due to a variety of reasons. Whenever the same *bandwidth* is used for transmission in both directions, echo is a concern. In all cases, some form of echo control must be used to compensate for these effects, which can be annoying for voice but devastating for data. Both the voice network and simple *modems* employ echo cancellation techniques.

Flat rate. A billing method where each user is charged a fixed monthly amount regardless of actual usage. Became popular during the Depression in the United States when telephone expenses formed a considerable percentage of a family's budget. Flat rate users tend to perceive usage as "free" after a certain point and always prefer flat rates to the alternative, known as metered service. Flat rates contribute to long holding times for data sessions.

Frame. (1) A variable length unit of information. Frames contain *packets* and are subject to varying *delays* as they make their way through a network. Neverthless, frames are the most popular way of transporting packets. Contrast with *cell*. (2) A fixed length unit used for the transport of bits over a physical link. This is technically a transmission frame and forms part of a framed transport. All xDSL technologies are framed transports. Not to be confused with the first meaning of the term *frame*.

Holding time. A amount of time that users interact with or over a network. Holding times vary widely, from hours on a cable TV network to minutes for a voice telephone call. Data sessions involving *clients* and *servers* on the *Internet* and *Web* typically have holding times of about one hour. The voice network becomes stressed because long holding time data sessions are run over voice links designed for voice calls lasting a few minutes.

Hypertext. The usual method of presenting information on the *Web*. Information is linked by a series of jumps from place to place. Today the term is misleading because much of the content is audio and video as well as text. However, the term *hypermedia* has been slow to catch on.

Information Highway. Sometimes a superhighway, this term is as meaningless as *cyberspace*. Most of the claims and actual abilities outlined for this system have been realized in the *Web* itself. The most glaring exception is lack of *broadband* access speeds.

Internet. A world-wide collection of networks forming one huge *internetwork* where almost any *client* can contact almost any *server*. From humble beginnings as an experimental US military network, the Internet is considered essential today mostly because a large portion of it houses the *Web*. If the Internet is used for corporate or other private purposes, this portion forms an *Intranet*. If a client on one Intranet is allowed to contact a server on another Intranet, this forms an *Extranet*.

ISDN. Integrated Services Digital Network. A system designed to deliver voice, video, and data information to everyone's home at an affordable price. In other words, everything that the *Web* is today.

ISDN is mainly used today as a faster form of access to the *Internet* and Web.

Loading coil.   A metallic, doughnut-shaped device used on *local loops* (or *access lines*) in North America, particularly within the United States, to extend their reach. Outside of the United States, loading coils are rare and extended distances are achieved by creative use of mixed wire gauges. Loading coils severely limit the bandwidth useful for telecommunications.

Local exchange.   Another term for *central office*, and officially used in ISDN. The term most often used outside of the United States to describe the basic public switched telephone network's *network node*. Used to attach *local loops* (or *access lines*) to users.

Local loop.   A term usually used for *access line*, which is in fact the more technically correct term. These are links used to allow users to access the *central office* (or *local exchange*) switch and have been optimized for voice use. Local loops have restricted *bandwidth*, known as the *passband*, that make it difficult to run at *broadband* speeds.

Modem.   The interface device required to carry *digital* information such as a text file on a computer across a link using *analog* signaling. Many of the devices used in xDSL technologies are basically advanced forms of modem, such as the ATU (ADSL Termination Unit). Contrast with *codec* and *DSU/CSU*.

Multiplexer.   Any one of a number of common devices used to combine and later split multiple telecommunications circuits into *channels* which are in turn sent over the larger *bandwidth* of another *circuit*. Allows for more efficient use of the capacity of the link. *Trunks* are usually multiplexed, but *access lines* (or *local loops*) are not.

Network node.   The heart of any network. The network node hooks the users together. In the voice network the links between *central office* (or *local exchange*) network nodes are called *trunks* and the links to the users are called *access lines*. On the *Internet* and *Web*, the network node is a *router* (or sometimes an ATM *switch*). DSL technologies in one sense try to get Internet traffic off the central office network node.

Network interface device.   The demarcation point where the public network ends and the "private" network within a home or office begins. All wiring and user devices inside the premises (such as a *modem*) are controlled and operated by the owner, not the network service provider. In the United States, in cases where the service provider wishes to provide equipment beyond the network demarcation, the provider must first have permission from any regulators involved and the customer must agree. Outside of the United States, users have less control over their wiring and equipment. Also called the *network termination unit* or *demarc*.

Packet.   The content of a *frame*, according to the first definition given above. In a packet-switched network, there is no fixed bandwidth (or usually a fixed path) assigned to users. Users are assigned bandwidth based on their needs of the moment. Developed for bursty data traffic, *trunks* between packet switching *network nodes* and even *access lines* do not typically use *channels* for the packets. Packets are efficient for data use. Contrast with *circuit* and *cell*.

Passband.   The *bandwidth* of a link is determined by the range of frequencies it can carry. *Local loops* (or *access lines*) will limit the range to the voice passband, around 300 to 3300 Hz. Any frequency outside this range is not carried on the link. This passband is adequate for voice, but severely limits the speed of digital information flow on the link.

Router.   The *network node* of the *Internet* (and *Web*). Routers forward packets to other routers until they reach their destination. Internet and Web routers use the IP protocol from *TCP/IP* as the format for *packets*. Nothing else is allowed. Routers are sometimes contrasted with *switches*, but the distinction becomes harder and harder to discern over time.

Server.   See *client/server*.

Splitter.   A device used in ADSL and perhaps some day in some other DSL technologies to allow users to continue to use their *analog* telephones while at the same time accessing the *Internet* and *Web* for *digital* information. Backward compatibility is the key idea here. The analog voice goes into the voice *switch* at the *central office* while data *packets* can be sent onto the Internet and Web through a *router*.

Switch.   The *network node* of the voice network and some data networks. In a data network, switches forward *packets* to other switches until they reach their destinations. The packets can take on a variety of formats, and usually follow a fixed path through the network, as opposed to packets in *router* networks. Data switches are sometimes contrasted with routers, but the distinction becomes harder and harder to discern over time.

TCP/IP.   Transmission Control Protocol/Internet Protocol. The official protocol of the *Internet* and *Web*. All *routers* on the Internet must be able to run TCP/IP. IP itself defines a *packet* structure which can be sent inside a variety of *frames*, or even **cells**. PCs must run TCP/IP to access the Internet and Web.

Trunk.   A link between *network nodes*. Most commonly used to refer to links between voice *central office* (or *local exchange*) *switches*, the term has been frequently applied to links between a telephone service provider and an Internet service provider, although technically these are just *local loops* (or *access lines*). However, modern usage employs the term *trunk* to indicate any links that are *not* specifically between a user and a network (e.g. voice switch network node to Internet router network node).

Web.   Sometimes World Wide Web, or less commonly WWW. That portion of the *Internet* which features multimedia content accessed through *hypertext* links. Wildly popular, the Web has been used for everything from marketing to commerce and usage is only expected to increase. The popularity of the Web has lead directly to proposed solutions like DSL technologies to ease some of the congestion on the voice network from dial-up *modem* connections.

ACRONYMS

2B1Q	Two Binary One Quaternary	CD-ROM	Compact Disc—Read-Only Memory
A/D	Analog to Digital		
AAL	ATM Adaptation Layer	CEBus	Consumer Electronics Bus
AC	Alternating Current	CERN	European Center for Nuclear Research
ADM	Add-Drop Multiplexer		
ADSL	Asymmetric Digital Subscriber Line	CLEC	Competitive Local Exchange Carrier
AIU	Access Interface Unit	CMIP	Common Management Interface Protocol
AM	Amplitude Modulation		
AMI	Alternate Mark Inversion	CO	Central Office
ANS	Advanced Network and Services	COT	Central Office Terminal
ANSI	American National Standards Institute	CPE	Customer Premises Equipment
AOL	America On-Line	CRC	Cyclical Redundancy Check
ARPA	Advanced Research Projects Agency	crc-i	cyclical redundancy check on the interleaved data
ARPANET	Advanced Research Projects Agency Network	CSA	Carrier Serving Area
		CSI	CompuServe, Inc.
ASCII	American Standard Code for Information Interchange	CSU	Channel Service Unit
		D/A	Digital to Analog
ASK	Amplitude Shift Keying	DAA	Data Access Arrangement
AT&T	American Telephone and Telegraph	DACS	Digital Access and Cross-connect System
ATM	Asynchronous Transfer Mode	DAML	Digital Added Main Line
ATU-C	ADSL Termination Unit— Central Office	DARPA	Defense Advanced Research Projects Agency
ATU-R	ADSL Termination Unit— Remote	DAVIC	Digital Audio-Visual Council
		DBS	Direct Broadcast Satellite
AUX	Auxiliary	DC	Direct Current
AWG	American Wire Gauge	DCE	Data Circuit-terminating Equipment
B-ISDN	Broadband Integrated Services Digital Network	DCS	Digital Cross-connect System
BER	Bit Error Rate	DCT	Discrete Cosine Transform
BERT	Bit Error Rate Test	DDD	Direct Distance Dialing
BOC	Bell Operating Company	DDN	Defense Data Network
BRA	Basic Rate Access	DDS	Digital Data Service
BRI	Basic Rate Interface	DLC	Digital Loop Carrier
BRITE	Basic Rate Interface Transmission Equipment	DMT	Discrete Multitone Technology
		DoD	Department of Defense
CAP	Carrierless Phase/Amplitude modulation	DoJ	Department of Justice
		DPSK	Differential Phase Shift Keying
CAP	Competitive Access Provider	DS	Digital Signal
CBR	Constant Bit Rate	DSE	Data Switching Equipment
CCITT	International Telegraph and Telephone Consultative Committee	DSL	Digital Subscriber Line
		DSLAM	Digital Subscriber Line Access Multiplexer
CCS	Hundred Call Seconds per Hour	DSP	Digital Signal Processing
CD	Compact Disc	DSS	Digital Satellite System

DSSC	Digital Subscriber Single Carrier	HTML	Hypertext Markup Language
DSU	Data Service Unit	HTTP	Hypertext Transfer Protocol
DSX	Digital System Cross-connect	HTU	HDSL Transmission Unit
DTE	Data Terminal Equipment	ib	indicator bits
DWMT	Discrete Wavelet Multitone	IBM	International Business Machines
EARN	European Academic Research Network	IDSL	ISDN Digital Subscriber Line
EC	European Community	IEC	Interexchange Carrier
EIA	Electronics Industry Association	IEEE	Institute of Electrical and Electronics Engineers
EO	End Office	ILEC	Incumbent Local Exchange Carrier
eoc	embedded operations channel	ILMI	Interim Local Management Interface
ES	End System		
ES-IS	End System-Intermediate System	IP	Internet Protocol
		IS	Intermediate System
ESP	Enhanced Service Provider	IS-IS	Intermediate System- Intermediate System
ESS	Electronic Switching System		
ETSI	European Telecommunications Standards Institute	ISDN	Integrated Services Digital Architecture
FCC	Federal Communications Commission	ISO	International Organization for Standardization
FDM	Frequency Division Multiplexing	ISP	Internet Service Provider
fe	forward error	ITFS	Instructional Television Fixed Service
FEBE	Far End Block Error		
febe-i	far-end block error on the interleaved data	ITU	International Telecommunications Union
febe-ni	far-end block error on the non-interleaved data	ITU-T	International Telecommunications Union—Telecommunications Standardization Sector
FEC	Forward Error Correction		
fecc-i	forward-error correction code on the interleaved data	IXC	Interexchange Carrier
		JPEG	Join Photographic Experts Group
fecc-ni	forward-error correction code on the non-interleaved data		
FITL	Fiber In The Loop	LADS	Local-Area Data Service
FM	Frequency Modulation	LAN	Local Area Network
FOT	Fiber Optic Terminal	LANE	Local Area Network Emulation
FRAD	Frame Relay Access Device	LAP-F	Link Access Procedure—Frame Relay
FSK	Frequency Shift Keying		
FT1	Fractional T1	LATA	Local Access and Transport Area
FTP	File Transfer Protocol	LE	Local Exchange
FTTC	Fiber To The Curb	LEC	Local Exchange Carrier
FTTH	Fiber To The Home	LEO	Low Earth Orbit
FTTN	Fiber To The Neighborhood	LEOS	Low Earth Orbit Satellite
GEO	Geosynchronous Earth Orbit	LEPA	Local Exchange Primary Access
GEOS	Geosynchronous Earth Orbit Satellite	LLC	Logical Link Control
		LMDS	Local Multipoint Distribution System
GUI	Graphical User Interface		
HDSL	High-bit rate Digital Subscriber Line	LOS	Line of Site
		los	loss of signal
HDTV	High Definition Television	LPF	Low Pass Filter
HFC	Hybrid Fiber-Coax	LSB	Least Significant Bit
HPF	High Pass Filter	LTU	Line Termination Unit
HPPI	High Performance Parallel Interface	MAC	Media Access Control
		MAE	Metropolitan Area Exchange

MB	Measured Business	PDN	Public Data Network
MBONE	Multicast Backbone	PDS	Premises Distribution System
MDF	Main Distribution Frame	PDU	Protocol Data Unit
MDSL	Multirate (or moderate) Digital Subscriber Line	PHY	Physical Layer
		PM	Phase Modulation
MEO	Medium Earth Orbit	PNNI	Private Network Node Interface
MEOS	Medium Earth Orbit Satellite	POP	Point of Presence
MFJ	Modification or Final Judgment	POTS	Plain Old Telephone Service
MILNET	Military Network	PPP	Point-to-point Protocol
MIT	Massachusetts Institute of Technology	PRA	Primary Rate Access
		PRI	Primary Rate Interface
MLHG	Multi-Line Hunt Group	PSDN	Public Switched Data Network
MMDS	Multichannel Multipoint Distribution System		
		PSK	Phase Shift Keying
MPEG	Motion Picture Experts Group	PSTN	Public Switched Telephone Network
MPOA	Multiprotocol over ATM		
MS	Microsoft	PUC	Public Utility Commission
MSB	Most Significant Bit	PVC	Permanent Virtual Circuit
NAP	Network Access Point	QAM	Quadrature Amplitude Modulation
NCSA	National Center for Supercomputer Applications		
		QoS	Quality of Service
NE	Network Element	QPSK	Quadrature Phase Shift Keying
NEXT	Near-End cross Talk	RADSL	Rate Adaptive Digital Subscriber Line
NGDLC	Next General Digital Loop Carrier		
		RAM	Random Access Memory
NI	Network Interface	RAM	Remote Access Multiplexer
NIC	Network Interface Card	RBOC	Regional Bell Operation Company
NID	Network Interface Device		
NII	National Information Infrastructure	rdi	remote defect indicator
		RDT	Remote Digital Terminal
NIU	Network Interface Unit	RF	Radio Frequency
NISDN	National ISDN	RFI	Radio Frequency Interference
NN	Network Node	RHC	Regional Holding Company
NNI	Network—Network Interface	RJ	Recommended Jack
NNI	Network Node Interface	ROM	Read-Only Memory
NSF	National Science Foundation	ROW	Rest of World
NSFnet	National Science Foundation Network	RT	Remote Terminal
		S/N	Signal-to-Noise ratio
NT	Network Termination	SAR	Segmentation and Reassembly
NTU	Network Termination Unit	sc	synchronization control
NVOD	Near Video On Demand	SCSI	Small Computer Systems Interface
OAM	Operations, Administration, and Maintenance		
		SDH	Synchronous Digital Hierarchy
OC	Optical Carrier	SDLC	Synchronous Data Link Control
occ	operations control channel		
OLEC	Other Local Exchange Carrier	SDN	Switched Digital Network
		SDSL	Symmetric (or Single-pair) Digital Subscriber Line
ONU	Optical Network Unit		
OSI	Open Systems Interconnect	SDU	Service Data Unit
OSI RM	Open Systems Interconnect Reference Model	SDV	Switched Digital Video
		sef	severely errored frame
PAM	Pulse Amplitude Modulation	SGML	Standard Generalized Markup Language
PBX	Private Branch Exchange		
PC	Personal Computer	SLC	Subscriber Loop Carrier

SLC	Simple Line Code
SM	Service Module
SMDS	Switched Multimegabit Data Service
SNMP	Simple Network Management Protocol
SOHO	Small Office/Home Office
SONET	Synchronous Optical Network
SPID	Service Profile Identifier
STDM	Statistical Time Division Multiplexing
STP	Shielded Twisted Pair
STS	Synchronous Transport Signal
SVC	Switched Virtual Circuit
TA	Terminal Adapter
TC	Transmission Convergence
TCP	Transmission Control Protocol
TCP/IP	Transmission Control Protocol/Internet Protocol
TDM	Time Division Multiplexing
TE	Terminal Equipment
TG	Trunk Group
TM	Terminal Multiplexer
TP	Twisted Pair
TU	Trunk Unit

UNI	User-to-Network Interface
USB	Universal Serial Bus
UTP	Unshielded Twisted Pair
VAR	Value-Added Reseller
vBNS	Very high speed Backbone Network Service
VBR	Variable Bit Rate
VC	Virtual Channel
VC	Virtual Circuit
VCI	Virtual Channel Identifier
VDSL	Very high-speed Digital Subscriber Line
VHS	Video Home System
VLAN	Virtual LAN
VOD	Video on Demand
VoIP	Voice over Internet Protocol
VP	Virtual Path
VPI	Virtual path Identifier
VTU	VDSL Termination Unit
WAN	Wide Area Network
WWW	World Wide Web
WYSIWYG	What You See Is What You Get
xDSL	x-Type Digital Subscriber Line

Index